Business Intelligence with SQL Server Reporting Services

Adam Aspin

Apress®

Business Intelligence with SQL Server Reporting Services

ISBN-13 (pbk): 978-1-4842-0533-4

ISBN-13 (electronic): 978-1-4842-0532-7

Managing Director: Welmoed Spahr
Lead Editor: Jonathan Gennick
Technical Reviewers: Rodney Landrum and Ian Rice
Editorial Board: Steve Anglin, Mark Beckner, Gary Cornell, Louise Corrigan, Jim DeWolf, Jonathan Gennick, Robert Hutchinson, Michelle Lowman, James Markham, Matthew Moodie, Jeff Olson, Jeffrey Pepper, Douglas Pundick, Ben Renow-Clarke, Gwenan Spearing, Matt Wade, Steve Weiss
Coordinating Editor: Jill Balzano
Copy Editor: Mary Behr
Compositor: SPi Global
Indexer: SPi Global
Artist: SPi Global
Cover Designer: Anna Ishchenko

Distributed to the book trade worldwide by Springer Science+Business Media New York, 233 Spring Street, 6th Floor, New York, NY 10013. Phone 1-800-SPRINGER, fax (201) 348-4505, e-mail orders-ny@springer-sbm.com, or visit www.springeronline.com. Apress Media, LLC is a California LLC and the sole member (owner) is Springer Science + Business Media Finance Inc (SSBM Finance Inc). SSBM Finance Inc is a Delaware corporation.

For information on translations, please e-mail rights@apress.com, or visit www.apress.com.

Apress and friends of ED books may be purchased in bulk for academic, corporate, or promotional use. eBook versions and licenses are also available for most titles. For more information, reference our Special Bulk Sales–eBook Licensing web page at www.apress.com/bulk-sales.

Any source code or other supplementary material referenced by the author in this text is available to readers at www.apress.com. For detailed information about how to locate your book's source code, go to www.apress.com/source-code/.

To Karine

Contents at a Glance

Contents

About the Author

Adam Aspin is an independent business intelligence consultant based in the United Kingdom. He has worked with SQL Server for nearly 20 years. During this time, he has developed several dozen reporting and analytical systems based on the Microsoft BI product suite.

A graduate of Oxford University, Adam began his career in publishing before moving into IT. Databases soon became a passion, and his experience in this arena ranges from dBase to Oracle, and Access to MySQL, with occasional sorties into the world of DB2. He is, however, most at home in the Microsoft universe, using SQL Server Analysis Services, SQL Server Reporting Services, and SQL Server Integration Services as well as Power BI.

Business intelligence has been his principal focus for the last 15 years. He has applied his skills for a range of clients including J.P. Morgan, The Organisation for Economic Co-operation and Development (OECD), Tesco, Centrica, Harrods, Vodafone, Crédit Agricole, Cartier, EMC Conchango, Alfred Dunhill, the RAC, and TNT Express.

Adam has been a frequent contributor to SQLServerCentral.com for several years. A fluent French speaker, Adam has worked in France and Switzerland for nearly two decades and has written numerous articles for various French IT publications. He is a regular speaker at user groups and conferences such as SQLBits.

He is the author of *SQL Server Data Integration Recipes* (Apress, 2012) and *High Impact Data Visualization with Power View, Power Map, and Power BI* (Apress, 2014).

About the Technical Reviewers

Rodney Landrum went to school to be a poet and a writer. And then he graduated, so that dream was crushed. He followed another path, which was to become a professional in the fun-filled world of information technology. He has worked as a systems engineer, UNIX and network admin, data analyst, client services director, and finally as a database administrator. The old hankering to put words on paper, while paper still existed, got the best of him, and in 2000, he began writing technical articles, some creative and humorous, some quite the opposite. In 2010 he wrote *The SQL Server Tacklebox*, a title his editor disdained, but a book closest to the true creative potential he sought; he wanted to do a full book without a single screen shot. He promises his next book will be fiction or a collection of poetry, but that has yet to transpire.

Ian Rice is a business intelligence and data warehouse consultant, specializing in the Microsoft suite of technologies. In his 12 years of consultancy, he has designed and delivered multiple data warehouses, including supporting report environments across finance, insurance, retail, and service sectors. He has also used his experience to assist organizations in recovering failed, previous data warehouse implementations. Ian is currently based in the UK, and accepts clients both nationally and internationally.

Acknowledgments

Writing a technical book can be a lonely occupation. So I am all the more grateful for the help and encouragement that I have received from so many friends and colleagues.

First, my heartfelt thanks go to Jonathan Gennick, the commissioning editor of this book. Throughout the publication process Jonathan has been an exemplary mentor. He has shared his knowledge and experience selflessly and courteously.

My deepest gratitude goes once again to Jill Balzano, the Apress coordinating editor, for managing this book through the publication process. She succeeded in the impossible task of making a potentially stress-filled trek into a journey filled with light and humor. Her team also deserve thanks for all their efforts. So thanks to Mary Behr for her tireless work editing and polishing the prose and to Dhaneesh Kumar for the hours spent formatting the text.

When lost in the depths of technical questions it is easy to lose sight of what should be the main objectives. Fortunately, technical reviewers Rodney Landrum and Ian Rice have worked unstintingly to remind me of these objectives. They have shared their considerable experience of SQL Server and Reporting Services and helped me considerably with their comments and suggestions.

However, my deepest gratitude must be reserved for the two people who have given the most to this book. They are my wife and son, who have always encouraged me to persevere while providing all the support and encouragement that anyone could want. I am very lucky to have both of them.

Introduction

Business intelligence (BI) means different things to different people. For some, it means advanced analytics. For others, it is self-service report creation. A few see it as a prelude to Big Data. For many users, however, BI is much more accessible than any of these things. It is quite simply the regular production of clear and meaningful reports and dashboards that translate data into actionable information.

In my experience, all that a majority of business people ask for is up-to-date data that they can access when they want in a ready-to-use format. They do not always have the time or the skills to develop their own flashy output. Most end users have no desire to delve deeply into vast realms of data to track processes, sales, and targets. They want only to identify anomalies and react swiftly to changing business conditions.

So, despite all the hype, there is still a place for good old corporate business intelligence. This is the mixture of data, technology, and IT skills that creates and distributes "canned" reports in all shapes and sizes to a multitude of users across an organization. Despite being out of the spotlight (thanks to the glare of self-service BI), corporate business intelligence still solves the vast majority of enterprise reporting needs. It delivers reliable and trusted data to users who have only to select a few parameters to get the results they need.

This book is about energizing this forgotten, but fundamental sector of the BI universe. More specifically, it aims to show you how you can capitalize on a technology that you probably already have-SQL Server Reporting Services-to push your corporate BI to the next level. After reading this book you will be able to produce designer dashboards that integrate gauges, maps, charts, and text to deliver targeted information. You will learn to exploit the subtleties of a variety of visualizations (KPIs, sparklines, trellis charts, bubble charts, and many others) to give your reports real pizzazz. Specifically, you will learn how to make important trends stand out, while providing clear visual alerts to draw your attention to significant thresholds and deviations from targets. All this can be done via a solid and reliable toolkit that is used regularly by a vast developer community around the world.

In this book you will also learn how to revamp the appearance of your reports to make them more user-friendly. You will learn how to add slicers and other interactive tweaks to make your output better adapted for delivery to mobile devices such as tablets and cellphones.

This book is not meant as an introduction to SSRS. It presumes that you already have some experience of producing basic reports. The aim here is to help you to build on your experience to develop more advanced scorecards and dashboards and then to adapt the output for mobile devices.

In the book you will see dozens of visualizations that can enhance the ways in which business intelligence is presented. All of them-and more-are available on the Apress web site in an SSDT project. Feel free to take them as the basis for your own reports and dashboards, and to adapt them to your specific needs.

CHAPTER 1

SQL Server Reporting Services as a Business Intelligence Platform

Not a month goes by without some shiny, new, all-singing, all-dancing business intelligence product being announced as the Next Big Thing. A few months later, no one can remember it ever existed, except your CEO, who wants to know what her organization is using to deliver corporate business intelligence (BI). She wants a solid, stable, and mature platform that can deliver flawless reports to users automatically and regularly. She does *not* want her staff wasting time developing their own reports and learning corporate data structures. Secretly, many users agree with her. She *does*, however, want a certain level of interactivity. Of course, she expects BI on mobile devices and a tangible level of "wow factor." At least, that is what you thought she said, before your eyes glazed over in fear as you wondered which product was going to be forced on your lovingly engineered SQL Server universe.

The product quite simply could be SQL Server Reporting Services.

SQL Server Reporting Services (SSRS) has been around for over 15 years. Longevity like this, fortunately, can have its advantages, and in the case of SSRS they are the following:

- SSRS is a mature, stable, and efficient platform.
- SSRS can accept data from many different sources, most often relational as well as dimensional and tabular.
- SSRS handles the details of scheduled report creation, snapshots, and automated report delivery with consummate ease.
- SSRS slots into a complete SQL Server BI stack efficiently and smoothly.
- SSRS reports can now also be made available in the cloud using Power BI.

Often underestimated, SQL Server Reporting Services is an extremely powerful environment for delivering clean, clear, and polished reports to users. If you add a final layer of BI polish to the mix (and that is what this book is all about), then you have an incredibly wide-ranging and effective business intelligence tool at your disposal.

SSRS is *not*, however, designed to be a self-service BI tool. It exists to deliver corporate BI, where prebuilt reports are what the user wants. Indeed, in this world the user specifically does *not* want to develop his or her own reports. They may want to change some core parameters or drill down or across through data, but they definitely do not want to spend (they might even say "waste") their time formatting their own reports. These users prefer to have an MI (Management Information) department prepare their reports for them.

So, if self-service BI is what you are looking for, then Power BI is probably a place to start, and my book *High Impact Data Visualization with Power View, Power Map, and Power BI* (Apress, 2014) should be your first port of call.

However, if you are looking at delivering pre-built, polished, and interactive business intelligence to your users, where timely, accurate, and attractive information delivery is the major requirement, then SSRS is probably the tool that you need. It can provide the following:

- KPIs
- Scorecards
- Dashboards

These features can be delivered to schedule or displayed on demand. The data can be updated according to your business requirements. Not only that, but SQL Server Reporting Services, if you are using SQL Server 2012 SP1 or later, can deliver your BI not only to browsers, but also to smartphones and tablets.

While users cannot generate reports, and slice and dice the data as they can in the self-service world, they can

- Set multiple parameters to define the data that will be displayed.
- Drill down into report hierarchies.
- Drill across to other reports

This book exists to show how you can use SSRS to provide corporate business intelligence that delivers all that is described above, plus add the "wow" factor that your boss wants. The trick is to learn how.

Business Intelligence Concepts

As this book is about delivering business intelligence using SQL Server Reporting Services, it will help to clarify a few basic concepts from the outset. When delivering corporate BI, you will mostly be developing key performance indicators, scorecards, and dashboards. Let's begin by defining these as succinctly as possible.

Key Performance Indicators

Key Performance Indicators (KPIs) have been an essential part of business intelligence for many years. They display a set of selected fundamental measurements that managers use to make productive decisions. A classic KPI will probably contain the following:

- A *value*: This is the figure returned from a business system.
- A *goal*: This is the figure that should have been attained, and comes from a budget document that has been loaded (somehow) into an accessible data source.
- A *status*: This is nearly always shown as a symbol that indicates how aligned the value is to the metric (in other words, how well you are doing). In some cases, it can be the color of a display element.
- A *trend*: This shows how the metric being measured has evolved over time and it can be displayed as a symbol or a small graph called a sparkline.

Put simply, a KPI will let you display a result, how it maps to target, and where it is going. It lets you see anything that can be measured that is of strategic or tactical importance.

Scorecards

Scorecards are collections of KPIs. They might all relate to the same area of business, or give a view across multiple business areas. Scorecards might themselves become part of a dashboard.

Dashboards

There are many overlapping–and sometimes conflicting–definitions of what makes up a dashboard. I have no intention of getting caught up in arcane theoretical disputes, so I will consider a dashboard to be a visual overview of essential corporate data. It can display many possible data snapshots of multiple aspects of a business, or show a highly specific set of metrics relating to a business area or department. I consider that it can include targets and objectives, but that these are not compulsory.

A dashboard can be made up of multiple elements, which many people also call "widgets" or visualizations. SSRS lets you use the following as the building blocks of your dashboards:

- KPIs
- Scorecards
- Tables
- Charts
- Gauges
- Sparklines
- Maps
- Images

These core elements can often be combined to produce multiple variations on a theme. How they can be added to reports and combined to deliver BI is the subject of the book.

SSRS for Business Intelligence, Practically

Now that you have had a rapid overview of the theory, it is time to get practical. You need to see more generally how SQL Server Reporting Services can deliver BI elements to your users. Specifically, you need to see what you will be doing when developing your BI solutions.

With SSRS for Business Intelligence, you will be developing the following:

- *Reports*: This is a more generic term that I use to describe an SSRS `.rdl` file, whatever the type of presentation it contains.
- *Visualizations*: BI dashboards, scorecards, and reports are normally made up of several elements. They could be charts, gauges, tables, maps, images, or combinations of these elements. I use the term *visualization* to describe a single element, even if it is composed of different elements from the SSRS toolbox. For instance, a KPI containing a sparkline, some metrics, and an indicator will make up a single visualization.
- *Widgets*: A widget is any element (or object) that can be stored independently and reused. In most cases it is really only a synonym for a visualization.

Delivering BI is not just a technical matter. It involves the following three basic threads that have to be melded into a final delivery:

- Data
- Design
- Platform

Let's take a look at these three.

Getting the Data Right

Making sure that the data is perfect is the key to delivering effective business intelligence. The data must be not "good enough," not "nearly right," but absolutely one hundred percent accurate. Obviously, this book is not the place to discuss data architecture, data cleansing, data quality, master data, or the definition of metrics and processing for BI. However, in the real world, you will have to deal with all of this.

As we are only dealing with the presentation layer in this book, I will simplify the learning curve by using a single database whose data we will presume is perfect (and thus ready for use in SSRS) to underpin the visualizations and dashboards that you will be creating. I will also make this single database the repository for all ancillary information that you may need when developing your outputs. By "ancillary" I mean data such as

- *Budget data*, used in KPIs, for instance
- *Geographical data*, used when creating maps or analyzing data by country or region
- Any other data that is *not* habitually sourced from an OLTP or OLAP system, such as certain types of reference data

However, having the data accurate and accessible is only the first part of the story. How you then present the data to SSRS is also fundamental. There are several aspects to this.

Prepare the Data

As you could well be sourcing data from multiple tables (for instance, you could be aggregating sales data from one table or view and comparing it with budget data from another set of tables), I generally advise that you mash up the data in the database *before* you present it to SSRS. There are several advantages to this approach, compared to filtering, joining and calculating data in SSRS itself:

- You are using data from a database, so it seems more logical to use the unrivalled power of the database language (T-SQL) rather than the reporting tool. SSRS is necessarily more limited where data manipulation is concerned.
- The heavy lifting takes place where the data is kept: in the database.
- SQL Server provides an amazing toolset to help you collect, rationalize, and output data. So do not be afraid to use temporary tables, CTEs, table variables, and all the other tools in the T-SQL toolkit to shape your data.

So what should your aims and objectives be when preparing datasets in T-SQL? Everyone has their own opinions, but I suggest that you take the following points into consideration when setting up your data for reports:

- Output only the precise data that is needed by your report. Extra rows and columns "in case they might be useful one day" slow down report processing and weigh on network transmission needlessly.
- Similarly, your T-SQL may include intermediate columns used for calculations. Often these are not displayed in the final report, so you may not need to output them to SSRS. They will likely only cause confusion when you (or your team) are developing or maintaining the report.
- Do not make code re-use into a target. Dashboards and reports change all the time, so trying to develop "one size fits all" stored procedures or scripts that underlie a group of reports can be a waste of time. Be prepared to create a stored procedure for each different visualization. This way, when the requirement for a minor change comes in, you will not have to worry about its ramifications across a whole range of dashboards. Alternatively, in the cases where you have a clear group of reports that are all based on the same data, reusable stored procedures can be a boon.
- Do document your code to remind yourself of any common elements across stored procedures. This way you know that if you change something for one chart (say), then you will need to modify other stored procedures too.
- Try where possible–and this may be a contentious point–to provide all the required data for a visualization in a single data set. This may seem obvious, so let me explain further. Some visualizations require data that is not a metric, such as the maximum for a gauge pointer or the thresholds used in ranges. These are only single, scalar figures. However, it can be easier sometimes to add them as columns to a dataset (even if the same figure is repeated identically in every record) and use them as the source data. This often only adds a few columns to a datasource, and has the advantage of keeping all the data in the same place for a visualization, and not cluttering up the report with added datasets containing only one or two values.
- Sometimes, however, you may want or need to reuse data such as thresholds or maximum values in different visualizations in the same report. In this case, do use separate datasets in a report to access these elements.
- Remember that in some cases (the second KPI in Chapter 2 is an example) you will have to create several datasets for a single visualization. Experience will tell you when you have to use multiple datasets and subsequently link them inside SSRS to produce the required result.

Of course, you may disagree strongly with these ideas. Feel free to do so; the essential thing is that you have thought through the reasons for how and why you have set up the data in the way that you have chosen. If you can justify them to yourself, then you can insist on them for your team–and defend your decisions to your boss.

Use Views and Stored Procedures

While it is perfectly possible to custom code every piece of SQL (or MDX or DAX) that feeds data into SSRS, it is frequently a much better practice to aim for some reusability. This nearly always means adding views and stored procedures to your data layer.

Views

It can often be worth the effort to prepare views that you can use and re-use as a conformed source of source data. You might be able to use views directly in certain reports, or (more likely) use them as the basis for specific processing and output using a stored procedure. If you will repeatedly be joining tables or aggregating data, then a view will save you a lot of repetitive coding. Of course, in most situations you will never really know what data you will be using before you start building visualizations. However, it can be worth refactoring your initial SQL queries into views earlier rather than later in the project lifecycle.

Stored Procedures

SSRS will let you use both T-SQL code blocks and stored procedures to assemble and deliver data as a dataset. However, I am a firm believer in using stored procedures rather than freeform code blocks in SSRS. My reasons include the following:

- Data manipulation is centralized in the database. You will not spend time opening multiple reports and then digging through datasets and scrolling through code in a tiny dialog in SQL Server Data Tools.
- Parameters are created automatically and correctly when you specify a stored procedure as a data source for a data set.
- Large and complex processes delivered as stored procedures will not exceed the character limit of the SSRS code box.
- You work using a suitable UI, rather than trying to code in a text box or (possibly worse) copying and pasting back and forth between SSRS and SQL Server Management Studio.
- You are a data professional and so probably feel happier using a more structured and coherent method of working.
- You are minimizing the risk of SQL injection attacks.

Some Ideas on Source Data Definition

I believe strongly that preparing the data for a visualization is a fundamental part of successful BI report creation with SSRS. This is why for every example in this book I always not only give the code used to deliver the required data, but show sample output and then explain the reasoning behind the T-SQL. I do this to draw the reader's attention to some of the many ways in which you can prepare data for BI reporting. As you will see, I tend to use temporary tables and CTEs quite regularly. I also tend to break down separate logical elements into small elements of code, rather than aim for monolithic structures. These are personal choices, and all I want to do is to provide some hints and suggestions for code use–and re-use. How you write your code is your choice. All I can suggest is that you try to make it as efficient as possible.

There are many design decisions that you will have to make when defining the datasets that feed into your reports. You may prefer to create many datasets–possibly not only one, but several for each widget. Alternatively, you might try to cram everything into a single datasource for all the elements on a dashboard, and then apply filters to apply only the relevant data to specify charts or gauges individually.

My approach is usually to find a common ground between these two extremes. In most cases, I find that a data source can be created (and subsequently filtered) for associated groups of elements–gauges, charts, or KPIs that make up a visualization. Sometimes an element may need several datasets. In yet other cases you may need multiple datasets because you are using disparate data sources.

Anyway, in this book you will find examples of most of these approaches. This will not mean that a certain way to prepare the data is best suited to a certain type of visualization, just that I am taking advantage of a scenario to show some of the ways that datasets can be used–and combined–in SSRS.

On a more general note, I find that spending time defining the data requirements at the start of a project–or even for a single BI element–can help me to focus on what each visualization is trying to achieve. It can also save a lot of development time as you will minimize the number of times that you modify your T-SQL code, and subsequently add, rename, and delete columns of data in SSRS datasets and objects. So I can only advise you to think through the data before you start building a chart, gauge, or map. To encourage good practice, therefore, I begin every example with the model that you are aiming to build and the code needed to produce the data that is required. This way you can see what is required for each visualization, why you need it, and what is, in my opinion, the best way to deliver it.

Of course, my SQL programming style may not be the same as yours. After all, we all have our own way of delivering BI to our users. So I am not saying "this is how you must do it," but merely "here are some ideas based on my experience." How you prepare the data for your dashboards and reports is completely up to you.

Real-World Data

In the real world, you will use data from many different sources in a variety of formats. SQL Server might not be the only relational database you are using. Relational data may not be the only data format; you could be using dimensional data from a SQL Server Analysis Services database or even data in the Analysis Services tabular format.

SSRS will allow you to use data from these varied platforms easily and simply. However, I decided not to use all these potential data sources in this book and stick to SQL Server relational data for the following reasons:

- Not everybody is using all the potential data sources that SSRS can handle.
- Many users are likely to be federating data in a SQL Server reporting layer (using techniques such as linked servers or ETL processes using SQL Server Integration Services).
- All readers are likely to know T-SQL, which makes it a comprehensible lingua franca for explaining how data is produced.
- Using SQL Server as the data layer means that readers can concentrate on SSRS rather than getting distracted by other languages such as MDX or DAX.

However, preparing the data is not the end of the process. You need to be able to reuse data in certain circumstances, and almost certainly you will need to cache both data and reports to ensure that your users enjoy a top-class BI experience. The final chapter of this book will outline some of the techniques that can help you enhance the user experience and deliver output smoothly–and quickly.

Designing SQL Server Reports for Business Intelligence

Once your data is in place, you can move on to designing the output. This is admittedly the fun part, and is often where the real challenge resides. You need to remember this, especially when faced with seemingly impossible demands from management or users, or apparently insane demands for stylistic pirouettes from the Corporate Style Police.

Presentation

Your first choices will inevitably concern the presentation approach that your reports will be taking. They could be

- Text-based
- Graphical
- Hybrid

You could be aiming for a traditional look and feel, or attempting something different. The choice is yours, as long as you remember that variety should not mean distraction, and novelty can wear off faster than you think.

Design

This is the area where your choice of backgrounds, borders, images, and text will come into play. Color and style are vast areas, and ones that are probably best left out of the hands of technical people like me. Nonetheless, a few comments are necessary.

The French have a saying that translates as "there is no discussing taste and color." This usually means that everyone knows that they (and only they) are right in matters of good taste-especially where report design is concerned. Moreover, any discussion with these people is pointless as it is clear (to them, at least) that *they* (whoever they are) are right. I agree: they (or even you) are always right. Or more probably, your boss, or another department somewhere in the organization, is inevitably right when it comes to the choice of colors and report presentation generally.

Consequently, to placate these people, I have adopted a color scheme that avoids garish colors and uses shades of gray for text and borders. This approach lets the information speak louder than the color scheme. You, and your internal clients, may prefer other color palettes and design styles, and this is up to you. The steps describing color and style are only there to indicate how to apply colors, not necessarily the color that you must use.

In some cases, I have thrown these principles out of the window, as style can also be a function of the delivery method. So, specifically when defining output for smartphones, I tend to use bold primary colors or even white-on-black displays. This is because on a phone a competing distraction is only a swipe away, so I want to grab the user's attention.

One thing that you will have to make abundantly clear to your users-and in-house style gurus-is that SSRS is not a design application. It cannot deliver the kinds of currently fashionable interfaces that certain other products can. What it can do (and this is the mantra you will need to repeat) is deliver BI in a fraction of the time and cost of many of the alternatives. Not only that, but it integrates perfectly into your SQL Server ecosystem, does not require expensive training, and is easy to maintain. Oh, and you will not need the licensing costs of third-party software. Or the extra consultants to get it up and running. Or to explain exactly why all of your reporting infrastructure needs to be replaced, yet again.

Layout

Having admitted that there are limitations to what SSRS can deliver visually, I now want to compensate by saying that it can nonetheless produce some really slick and impressive output. To get the most out of SQL Server Reporting Services to deliver visually compelling BI, you need to know a set of techniques-and a few tricks-that help you get the job done. These include

- Aligning visualizations
- Enclosing objects in other objects

- Controlling the size and growth of objects
- Applying images as backgrounds to many objects

Ideas for these aspects of report creation are discussed in Chapter 7.

Interface and Interactivity

You probably do not need me to remind you that the SSRS web interface (Report Manager) will never win any prizes for ergonomics. With a little effort, however, you can revamp and even replace this tired old UI to give the user a more pleasant experience. This revamping also covers ways to enhance the (admittedly limited) interactivity that is on offer.

This revamping is mainly based on using the following features:

- *"Postback" and hidden variables*: This means using the `Action` property of an item on a report to re-display the same report where some parameters are changed and others kept the same.
- *SSRS Expressions*: Expressions are the key to enhancing most aspects of SSRS, and interactivity is no exception. You will see many examples of expressions in this book. Fortunately, you do not have to learn an entire programming language, as only a handful of Visual Basic keywords are essential when revamping reports. So, if you are prepared to learn to use `If()`, `Switch()`, `Lookup()` and one or two others, then you can bring the SSRS interface into the 21st century.

Chapters 8, 9, and 10 will show you some of the techniques that can be applied when revamping the report interface.

Multi-Purposing

Many of the techniques and most of the widgets that you learn to build in this book can be used in a plethora of circumstances. Just because you learn to apply a technique to tablet BI does not mean that it cannot be used effectively in reports that are read on a laptop. Certain techniques explained in the context of output for smartphones can be used on a PC. So please do not think that just because I may explain an approach to business intelligence delivery in the context of a certain output device that you can only use it on that specific device. Indeed, I encourage you to experiment and to discover which techniques work best in which circumstances and on which devices.

The Sample Database

The sample database that is used throughout this book as the basis for all reports contains the sales data for a small English car reseller called Brilliant British Cars Ltd. It has been going since 2012 and now exports a small set of English luxury and sports cars to Europe and North America.

Preparing Your Environment

If you intend to follow the examples in this book, you will need to set up a SQL Server Data Tools (SSDT) environment that is correctly configured so that you do not waste any time and can create your own version of the reports that are detailed in the various chapters. If you have copied the CarSalesReports solution from the Apress web site and installed it as described in Appendix A, then you already have an environment that contains all the examples. However, as it can be fruitful to build your own reports from scratch, here is how to set up and configure an empty SSDT project.

Creating an SSDT Project

The first thing to do is to create an SSDT project.

1. Run SQL Server Data Tools (Start ➤ All Programs ➤ SQL Server 2014 ➤ SQL Server Data Tools).
2. Click New Project.
3. Click Report Server Project. Set the name and select a location such as `C:\BIWithSSRS`. Uncheck Create directory for solution.
4. Click OK.

■ **Note** I am presuming that you are using the 2014 version of SQL Server. However, 2012 will work just as well. Indeed, for everything except output to mobile devices, SQL Server 2008 R2 will work just as well.

Adding a Shared Datasource

You now need to add a shared datasource that connects to the source database. I am presuming that you have downloaded and restored the `CarSales_Reports` database as described in Appendix A.

1. Display the Solution Explorer window (Ctrl+Alt+L).
2. Right-click Shared Data Sources and select Add New Data Source from the context menu.
3. In the Shared Data Source Properties dialog, enter the following:
    a. *Name*: CarSales_Reports.rds
    b. *Type*: Microsoft SQL Server
    c. *Connection string*: `Data Source=YourServer\YourInstance; Initial Catalog= CarSales_Reports`
4. Click OK.

■ **Note** You can, of course, click Edit and configure the data source to point to another database using a different security configuration.

Add Shared Datasets

You now need to add the shared datasets that are used in the examples in this book. Before adding them, it is best to know what they are and what they do.

- `CurrentMonth`: This gives the current month that is passed as the default to the `ReportingMonth` parameter of most reports.
- `CurrentYear`: This gives the current year that is passed as the default to the `ReportingYear` parameter of most reports.
- `ReportingFullMonth`: This gives the month names and month number. It is used in a couple of reports to display the full month name as a report parameter.
- `ReportingMonth`: This gives the months in the year. It is used in the `ReportingMonth` parameter (found in most reports) to allow the user to choose the month for which data will be displayed.
- `ReportingYear`: This gives all the years for which there are data in the database. It is used in the `ReportingYear` parameter of most reports to allow the user to choose the year for which data will be displayed.

You add a shared dataset like this, using `ReportingYear` as an example.

1. Right-click Shared Datasets (in the Solution Explorer window) and select Add New Dataset from the context menu.
2. Enter a dataset name (`CurrentMonth` in this example) and set the Data source to `CarSales_Reports`.
3. Set the Query type to Stored Procedure.
4. Select (or paste) the stored procedure `Code.pr_ReportingYear` as the stored procedure name.
5. Click OK.

You can now create the five other shared datasets, using the following properties:

Dataset	Property	Value
`CurrentMonth`	Query type	Stored Procedure
	Stored procedure name	`Code.pr_CurrentMonth`
`CurrentYear`	Query type	Stored Procedure
	Stored procedure name	`Code.pr_CurrentYear`
`ReportingYear`	Query type	Stored Procedure
	Stored procedure name	`Code.pr_ReportingYear`
`ReportingFullMonth`	Query type	`Text`

(continued)

Dataset	Property	Value
	Query	SELECT 1 AS ReportingMonth, 'January' AS MonthName UNION SELECT 2 AS Expr1, 'February' AS Expr2 UNION SELECT 3 AS Expr1, 'March' AS Expr2 UNION SELECT 4 AS Expr1, 'April' AS Expr2 UNION SELECT 5 AS Expr1, 'May' AS Expr2 UNION SELECT 6 AS Expr1, 'June' AS Expr2 UNION SELECT 7 AS Expr1, 'July' AS Expr2 UNION SELECT 8 AS Expr1, 'August' AS Expr2 UNION SELECT 9 AS Expr1, 'September' AS Expr2 UNION SELECT 10 AS Expr1, 'October' AS Expr2 UNION SELECT 11 AS Expr1, 'November' AS Expr2 UNION SELECT 12 AS Expr1, 'December' AS Expr2
ReportingMonth	Query type	Stored Procedure
	Stored procedure name	Code.pr_ReportingMonth

Configuring Parameters

Nearly every report used in this book will use two parameters to choose the month and year for which data is displayed. These parameters are

- ReportingYear
- ReportingMonth

Here is how to add them to a report.

1. With a report open (or freshly created) right-click Parameters in the Report Data window and select Add Parameter from the context menu.
2. In the General tab, add the parameter name (ReportingYear, in this example).

3. Set the data type to Integer.
4. Click Available Values on the left.
5. Select Get values from a query.
6. Set the following:
    a. Dataset: ReportingYear
    b. Value field: ReportingYear
    c. Label field: ReportingYear
7. Click Default Values on the left and set the following:
    a. Dataset: CurrentYear
    b. Value field: CurrentYear
8. Click OK.

You can now do exactly the same for a second parameter named CurrentMonth. The only differences are that the Dataset, Value field, and Label field will use ReportingMonth for the available values, and the default values will use CurrentMonth for the dataset and Value field.

■ **Note** Even if you do not add these parameters to a report that uses them, they will be added by the stored procedures that return the data (in most cases). However, you will still have to configure them as described above for the reports to function correctly.

You can now use this project to build the examples from this book without any risk of overwriting the examples in the sample project. Indeed, you can then compare your work with the sample report files in the sample project, or even copy and paste items between projects if you want to speed up the learning curve.

When you are following the examples in this book, you will see that I am giving each SSRS report the name of the example that you can find in the CarSalesReports project. This way you can compare your work with the model that was used to produce each image that you can see for each example. You can name your reports anything you want, of course.

Code and Stored Procedures

All the code examples in this book are shown as stand-alone code snippets that you can run as-is in an SSDT query window (assuming that you have downloaded the `CarSales_Reports` database from the Apress web site, restored it to your computer, and are running the query against this database). When it comes to creating the actual reports, however, I will presume that you have either made the code into a stored procedure, or are using the referenced stored procedure from the `CarSales_Reports` database. As I am assuming that you know how to create and use stored procedures I will not be explaining how to tweak the code to add a 'sproc header.

Reusability

BI reports and dashboards can be immensely varied. They can use a multitude of different types of visualization. Not only that, but each type can vary in a myriad of subtle ways.

One of the objectives of this book is to provide you with a starter kit of reusable widgets that you can then tailor to suit your specific requirements. These widgets are not, however, designed to be generic. They are designed to illustrate the techniques that can be brought to bear to solve BI reporting challenges using SSRS.

However, you should be able to adapt most of the widgets described in this book, or taken from the source code download page on the Apress web site, to your own requirements. One hint that I can give you is that if you need to change, for instance, field names in a widget to suit your data, the best approach is to follow these steps.

1. Display the Solution Explorer window. Ctrl+Alt+L is one way of doing this.
2. Right-click the report file that you want to modify and select View code from the context menu.
3. Select Edit ➤ Find and Replace ➤ Quick Replace.
4. Enter the text that you want to find and the replacement text. This could be a file name, for instance.
5. Click Replace All.
6. Save the file and close it.
7. Reopen the report normally.
8. Preview and test the report.

If all has gone well, you will have taken a major step towards adapting the widget to your specific requirements.

I hope also that the code samples will be reusable and customizable to some extent. Obviously, they cannot cover more than a fraction of the needs of a reporting system, yet many of the overall approaches that are taken can hopefully provide you with some ideas and prototypes.

The CarSales_Reports Database

If you are following the examples in this book, you must install the `CarSales_Reports` database as described in Appendix A. This database is deliberately simple, so that you can concentrate on building dashboards and BI reports rather than struggle to understand an overly complex mound of data.

As a nod to the real world, the source data is in a series of relational tables, but the reports source their sales data from a view over the OLTP tables. This avoids you having to create the same joins again and again, as well as reminding you that reporting data is often prepared in ways like this.

The sample database also contains three user-defined functions that round up output. They are used for charts and gauges to set the upper limit of scales in many cases. The reason for this is that if you set a scale limit to a "rounded" figure, the scale increments (and gridlines and tick marks, if appropriate) will be much easier to read. If you need to adapt these functions to set other increments to round up by, you can use them as a starting point.

Book Audience

This is not a book for total SSRS novices. If you are using this book, then I am presuming that you can already create reports and are at ease with the concepts and practice of

- Data sources
- Datasets

- Tables
- Matrixes
- Basic charts
- Images

You will need a basic familiarity with these elements because BI with SSRS requires you to build on these foundations and to add

- Gauges
- Complex charts
- Sparklines
- Indicators
- Data bars

Not only will you be using these more "BI-oriented" elements, but you will be learning to combine most of the tools that SSRS has to offer. Moreover, you will be using all these tools together to create dashboards and output for mobile devices.

This book has no pretentions about being exhaustive. There are simply too many ways to use SSRS for business intelligence to cover in one small volume. Consequently, you will find only a selection of techniques and approaches that my experience has shown to be useful. There are plenty more ways to develop and use gauges and charts, for instance, than are explained in this book. Indeed, I hope that you will want to take the models that you find here and further enhance and develop them to create your own visualizations.

You are welcome to begin your BI development using this book as a starting point if you wish. However, if you feel that you need to revise some core knowledge, then *Pro SQL Server 2012 Reporting Services* by Brian McDonald, Shawn McGehee, and Rodney Landrum (Apress, 2012) is a perfect place to get the information that you might need.

As I am assuming that you are an intermediate-to-advanced SSRS user, I have had to make some presumptions about your level of knowledge. Nonetheless, I do not want to leave near beginners completely out in the cold, as there are simply too many cool things that can be done even if your knowledge of SSRS is only rudimentary. My approach is to start many chapters with an initial example that explains many of the core techniques and tools that you will be using in greater detail further on in the chapter. This first item isolates and explains each step in the process of developing a BI visualization. I hope that this approach will give novices a rapid introduction to the relevant techniques. From then on in the chapter, all other items are explained a little more succinctly. Specifically, the multiple properties that most elements require to be set are given as a table of properties that you have to set from one of the requisite dialogs.

SSRS allows you to set many properties in two or even three ways. I will try always to explain how to set properties using the dialogs that appear after you right-click an option in the context menu. This does not mean that you cannot use the properties window (or any other available method) if you prefer.

Some SSRS objects–charts spring to mind–are quite complex to configure. It follows that defining all the required properties for an arresting visual element can take quite a few clicks. This can also mean traversing a multitude of dialogs, panes, and windows full of settings.

It can also be difficult to select the appropriate part of a gauge or chart when you want to modify it. Admittedly, this becomes second nature after a little time, but until then you may need a little practice to get the job done.

Nearly all the examples in this book tend to use virtually the same basic template where the only parameters are the year and month. Inevitably, your dashboards and visualizations will be more complex than this, but I prefer to stick to a simple core report rather than cause confusion by altering the parameters for each report. This way you can concentrate on the elements that make up a dashboard rather than the mechanics of building and configuring different parameter sets. The report `__BaseReport.rdl` can be used as a template for nearly all reports if you wish to save time when it comes to adding the shared data source and shared datasets, and configuring the core parameters. Just remember to make a copy of the file first, so that you can reuse it easily for other reports.

How Best to Use This Book

This book is designed to be read in a linear fashion, from start to end. Indeed, some later chapters build on visualizations developed in previous chapters. To avoid this becoming a constraint there is nothing to stop you from merely taking the completed report or widget from the sample application and using it as a basis for development if you are jumping in at a later chapter. Equally (and if you are already a proficient SSRS developer) you can certainly leap in to any chapter or example to glean some useful information and tips.

Chapters 2-7 are designed to be read as a coherent unit. They explain how to build dashboards from component units. The basic components are described in Chapters 2-5, as follows:

- *Chapter 2* is an introduction to KPIs and some of the ways (from classic to more adventurous) of presenting your Key Performance Indicators.
- *Chapter 3* introduces gauges in business intelligence for SSRS. You will see some fairly standard gauges alongside some of the more interesting things that you can do with gauges.
- *Chapter 4* looks at some of the chart types that my experience has led me to believe are suited to BI. This is not a complete introduction to all the chart types that SSRS offers, but a trip into some of the less traditional aspects of chart design and creation.
- *Chapter 5* introduces maps and geographical data in SSRS. In a break with the format of this book, it provides a more complete introduction and does not presume prior experience with maps in SSRS.

Chapters 6 and 7 show how to assemble these components to build sample dashboards.

- *Chapter 6* shows how to use and manipulate images to standardize presentation in dashboards.
- *Chapter 7* shows how to assemble the composite parts of a dashboard into a coherent and attractive report.

Chapters 8 and 9 show how to customize and revamp the SSRS user interface and add interactivity.

- *Chapter 8* shows how to extend interactivity in SSRS using slicers and highlighting. You will also see how to create a kind of popup menu to enhance the user experience.
- *Chapter 9* shows how to enhance the user experience through adding tiles to the SSRS report interface. You will even see how to imitate Power BI carousels to scroll through lists of data as well as developing controlled paged datasets. Finally, you will see how to create "tabbed" reports.

The final three chapters cover mobile output and a series of techniques to deliver BI efficiently and quickly:

- *Chapter 10* gives you a series of tips and tricks that you will find useful when preparing reports that are destined to be viewed on tablets and smartphones.
- *Chapter 11* explains how to work more efficiently through standardizing certain methods and techniques.
- *Chapter 12* concludes the book with approaches that help you deliver data faster to the output devices and consequently enhance the user experience.

■ **Note** In this book I intend to be agnostic as to the publishing platform, so you can apply nearly every technique explained in the book to native Reporting Services just as you can to SSRS in SharePoint.

Conclusion

Now that you have seen what this book is about, it is time to get practical and to start using SSRS to create and deliver business intelligence to your users. The first stop on the road to the next level of SSRS is producing KPIs. This is the subject of the next chapter.

CHAPTER 2

KPIs and Scorecards

Key Performance Indicators, or KPIs, are a core element in most business intelligence. SSRS has been able to display KPIs for some time now, and they can be invaluable when it comes to delivering essential performance data in a succinct and meaningful way.

A set of KPIs is often referred to as a scorecard. Here I will not be delving into the management theory of what makes up a "balanced" scorecard, but will only refer to a scorecard as being a collection of KPIs.

Nonetheless, there are many ways to display KPIs. You may be used to a more traditional style of presentation where each KPI is on a separate row in a table, with visual elements to alert the user to deviations from an expected result. However, there are many other ways to deliver KPIs that need not involve tabular data. These can include using gauges to present the information. They may also include using text and colors to highlight changes in status and showing trends with the aid of sparklines. This chapter will begin with traditional KPIs, both to explain the concepts and to demonstrate the techniques. Then I will move on to some examples of less traditional KPIs. The underlying principle will always be the same, nevertheless. I will be comparing a metric to a target and highlighting any deviation from the objective, be it positive or negative. Similarly, I will display a trend over time.

In any case, I hope that you will discover that SSRS can deliver visually arresting KPIs in many ways, and I also hope that the examples in this chapter will be of use in your organization.

What Are Key Performance Indicators?

Before you can start using SSRS to display KPIs, you need to define what they are. Without wishing to get lost in layers of management-speak, let's say that Key Performance Indicators are a way of measuring progress towards a defined organizational objective. In practical terms, this usually means displaying the following elements:

- *A goal*: This is the target you are measuring an outcome against. This will be a figure, perhaps a budget or a sales target.
- *A value*: This is the actual data that will be compared to the target.

These two elements are the core of any KPI. However, they can be extended with one or both of the following elements:

- The *status* of the value compared to the goal. This indicates how well you are doing.
- The *trend* (over time, inevitably) of how well you are doing.

What you want to display in a KPI is up to you, or rather, up to the business you are working in. I consider that as long as you have a target figure and valid data to compare to the target, along with some indication of progress, then you have a KPI that you can use to convey meaningful information to your audience.

KPI Value

These metrics are often the easiest to understand. They are the figures that represent the business reality. In practice, however, delivering high-level data can require a lot of work preparing the source data. So you must be prepared to aggregate, calculate, filter, and rationalize your data in order to produce a meaningful set of KPI values.

KPI Goal

It probably sounds a little obvious, but you need to have goals (or targets) for a KPI to function. This means getting your business users to specify exactly what the targets are; they can be simple values (such as budgetary data) or they can be calculated values such as a defined percentage increase or decrease. In any case, you need to store and/or calculate the goal data for each metric. In practice, this can be tougher than it sounds. The difficulties are often more operational than technical, because

- Business users do not always have clearly defined quantifiable goals.
- Budgetary values, where they exist, are often in separate systems, or (worse in some cases) in spaghetti-like spreadsheets, which present difficulties such as:
    - Loading and updating the data can be painful, as it is not structured.
    - Mapping the budget data to the data from line of business systems is difficult.

Resolving these problems is outside the scope of this book. In the following examples, you will be using a set of budgetary data from the sample database that has been pre-crunched to map to the business data and that allows you to create meaningful KPIs.

KPI Status

Status is probably best thought of as a visual signal of how well you are doing. Frequently a status indicator is a symbol like a traffic light. Alternatively, it can be an image that indicates success or failure. It may even be a symbol that changes color to indicate good, average, or bad (or even under-achievement, on target, or over-achievement). As these examples imply, most status indicators will only display three values. You can extend this to five or even more values if you want. Just remember that you can easily lose the visual effect and clarity if you make it too complex for instant and intuitive understanding by the user.

What you need to retain is that the KPI status is essentially a flag indicator that has to be calculated. So you will need to know three things:

- The value (as for any KPI)
- The target (also as for any KPI)
- The thresholds at which a status flag switches from one value to another

KPI Status Thresholds

Let's take a simple example of a status flag in a KPI. Suppose that you have a target of 100. The business has said that any result under 80 is bad news, whereas anything over 125 will win a vacation. This means that you have the target and the thresholds. Once you have the value, you can then divide the result by the target. Under 80 percent is bad news, between 80 percent and 125 percent is expected, and anything over 125 percent is good news indeed. It is that easy. (In the real world, threshold calculations can be more complex than this, but I want to demonstrate the principle.) Of course, the thresholds can be absolute values rather than percentages; it will all depend on the business requirements. In the case that I just mentioned, the example only has three status flags; you can have as many status flags as you wish in your visualizations. However, I will never exceed three status indicators in the examples in this book as I firmly believe that using more than three status flags obscures the information rather than clarifies it.

Threshold Flags

If you are carrying out any status calculations in a database (relational or dimensional), I generally advise that you convert the status to a simple indicator, such as 1 for bad, 2 for OK, and 3 for good. This is because

- It is often easier to have the business logic in one place.
- You can standardize the threshold flags across your entire reporting suite, or better still across the enterprise.
- It is simpler to maintain reports if the business logic is in a structured (and hopefully annotated) environment.
- It is generally easier to concentrate on the way that an indicator value is *displayed* in SSRS (the choice of symbols or images and the color selection, for instance) if you are not having to mix this in with the calculation and the business logic in the report at the same time.

This threshold value will then be used by SSRS to display a more visual and meaningful indicator.

You can, of course, perform calculations and apply business logic in SSRS, and there are many valid reasons for this approach, too. However, as I mentioned in Chapter 1, in this book I will always place business logic at a lower layer of the solution.

KPI Trends

KPIs can also give a visual indication of how the current results compare to past outcomes. Here again, exactly how you define the time-based comparison will depend on the business. For example, you may need to

- Compare the current month's sales with the previous month's sales
- Compare the current month's sales with the same month of the previous year
- Visualize the last few months or year's data as a sparkline

KPI Trend Thresholds and Flags

The principles that I outlined for status thresholds and flags apply in a very similar way to trend thresholds and flags. You need the current value and the value you are comparing it with. Then you need to compare the two and indicate what the trend is on a simple scale. Interestingly, trend indicators in KPIs tend to show at least five values, and can show more; though, once again, clarity is essential, and a trend indicator that has nine different levels can be hard to read. So I will stick to a maximum of five in this book. Later I will show you how to use sparklines to display trends in a different type of detail.

As for the indicators, I feel that trend indicators are best calculated at the database layer and a simple flag (say 1 through 5) is sent to the report to be displayed in a more visual and intuitive way. This is what you will do in the next few examples.

A Simple KPI

I suspect that you have had enough theory and want to get down to delivering KPIs with SSRS. So let's move on to a simple KPI that will let you see how the principles can be applied in practice. This KPI takes sales data for the selected year (up to the selected month) and does the following:

- Displays the sales data.
- Displays the forecast data.
- Compares the sales data with the sales forecast and indicates the status on a scale of 1 to 3. The status is displayed as a traffic light.
- Compares the current sales up to the selected month with the sales for the same period in the previous year and returns a trend flag on a scale of 1 to 5. The trend is displayed as five pointed star that ranges from empty (a threshold indicator of 1) to full (an indicator of 5).

You will use a single stored procedure to gather the data and apply business logic. Status and trend indicators will be delivered in the same dataset as the business data and target metrics. At the risk of belaboring the point, I prefer to centralize the business logic and source data in a single place (a stored procedure in a reporting database) wherever possible. Of course, the reporting suite that you develop may take another approach, and there is nothing to prevent you carrying out the calculations and applying business logic in SSRS.

As this is the first example in this chapter (indeed in the whole book) I will be explaining it in more detail than the subsequent KPIs in this chapter. So this is an excellent place to start if you have a basic grasp of SQL Server Reporting Services and need to consolidate your knowledge before moving on to some of the more advanced techniques that you will find later in this chapter.

Figure 2-1 shows the KPI that you are trying to create. It displays sales and budget figures alongside the status and trend indicators for all the makes of car sold in 2014 up to the end of June.

Make	Sales	Sales Budget	Status	Trend
Aston Martin	696,500	729,000		
Bentley	722,500	729,000		
Jaguar	545,000	729,000		
Rolls Royce	765,000	729,000		

Figure 2-1. *A KPI showing sales by make for the first half of 2014*

The Source Data

First, you need some data to work with. The following code snippet (available as `Code.pr_CarSalesYearToDateKPISimple` in the `CarSales_Reports` database) gives you the data that you need for June 2014:

```
DECLARE @ReportingYear INT = 2014
DECLARE @ReportingMonth TINYINT = 6

IF OBJECT_ID('Tempdb..#Tmp_KPIOutput') IS NOT NULL DROP TABLE Tempdb..#Tmp_KPIOutput
```

```
CREATE TABLE #Tmp_KPIOutput
(
ReportingYear INT
,Make NVARCHAR(80) COLLATE DATABASE_DEFAULT
,Sales NUMERIC(18,6)
,SalesBudget NUMERIC(18,6)
,PreviousYear NUMERIC(18,6)
,StatusIndicator SMALLINT
,TrendIndicator SMALLINT
)

INSERT INTO #Tmp_KPIOutput
(
ReportingYear
,Make
,Sales
)

SELECT     ReportingYear
          ,Make
          ,SUM(SalePrice)
FROM      Reports.CarSalesData
WHERE     ReportingYear = @ReportingYear
          AND ReportingMonth <= @ReportingMonth
GROUP BY  ReportingYear
          ,Make
-- Previous Year Sales
;
WITH SalesPrev_CTE
AS
(
SELECT     ReportingYear
          ,Make
          ,SUM(SalePrice) AS Sales
FROM      Reports.CarSalesData
WHERE     ReportingYear = @ReportingYear - 1
          AND ReportingMonth <= @ReportingMonth
GROUP BY  ReportingYear
          ,Make
)

UPDATE    Tmp
SET       Tmp.PreviousYear = CTE.Sales
FROM      #Tmp_KPIOutput Tmp
          INNER JOIN SalesPrev_CTE CTE
          ON Tmp.Make = CTE.Make

;
WITH Budget_CTE
AS
```

```
(
SELECT     SUM(BudgetValue) AS BudgetValue
          ,BudgetDetail
          ,Year
FROM      Reference.Budget
WHERE     BudgetElement = 'Sales'
          AND Year = @ReportingYear
          AND Month <= @ReportingMonth
GROUP BY  BudgetDetail
          ,Year
)

UPDATE    Tmp
SET       Tmp.SalesBudget = CTE.BudgetValue
FROM      #Tmp_KPIOutput Tmp
          INNER JOIN Budget_CTE CTE
          ON Tmp.Make = CTE.BudgetDetail
          AND Tmp.ReportingYear = CTE.Year

-- Internal Calculations
-- Year on Year Delta
-- TrendIndicator

UPDATE  #Tmp_KPIOutput
SET     TrendIndicator =
           CASE
               WHEN ((Sales - PreviousYear) / Sales) + 1 <= 0.7 THEN 1
               WHEN ((Sales - PreviousYear) / Sales) + 1  > 1.3 THEN 5
               WHEN ((Sales - PreviousYear) / Sales) + 1  > 0.7
                              AND ((Sales - PreviousYear) / PreviousYear) <= 0.9 THEN 2
               WHEN ((Sales - PreviousYear) / Sales) + 1  > 1.1
                              AND ((Sales - PreviousYear) / PreviousYear) <= 1.3 THEN 4
               WHEN ((Sales - PreviousYear) / Sales) + 1  > 0.9
                              AND ((Sales - PreviousYear) / PreviousYear) <= 1.1 THEN 3
               ELSE 0
           END

-- StatusIndicator

UPDATE    #Tmp_KPIOutput
SET       StatusIndicator =
              CASE
                WHEN ((Sales - SalesBudget) / Sales) + 1 <= 0.8 THEN 1
                WHEN ((Sales - SalesBudget) / Sales) + 1  > 1.2 THEN 3
                WHEN ((Sales - SalesBudget) / Sales) + 1  > 0.8
                     AND ((Sales - SalesBudget) / SalesBudget) <= 1.2 THEN 2
                ELSE 0
              END
-- Output

SELECT   Make, Sales, SalesBudget, StatusIndicator, TrendIndicator
FROM     #Tmp_KPIOutput
```

Running this snippet gives the output in Figure 2-2.

	Make	Sales	SalesBudget	StatusIndicator	TrendIndicator
1	Aston Martin	696500.000000	729000.000000	2	1
2	Bentley	722500.000000	729000.000000	2	5
3	Jaguar	545000.000000	729000.000000	1	1
4	Rolls Royce	765000.000000	729000.000000	2	2

Figure 2-2. *The data for a simple KPI*

How the Code Works

This T-SQL snippet is quite simple, really; it does the following:

- Prepares a temporary table to hold the output that will be sent to SSRS. This table contains only the essential data used by the KPI. This means the make, the (sales) value, the (budgetary) target, and the status indicator are on a scale of 1-3 and the trend indicator is on a scale of 1-5.
- Groups the sales per make of car for the selected year (2014) up to and including the selected month (June) and adds them to the output table.
- Updates the table with the corresponding figures for the previous year.
- Updates the table with the budget figures for the current year.
- Calculates the status and trend indicators using a predefined and hard-coded logic. This sets the status from 1 through 3 and the trend from 1 through 5 using a predefined business rule that we will imagine has been chosen by the CFO.

■ **Note** I have got into the habit of defining status indicators as 1-3 and trend indicators as 1-5. This is far from the only way of doing this, and many other approaches exist. You could (for status) use -1, 0, and 1, for instance. My only advice is to set a standard and stick to it across all of your KPIs.

Building the KPI

Now that you have your source data, you can get into the fun bit: building the KPI itself. As this is not only the first example in this chapter, but the first in the entire book, I will try to make the explanation reasonably comprehensive so that SSRS novices are not left floundering. In any case (and whatever your level of SSRS knowledge), remember that you can refer back to Figure 2-1 at any time when you are building this KPI if you need to check that you are doing things the right way.

1. Create a new SSRS report named `KPI_Basic.rdl`.
2. Add the shared data source `CarSales_Reports`. Name it `CarSales_Reports` (and not `Datasource1`, which is the default).

3. Add the following four shared datasets (ensuring that you also use the same name for the dataset in the report):

    a. CurrentYear

    b. CurrentMonth

    c. ReportingYear

    d. ReportingMonth

4. Add the parameters ReportingYear and ReportingMonth, and set their properties as defined in Chapter 1.

5. Add a new stand-alone (i.e., not shared) dataset named CarSalesYearToDateKPISimple. Have it use the CarSales_Reports data source and the stored procedure Code.pr_CarSalesYearToDateKPISimple that you saw earlier.

6. Drag a table from the SSRS toolbox onto the report body.

7. Select the table and display the properties window unless it is already visible (pressing F4 is one way to do this; another is to select View ➤ Properties Window).

8. Find the DataSetName property and click the pop-up list to the right. Select the dataset CarSalesYearToDateKPISimple.

9. Select all the textboxes (or cells if you prefer) in the table. Expand BorderStyle in the Properties window and set the Default property to None. This will ensure that the table will only display the borders that you have specifically added.

10. Add two new columns to the table. One way to do this is to right-click a text box anywhere inside a table. You then select Insert Column ➤ Right (or left) from the context menu. You can add these columns anywhere. Your table will now have five columns.

11. Add the following three fields to the Details row of the leftmost three columns, in this order: Make, Sales, SalesBudget. You can either drag the fields into the detail (second) row from the dataset CarSalesYearToDateKPISimple, or click the tiny table symbol that appears when you hover the mouse pointer over a textbox and select the field from the pop-up list.

12. Drag an indicator (from the SSRS toolbox) to the fourth column of the details row. The Select Indicator dialog will appear. Select the three-gray-arrows indicator shown in Figure 2-3.

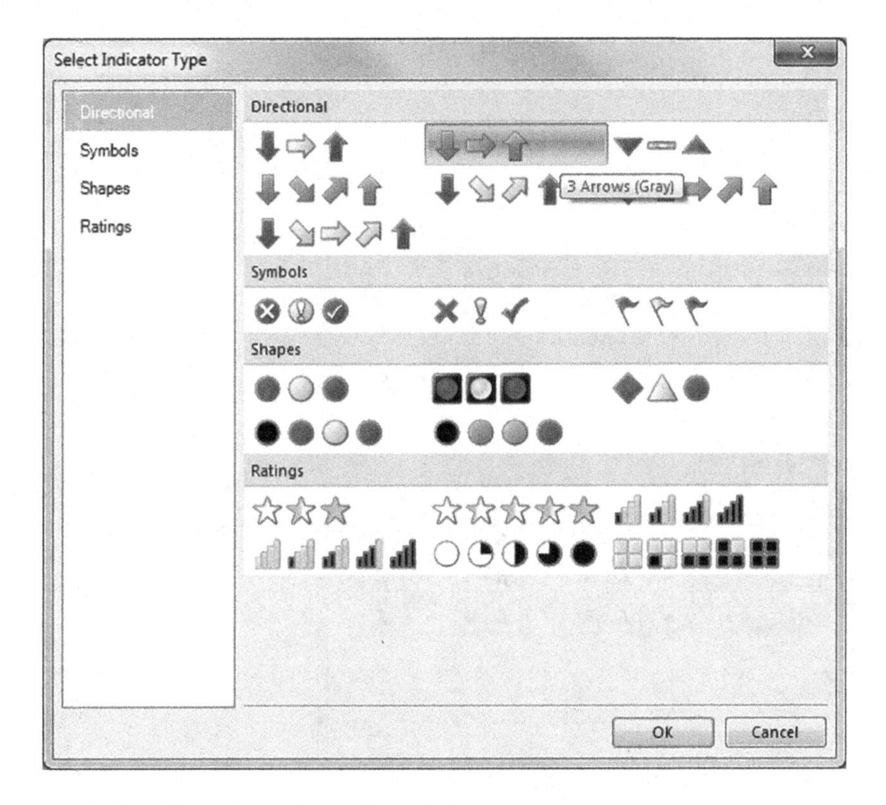

Figure 2-3. *The Select Indicator Type dialog*

13. Click OK.
14. Right-click the indicator that you can now see in the fourth column and select Indicator Properties from the context menu.
15. Click the Values and States option on the left to display the Change Indicator Value pane of the dialog.
16. Set the Value to `[Sum(StatusIndicator)]`.
17. Select Numeric as the States measurement unit.
18. Leaving the icon images as they are, set the following start and end attributes for the three icons (in this order from top to bottom):
    a. Down-facing arrow: Color: Gainsboro, Start and End: 1.
    b. Right-facing arrow: Color: Silver, Start and End: 2.
    c. Up-facing arrow: Color: Dim Gray, Start and End: 3.

The dialog should look like Figure 2-4.

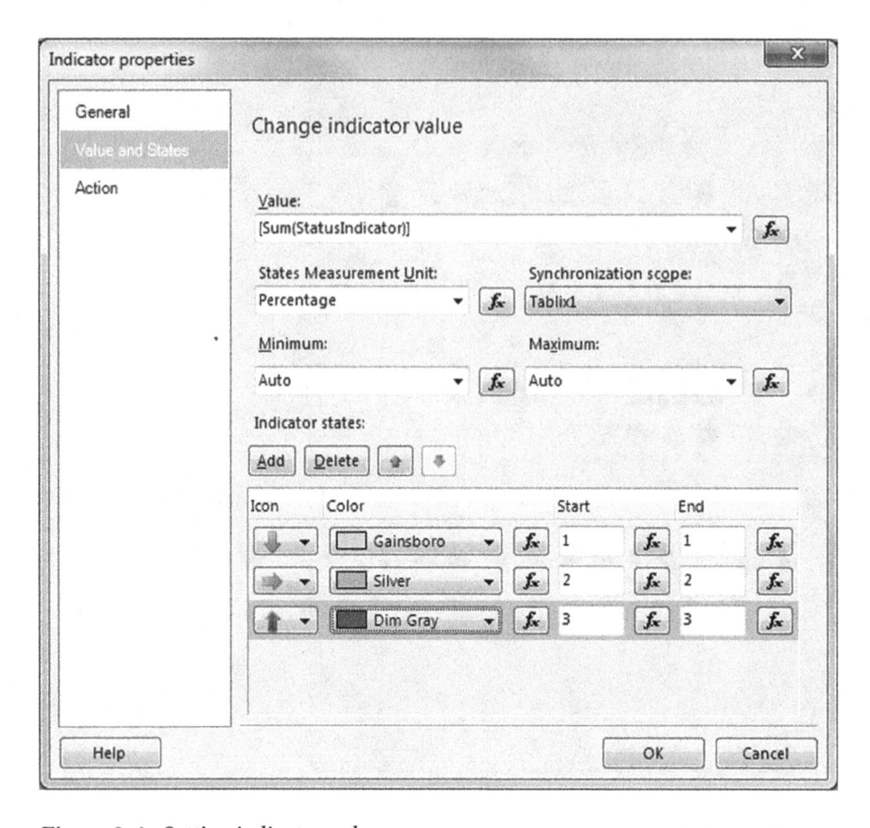

Figure 2-4. *Setting indicator values*

19. Click OK.
20. Drag a second indicator into the fifth column of the detail row. Select the Five Stars rating (it is on the top row of the ratings section in the center) and click OK.
21. Right-click the indicator that you can now see in the fifth column and select Indicator Properties from the context menu. In the Values and States pane, set the following:
    a. Value: `[Sum(TrendIndicator)]`.
    b. States measurement Unit: Numeric.
    c. First Star: Color: Gainsboro, Start and End: 1.
    d. Second Star: Color: Light Gray, Start and End: 2.
    e. Third Star : Color: Silver, Start and End: 3.
    f. Fourth Star: Color: Dark Gray, Start and End: 4.
    g. Fifth Star: Color: Dim Gray, Start and End: 5.
22. Click OK.

23. Add titles to the first row for the fourth and fifth columns of Status and Trend, respectively.

24. Set all the text boxes to the following font attributes. To save time, I suggest that you Ctrl-click on the text boxes one by one (to select a group of text boxes) and then use the Properties window to set the color and font.

    a. Font: Arial

    b. Font Size: 10 pt.

    c. Color: Gray

25. Set the number format the Sales and SalesBudget text boxes in the `Detail` row to Number using the thousands separator and with no decimals (right-click each text box, select Text Box Properties, then the Number pane to do this). Right-align these numbers.

26. Center all the titles and set them to use boldface. Using the toolbar is the quickest way to do this, but you can use the properties window or the Text Box Properties dialog (from the context menu) if you prefer.

27. Italicize the Make textbox in the details row.

28. Add a 2pt gray border to the bottom of the header row.

This is your first KPI. If you select 2014 and 6 (for the year and month parameters) it should look like Figure 2-1 (which you saw at the start of this example) and can be delivered as it is (on a tablet or smartphone if you want) or it can become part of a scorecard containing other KPIs.

■ **Note** You can replace steps 1-4 by using a copy of the report `_BaseReport.rdl`, which is in the sample SSRS project CarSalesReports on the Apress web site. This report contains the datasource, datasets, and parameters that you need for most of the examples in this book.

How It Works

Once the data has been defined, this KPI is largely a question of using the right datasource field in the right way. This means applying the numeric fields to certain text boxes, and then adding indicators to the remaining fields. Then you apply the fields that contain the precalculated status and trend flags to these indicators. As you saw in steps 18 and 21, you finish by telling SSRS which status indicator translates into which indicator shape and color.

The indicators for status and trend were defined in the code, so all SSRS had to do was to attribute the value returned in the data to the corresponding icon for that value. This is why the States measurement unit was set to numeric, so that it would recognize the value in the data. If your data contains a range of values for each status or trend element, then you need to set the corresponding start and end values, and not the same values for each icon as you did here.

I formatted this KPI in a slightly subdued way using grays for borders and text, as this lets the figures and indicators speak without shouting. You are, of course, free to choose the formatting style that you prefer. Explaining the formatting was more a quick revision of formatting techniques than any attempt at imposing a choice of presentation style.

A More Complex KPI Using Sparklines

Now that you have seen how to create a "starter" KPI, let's move on to a more complex example. Specifically, I want now to replace the trend indicator with a sparkline so that you can see the evolution of sales over the last 12 months. Also, I want to make the status indicator into a data bar that indicates the extent of sales and uses color to indicate the status relative to the target.

Figure 2-5 show the kind of visualization that you are trying to create, for August 2013. It shows sales by color of vehicle (let's imagine that the head of marketing requested this).

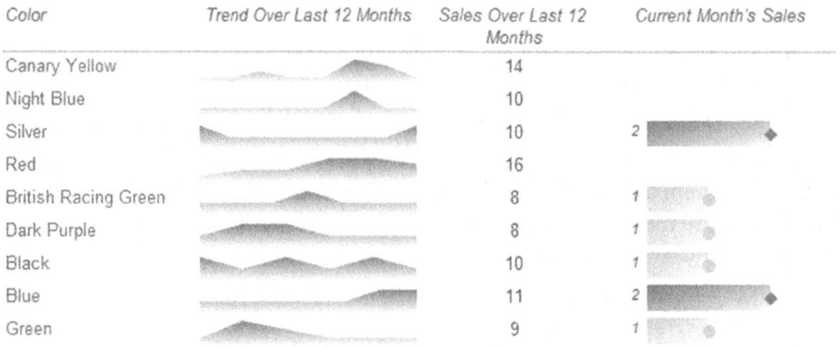

Color	Trend Over Last 12 Months	Sales Over Last 12 Months	Current Month's Sales
Canary Yellow		14	
Night Blue		10	
Silver		10	2
Red		16	
British Racing Green		8	1
Dark Purple		8	1
Black		10	1
Blue		11	2
Green		9	1

Figure 2-5. *A complex KPI using sparklines and data bars*

The Source Data

This KPI cannot really be delivered using a single stored procedure. I found it much easier to use separate stored procedures to feed the necessary data into the KPI. The four 'sprocs are the following:

```
DECLARE @ReportingYear INT = 2013
DECLARE @ReportingMonth TINYINT = 8

-- Code.pr_WD_MonthlyColorSales

IF OBJECT_ID('tempdb..#Tmp_Output') IS NOT NULL DROP TABLE tempdb..#Tmp_Output

CREATE TABLE #Tmp_Output
(
NoSales INT NULL
,Color VARCHAR(50) COLLATE DATABASE_DEFAULT NULL
,SortDate DATE NULL
)

INSERT INTO #Tmp_Output
(
NoSales
,Color
,SortDate
)
```

```
SELECT  NoSales, Color, SortDate
FROM     (
          SELECT     COUNT(Color) AS NoSales
                    ,Color
                    ,CAST(CAST(YEAR(InvoiceDate) AS CHAR(4)) + '-' + RIGHT('0'
                    + CAST(MONTH(InvoiceDate) AS VARCHAR(2)),2) + '-01' AS DATE) AS SortDate
          FROM      Reports.CarSalesData
          GROUP BY  Color
                    ,CAST(CAST(YEAR(InvoiceDate) AS CHAR(4)) + '-' + RIGHT('0'
                    + CAST(MONTH(InvoiceDate) AS VARCHAR(2)),2) + '-01' AS DATE)
                    ,YEAR(InvoiceDate)
                    ,MONTH(InvoiceDate)
        ) A

WHERE  SortDate > DATEADD(mm,-12,CAST(CAST(@ReportingYear AS CHAR(4)) + '-' + RIGHT('0'
       + CAST(@ReportingMonth AS VARCHAR(2)),2) + '-01' AS DATE))
       AND SortDate <= CAST(CAST(@ReportingYear AS CHAR(4)) + '-' + RIGHT('0'
       + CAST(@ReportingMonth AS VARCHAR(2)),2) + '-01' AS DATE)

SELECT * FROM #Tmp_Output ORDER BY SortDate

-- Code.pr_WD_ColorSalesCurrentMonth

CREATE TABLE #Tmp_KPIOutput
(
 ReportingYear INT
,Color NVARCHAR(80) COLLATE DATABASE_DEFAULT
,NoSales NUMERIC(18,6)
,SalesBudget NUMERIC(18,6)
,StatusIndicator SMALLINT
)

INSERT INTO #Tmp_KPIOutput
(
 ReportingYear
,NoSales
,Color
)

SELECT     ReportingYear
          ,COUNT(Color) AS NoSales
          ,Color
FROM      Reports.CarSalesData
WHERE     YEAR(InvoiceDate) = @ReportingYear
          AND MONTH(InvoiceDate) = @ReportingMonth
GROUP BY  ReportingYear, Color

;
WITH Budget_CTE
AS
(
```

```
SELECT      SUM(BudgetValue) AS BudgetValue
           ,BudgetDetail
           ,Year
FROM       Reference.Budget
WHERE      BudgetElement = 'Color'
           AND Year = @ReportingYear
           AND Month = @ReportingMonth
GROUP BY   BudgetDetail
           ,Year
)

UPDATE     Tmp
SET        Tmp.SalesBudget = CTE.BudgetValue
FROM       #Tmp_KPIOutput Tmp
           INNER JOIN Budget_CTE CTE
           ON Tmp.Color = CTE.BudgetDetail
           AND Tmp.ReportingYear = CTE.Year

-- StatusIndicator

UPDATE     #Tmp_KPIOutput
SET        StatusIndicator =
               CASE
                  WHEN ((NoSales - SalesBudget) / NoSales) + 1 <= 0.8 THEN 1
                  WHEN ((NoSales - SalesBudget) / NoSales) + 1  > 1.2 THEN 3
                  WHEN ((NoSales - SalesBudget) / NoSales) + 1  > 0.8
                  AND ((NoSales - SalesBudget) / SalesBudget) <= 1.2 THEN 2
                  ELSE 0
               END

SELECT Color, NoSales, StatusIndicator FROM  #Tmp_KPIOutput

-- Code.pr_WD_ColorSalesInLast12Months

IF OBJECT_ID('tempdb..#Tmp_Output') IS NOT NULL DROP TABLE tempdb..#Tmp_Output

CREATE TABLE #Tmp_Output
(
NoSales INT NULL
,Color VARCHAR(50) COLLATE DATABASE_DEFAULT NULL
)

INSERT INTO #Tmp_Output
(
NoSales
,Color
)
```

```
SELECT   SUM(NoSales) AS NoSales, Color
FROM     (
          SELECT      COUNT(Color) AS NoSales
                      ,Color
                      ,CAST(CAST(YEAR(InvoiceDate) AS CHAR(4)) + '-' + RIGHT('0'
                      + CAST(MONTH(InvoiceDate) AS VARCHAR(2)),2) + '-01' AS DATE) AS
SortDate
          FROM        Reports.CarSalesData

          GROUP BY    Color
                      ,CAST(CAST(YEAR(InvoiceDate) AS CHAR(4)) + '-' + RIGHT('0'
                      + CAST(MONTH(InvoiceDate) AS VARCHAR(2)),2) + '-01' AS DATE)
                      ,YEAR(InvoiceDate)
                      ,MONTH(InvoiceDate)
         ) A

WHERE     SortDate >= DATEADD(mm,-12,CAST(CAST(@ReportingYear AS CHAR(4)) + '-' + RIGHT('0'
          + CAST(@ReportingMonth AS VARCHAR(2)),2) + '-01' AS DATE))
          AND SortDate <= CAST(CAST(@ReportingYear AS CHAR(4)) + '-' + RIGHT('0'
          + CAST(@ReportingMonth AS VARCHAR(2)),2) + '-01' AS DATE)
GROUP BY  Color

SELECT * FROM #Tmp_Output

-- Code.pr_WD_ColorSalesCurrentMonthMAX

SELECT MAX(NoSales) + 1 AS MaxSales
FROM
     (
          SELECT      COUNT(Color) AS NoSales
                     ,Color
          FROM       Reports.CarSalesData
          WHERE      YEAR(InvoiceDate) = @ReportingYear
                     AND MONTH(InvoiceDate) = @ReportingMonth
          GROUP BY    Color
     ) A
```

These stored procedures return the four datasets shown in Figure 2-6 (the monthly color sales are not shown in full):

pr_WD_MonthlyColorSales

	NoSales	Color	SortDate
1	1	Canary Yellow	2012-09-01
2	1	Night Blue	2012-09-01
3	2	Silver	2012-09-01
4	1	Red	2012-10-01
5	1	British Racing Green	2012-10-01
6	1	Canary Yellow	2012-10-01
7	2	Canary Yellow	2012-11-01
8	1	Dark Purple	2012-11-01
9	1	Night Blue	2012-11-01
10	1	Night Blue	2012-12-01
11	1	British Racing Green	2012-12-01
12	2	Black	2012-12-01
13	1	Black	2013-01-01
14	1	Blue	2013-01-01
15	1	Canary Yellow	2013-01-01
16	1	British Racing Green	2013-01-01

pr_WD_ColorSalesInLast12Months

	NoSales	Color
1	10	Black
2	11	Blue
3	8	British Racing Green
4	14	Canary Yellow
5	8	Dark Purple
6	9	Green
7	10	Night Blue
8	16	Red
9	10	Silver

pr_WD_ColorSalesCurrentMonth

	Color	NoSales	StatusIndicator
1	Black	1.000000	2
2	Blue	2.000000	3
3	British Racing Green	1.000000	2
4	Dark Purple	1.000000	2
5	Green	1.000000	2
6	Silver	2.000000	3

pr_WD_ColorSalesCurrentMonthMAX

	MaxSales
1	3

Figure 2-6. *The four datasets needed to produce a complex KPI*

How the Code Works

The four datasets to produce this KPI are the following:

- `Code.pr_WD_MonthlyColorSales`: This is used to generate the sparkline of the sales trend over the last 12 months.
- `Code.pr_WD_ColorSalesCurrentMonth`: This stored procedure gives you the sales for the selected month as well as the trend indicator for the month.
- `Code.pr_WD_ColorSalesInLast12Months`: This data source gives you the cumulated sales for the twelve months up to and including the selected month.
- `Code.pr_WD_ColorSalesCurrentMonthMAX`: This dataset returns the value of sales for the color with the most sales for the month, plus one. This is used to set the scale maximum for the data bar of sales for the month.

It might be perfectly possible to coerce all the data you are using into a single dataset for this KPI. However, reality frequently dictates that it is easiest to use separate datasets and combine them in a single visualization. So I want here to give an idea of how such an approach can be used in practice.

Building the KPI

As this is the second KPI that you are creating in this chapter, I will describe the process a little more succinctly, especially where there are common features shared with the previous example. If you are relatively new to SSRS, you may want to build the previous example before beginning this one.

1. Create a new SSRS report named `_ColorSales.rdl`, and add the shared data source `CarSales_Reports`. Name it `CarSales_Reports`.

2. Add the following four shared datasets (ensuring that you also use the same name for the dataset in the report): `CurrentYear`, `ReportingYear`, `CurrentMonth`, `ReportingMonth`. Set the parameter properties for `ReportingYear` and `ReportingMonth` as defined at the start of Chapter 1.

3. Add four datasets corresponding to the four stored procedures shown above. Name them `MonthlyColorSales` (using `Code.pr_WD_MonthlyColorSales`), `ColorSalesCurrentMonth` (using `Code.pr_WD_ColorSalesCurrentMonth`), `ColorSalesInLast12Months` (using `Code.pr_WD_ColorSalesInLast12Months`) and `ColorSalesCurrentMonthMAX` (using `Code.pr_WD_ColorSalesCurrentMonthMAX`).

4. Add a table from the SSRS toolbox. Set it to use the dataset `MonthlyColorSales`, and add two more columns so that there are five in total. Set the text box border default to None for all cells in the table (as described in the first example).

5. Add the field Color to the first column of the detail row.

6. Add the expressions given below to the Detail row. If you are new to expressions, you can do this by right-clicking the appropriate text box and selecting Expression from the context menu.

    a. Column 3: `=Lookup(Fields!Color.Value, Fields!Color.Value, Fields!NoSales.Value, "ColorSalesInLast12Months")`

    b. Column 4: `=Lookup(Fields!Color.Value, Fields!Color.Value, Fields!NoSales.Value, "ColorSalesCurrentMonth")`

7. Click a text box inside the detail row and then right-click the row selector (the grey square to the left). Select Row Group ➤ Group Properties from the context menu. In the General pane of the Group Properties dialog, click Add to add a group expression and select Color as the field to group on. Click OK to finish setting this property.

8. Drag a sparkline from the SSRS toolbox into the second column of the detail row. Select Area as the sparkline type. This is the leftmost of the area sparklines. Click OK.

9. Click twice on the sparkline to display the Chart Data pane. Add NoSales as the ∑ Values.

10. Click twice on the sparkline to ensure that it is selected (and not the text box that contains it). Right-click the Sparkline and select Series Properties. In the Series Properties dialog, select Fill on the left. In the Fill pane, set the following and then confirm with OK:

    a. Fill style: Gradient

    b. Color: Cornflower Blue

    c. Secondary color: White

    d. Gradient style: Top bottom

11. Drag a data bar into the fifth column of the detail row. Select Bar (this is the leftmost of the data bars) as the sparkline type and click OK.

12. Click twice on the data bar to display the Chart Data pane. Click the plus symbol to the right of the ∑ Values and select Expression from the pop-up. Add the following expression:

```
=Lookup(Fields!Color.Value, Fields!Color.Value, Fields!NoSales.Value, "ColorSalesCurrentMonth")
```

13. Click twice on the data bar, then right-click and select Series Properties from the context menu. Set the following properties:

Section	Property	Value
Markers	Marker type (expression)	=IIF(Lookup(Fields!Color.Value, Fields!Color.Value, Fields!StatusIndicator.Value, "ColorSalesCurrentMonth") = 1, "Square",IIF(Lookup(Fields!Color.Value, Fields!Color.Value, Fields!StatusIndicator.Value, "ColorSalesCurrentMonth") = 2, "Circle","Diamond"))
	Marker size	7pt
	Marker color (expression)	=IIF(Lookup(Fields!Color.Value, Fields!Color.Value, Fields!StatusIndicator.Value, "ColorSalesCurrentMonth") = 1, "LightCoral",IIF(Lookup(Fields!Color.Value, Fields!Color.Value, Fields!StatusIndicator.Value, "ColorSalesCurrentMonth") = 2, "LightGreen","CornflowerBlue"))
	Marker border color (expression)	=IIF(Lookup(Fields!Color.Value, Fields!Color.Value, Fields!StatusIndicator.Value, "ColorSalesCurrentMonth") = 1, "LightCoral",IIF(Lookup(Fields!Color.Value, Fields!Color.Value, Fields!StatusIndicator.Value, "ColorSalesCurrentMonth") = 2, "LightGreen","CornflowerBlue"))
Fill	Fill style	Gradient
	Color	=IIF(Lookup(Fields!Color.Value, Fields!Color.Value, Fields!StatusIndicator.Value, "ColorSalesCurrentMonth") = 1, "Red",IIF(Lookup(Fields!Color.Value, Fields!Color.Value, Fields!StatusIndicator.Value, "ColorSalesCurrentMonth") = 2, "LightGreen","CornflowerBlue"))
	Secondary color	White
	Gradient Style	Diagonal left

14. Add a 2-point light gray border under the header row.
15. Set the following column widths (or something near to this):
    a. Column 1: 1.5in.
    b. Column 2: 1.85in.
    c. Column 4: 0.4in.
16. Select the fourth and fifth cells on the header row and then right-click and select Merge Cells from the context menu. Add the titles you see in Figure 2-5. Set the titles to be in italics.
17. Set all the text boxes to Dim Gray. Center the figures in column three and right-align the figures in column 4. Center the titles in columns 2-4.
18. Right-click the fourth data column and select Text Box Properties. Click Number on the left and choose Custom as the category. Then set the custom format to `#,#`. This will remove decimals and add a thousands separator.

That is your KPI completed. If you preview it, it should look like Figure 2-5.

How It Works

The first point concerning this KPI is that it is centered on the sparkline. This is why the dataset ColorSales is used for the table; it contains the sales per month per color, which are needed to display the sparklines for each color of car sold over the 12-month period. However, this dataset has to group the data by color for the sparkline to display correctly. Hence the need to group the detail row using the Color field. Indeed, if you try to add the sparkline without the grouping element you will get an error message.

With the sparkline in place you can add the figures for the current month's sales and the sales over the last twelve months. However, there is a trick here, too. As the data for each of these metrics does not come from the dataset used by the table, you need to link the core dataset to the dataset used for these metrics. This is done using the `Lookup` function. This function is in four parts, and indicates the following:

a. The field in the dataset used by the table that you want to use as a filter in the other dataset. This is the field `Color`.

b. The field in the other dataset which will map to the field in the table (`Color` again).

c. The field that will be returned from the other dataset.

d. The name of the other dataset.

Lookup fields can be used in many ways, and this KPI shows how you can use them to return values and status indicators. The status indicator is then used to set the color of the data bar for the month's sales and the shape of the marker as well. This is not just a stylistic flourish. It is always recommended to use shape as well as color in status indicators to cater for any color-blind readers of your KPIs.

■ **Note** Using color and shape to indicate status is less traditional (and consequently less soporific) than using classic indicators. However, you may have to educate your users to their meaning-and use this approach across all your BI delivery.

Above all, this KPI has demonstrated that you can use multiple datasets for a single visualization. However, this can require some careful preparation of the source data so that the various datasets can be joined in the KPI.

Gauge-Based KPIs

The aim of a KPI is to show how well your business area is doing. In the previous example, you looked at a tabular KPI, where the figures delivered the information. However, figures are not the only way of presenting high-level information. SSRS includes a wealth of gauges to display data, and gauges can be adapted extremely easily to show values, target status, and trend.

If you are using gauges to present your KPIs, then it is probably worth seeing the advantages and drawbacks of each approach. Table 2-1 resumes the essential aspects of both tabular and gauge-based KPIs.

Table 2-1. *Tabular vs. Gauge-Based KPIs*

Tabular KPIs	Gauge-Based KPIs
Multiple elements per page	Intuitive comparison over multiple elements
Can present lots of information	Can take longer to build
Can be rolled up and drilled down like any table or matrix	Nearly always take up more space
Easy to overload and obscure key data	Fewer gauges than rows in a table
	Harder to maintain
	Greater "wow" factor

Suppose you want to create a series of gauge-based KPIs that look like Figure 2-7.

Figure 2-7. Gauge-based KPIs

Each gauge shows the sales (the bar) and budget (the marker) for a make of car as well as the status (using the bar color). They also have a trend indicator in the center of the gauge. Once an initial gauge has been created, it is simply copied and pasted (and the country filter tweaked) to show the data for several different makes of car.

There are a couple of subtleties that I want to make clear before you start configuring these gauge-bases KPIs:

- The gauge uses two pointer types, a bar for the value and a marker for the target.
- Status is displayed using the color of the bar pointer.

So core KPI elements are nonetheless present. They are, however, presented differently from a more traditional KPI.

The Source Data

These gauges only need a single source dataset. It is on the source database as Code.pr_YearlyCarSalesKPIGauges. Although the T-SQL is nearly identical to that used in the first example in this chapter, I will nonetheless reproduce it all here as there are enough differences to justify this. The code output and the gauges show the data for May 2013.

```
IF OBJECT_ID('Tempdb..#Tmp_KPIOutput') IS NOT NULL DROP TABLE Tempdb..#Tmp_KPIOutput

CREATE TABLE #Tmp_KPIOutput
(
ReportingYear INT
,Make NVARCHAR(80) COLLATE DATABASE_DEFAULT
,Sales NUMERIC(18,6)
,SalesBudget NUMERIC(18,6)
,PreviousYear NUMERIC(18,6)
,StatusIndicator SMALLINT
,TrendIndicator SMALLINT
,ScaleMax INT
)

INSERT INTO #Tmp_KPIOutput
(
ReportingYear
,Make
,Sales
)

SELECT          ReportingYear
               ,Make
               ,SUM(SalePrice)
FROM           Reports.CarSalesData
WHERE          ReportingYear = @ReportingYear
               AND ReportingMonth <= @ReportingMonth
GROUP BY        ReportingYear
               ,Make

-- Previous Year Sales
;
WITH SalesPrev_CTE
AS
(
SELECT         ReportingYear
              ,Make
              ,SUM(SalePrice) AS Sales
```

```
FROM         Reports.CarSalesData
WHERE        ReportingYear = @ReportingYear - 1
             AND ReportingMonth <= @ReportingMonth
GROUP BY     ReportingYear
             ,Make
)

UPDATE      Tmp
SET         Tmp.PreviousYear = CTE.Sales
FROM        #Tmp_KPIOutput Tmp
            INNER JOIN SalesPrev_CTE CTE
            ON Tmp.Make = CTE.Make

;
WITH Budget_CTE
AS
(
SELECT      SUM(BudgetValue) AS BudgetValue
            ,BudgetDetail
            ,Year
FROM        Reference.Budget
WHERE       BudgetElement = 'Sales'
            AND Year = @ReportingYear
            AND Month <= @ReportingMonth
GROUP BY    BudgetDetail
            ,Year
)

UPDATE      Tmp
SET         Tmp.SalesBudget = CTE.BudgetValue
FROM        #Tmp_KPIOutput Tmp
            INNER JOIN Budget_CTE CTE
            ON Tmp.Make = CTE.BudgetDetail
            AND Tmp.ReportingYear = CTE.Year

-- Scale maximum

UPDATE   #Tmp_KPIOutput
SET      ScaleMax =
                   CASE
                        WHEN Sales >= SalesBudget
                        THEN (SELECT CarSales_Reports.Code.fn_ScaleDecile (Sales))
                        ELSE (SELECT CarSales_Reports.Code.fn_ScaleDecile (SalesBudget))
                   END

-- Internal Calculations
-- Year on Year Delta

-- TrendIndicator
```

```
UPDATE   #Tmp_KPIOutput
SET      TrendIndicator = CASE
                          WHEN ((Sales - PreviousYear) / Sales) + 1 <= 0.7 THEN 1
                          WHEN ((Sales - PreviousYear) / Sales) + 1  > 1.3 THEN 5
                          WHEN ((Sales - PreviousYear) / Sales) + 1  > 0.7
                          AND ((Sales - PreviousYear) / PreviousYear) <= 0.9 THEN 2
                          WHEN ((Sales - PreviousYear) / Sales) + 1  > 1.1
                          AND ((Sales - PreviousYear) / PreviousYear) <= 1.3 THEN 4
                          WHEN ((Sales - PreviousYear) / Sales) + 1  > 0.9
                          AND ((Sales - PreviousYear) / PreviousYear) <= 1.1 THEN 3
                          ELSE 0
                          END

-- StatusIndicator

UPDATE    #Tmp_KPIOutput
SET       StatusIndicator = CASE
                            WHEN ((Sales - SalesBudget) / Sales) + 1 <= 0.8 THEN 1
                            WHEN ((Sales - SalesBudget) / Sales) + 1  > 1.2 THEN 3
                            WHEN ((Sales - SalesBudget) / Sales) + 1  > 0.8
                            AND ((Sales - SalesBudget) / SalesBudget) <= 1.2 THEN 2
                            ELSE 0
                            END

-- Output

SELECT    Make, Sales, SalesBudget, StatusIndicator, TrendIndicator, ScaleMax
FROM      #Tmp_KPIOutput
```

The T-SQL used in this example returns the data shown in Figure 2-8.

	Make	Sales	SalesBudget	StatusIndicator	TrendIndicator	ScaleMax
1	Aston Martin	1095090.000000	600000.000000	3	5	2000000
2	Bentley	410750.000000	600000.000000	1	2	600000
3	Jaguar	670000.000000	600000.000000	2	5	700000
4	MGB	149250.000000	NULL	0	0	NULL
5	Rolls Royce	824250.000000	600000.000000	3	5	900000
6	Triumph	115250.000000	NULL	0	0	NULL
7	TVR	71000.000000	NULL	0	0	NULL

Figure 2-8. *The data for a collection of KPI gauges*

How the Code Works

As I said previously, this code is largely similar to the first code block at the start of this chapter. It uses a temporary table to hold the sales by make for a specific period, adds the budget figures and previous year's figures, and then calculates the status and trend calculations.

This addition to the code uses a tiny "helper" function that you can find in the `CarSales_Reports` database. It rounds up the maximum value of the gauge scale to the nearest decile, which makes the scale easier and more pleasant to read.

Building the Gauge-Based KPI

Now that you have the data you need, you can start by building the first gauge that will be the model for all the gauges in this composite KPI. If this example seems a little complex, feel free to look ahead at Chapter 3 first, as it explains more of the techniques used when creating gauge-based KPIs.

1. Create a new SSRS report named `KPI_Gauges.rdl`, and add the shared data source `CarSales_Reports`. Name it `CarSales_Reports`.
2. Create a dataset named `BudgetComparisons`. Set it to use the `CarSales_Reports` data source and the query `Code.pr_ScorecardCostsGauges`. Yes, I am making the dataset name different from the stored procedure name for once.
3. Create a dataset named `YearlyCarSalesKPIGauges`. Set it to use the `CarSales_Reports` data source and the query `Code.pr_YearlyCarSalesKPIGauges`.
4. Add the following four shared datasets (ensuring that you also use the same name for the dataset in the report): `CurrentYear`, `ReportingYear`, `CurrentMonth`, `ReportingMonth`. Set the parameter properties for `ReportingYear` and `ReportingMonth` as defined at the start of Chapter 1.
5. Drag a gauge from the toolbox onto the report surface. Select Radial (the top left gauge) as the gauge type.
6. Right-click the range (the red area) on the right of the gauge and select Delete Range from the context menu.
7. Right-click just outside the gauge-but inside the gauge bounding box-and select Gauge Panel Properties from the context menu. Select `YearlyCarSalesKPIGauges` as the dataset name.
8. Select Filters on the left of the Gauge Panel Properties dialog and click Add to add a filter. Set the Expression to Make the operator to = (equals) and the Value to Rolls Royce, then click OK.
9. Right-click the scale, select Scale Properties from the context menu, and set the following properties:

Section	Property	Value
General	Minimum	0
	Maximum	`Sum([ScaleMax])`
Layout	Scale Radius (percent)	39
	Start angle (degree)	20
	Sweep angle (degree)	320
	Scale bar width (percent)	0
Labels	Placement (relative to scale)	Cross
	Distance from scale (percent)	0
Label Font	Font	Arial
	Size	11 point
	Color	Dark Blue
	Bold	Checked
Number	Category	Number
	Use 1000 separator (,)	Checked
	Decimal places	2
Major Tick Marks		Check Hide major tick marks
Minor Tick Marks		Uncheck Hide minor tick marks
	Minor tick mark shape	Diamond
	Width (percent)	1
	Minor tick mark placement	Cross
	Length (percent)	2

10. Right-click the pointer (the needle) and select Pointer Properties from the context menu. Set the following properties:

Section	Property	Value
Pointer Options	Value	`[Sum(Sales)]`
	Pointer type	Bar
	Placement (relative to scale)	Inside
	Distance from scale (percent)	15
	Width (percent)	12
Pointer Fill	Fill style	Gradient

	Color	`=Switch(`
		`Sum(Fields!StatusIndicator.Value) = 1, "Red"`
		`,Sum(Fields!StatusIndicator.Value) = 2, "DarkGray"`
		`,Sum(Fields!StatusIndicator.Value) = 3, "Blue"`
		`)`

	Secondary color	White
	Gradient style	Vertical center
Cap Options	Hide pointer cap	Checked
Shadow	Shadow Offset	1 Point
	Shadow intensity	25

11. Right-click inside the gauge and select Add Pointer from the context menu. Right-click the newly added pointer (RadialPointer2) and set the new pointer properties to the following values:

Section	Property	Value
Pointer Options	Value	`[Sum(SalesBudget)]`
	Pointer type	Marker
	Marker style	Diamond
	Placement (relative to scale)	Cross
	Distance from scale (percent)	15
	Width (percent)	15
	Length	25
Pointer Fill	Fill style	Gradient
	Color	Black
	Secondary color	Silver
	Gradient style	Left right
Shadow	Shadow Offset	1 point
	Shadow intensity	25

12. Right-click inside the gauge and select Add Indicator ➤ Child from the context menu. Select 5 Ratings (the bottom left collection of icons) and click OK.

13. Right-click the indicator and select Indicator Properties from the context menu. Set the following properties:

Section	Property	Value
General	X position (percent)	40
	Y position (percent)	36
	Width (percent)	25
	Height (percent)	25
Values and states	Value	[Sum(TrendIndicator)]
	States Measurement Unit	Numeric
	Icon 1	Color: Black, Start and End: 1
	Icon 2	Color: Black, Start and End: 2
	Icon 3	Color: Black, Start and End: 3
	Icon 4	Color: Black, Start and End: 4
	Icon 5	Color: Black, Start and End: 5

14. Right-click the gauge and select Gauge Properties from the context menu. Set the following properties:

Section	Property	Value
Back Fill	Color	White
Frame	Style	Edged
	Shape	Circular 1
Frame Fill	Color	WhiteSmoke

Your gauge is now complete, and should resemble the top left gauge in Figure 2-7. It is now easy to copy the gauge as many times as there are makes of cars in the dataset and set the filter for each gauge to a different make of car to produce a visualization that covers the entire range of products sold. You can add text boxes containing the names of the car makes as a final flourish.

How It Works

This KPI uses two gauge pointers to display the sales figure and the corresponding budget. The visual trick is to use two different pointer types so that the marker (the diamond) indicates the target value. The status is indicated by the color of the bar pointer using the Switch code shown in step 10. The trend indicator is just another trend indicator chosen from those available in SSRS-but it is added as a child indicator of the main gauge.

In this example, the maximum value for each scale varies according to the make displayed in the gauge. This means that the gauges are not standardized on a single scale-and consequently that the data in the gauges is not strictly comparable. In practice, this may not be what your users are expecting, and so you may have to indicate this to them in some way.

Hints and Tips

Here are some tips and tricks to keep in mind.

- Once again, the choice of colors is up to you. I suggest that you try to let the status colors stand out against the rest of the gauge or the visual indication will be drowned out.
- You have a wide choice of pointer types in gauges, so try them out until you find the one that projects the effect you want.
- Ensure that you place the trend indicator to avoid clashing with the pointer and obscuring the visibility of one-or both.

Text-Based KPIs

The previous sections showed you what a KPI is and what it can deliver. Sometimes, however, you may need to produce more complex KPIs, where several pieces of information are presented to the user as a coherent whole. This is where a well-constructed text-based KPI can be truly useful. Through the use of font size, color, and position, you can present several essential pieces of data to a user, as well as provide an idea of how the information is to be prioritized.

A Simple Text-Based KPI

To show you that KPIs that do *not* have indicators in them can be just as powerful as "classic" KPIs, let's look at a first example of a text-based KPI. Suppose that you want to display the value for a collection of key enterprise metrics. For each one you want to show the percentage attainment of the target-which will display the metric's status through changing the color of the percentage. You then want to display the trend as a sparkline. Here is how it can be done.

Figure 2-9 shows a text-based KPI. This KPI relies on color to convey the status of certain metrics; this is less obvious in a black and white book than on a color screen, so you may need to preview the report in the sample application to get the full effect.

Figure 2-9. *A simple text-based KPI*

The Source Data

This particular KPI requires six datasets: one for each sparkline (though they could, admittedly, be refactored as a single dataset) and another dataset for the grouped metrics and targets for each cost area. This dataset will then be filtered to use the appropriate figures for each component of the visualization. The stored procedures in the sample database that produce the data are the following (for May 2015):

```
DECLARE @ReportingYear INT = 2015
DECLARE @ReportingMonth TINYINT = 5

-- Code.pr_WD_KeyFiguresForMonthLaborSparkline
```

```
SELECT
ReportingMonth
,SUM(LaborCost) AS MetricValue
FROM            Reports.CarSalesData
WHERE           ReportingYear = @ReportingYear
                AND ReportingMonth <= @ReportingMonth
GROUP BY        ReportingMonth

-- Code.pr_WD_KeyFiguresForMonthDeliverySparkline

SELECT
ReportingMonth
,SUM(DeliveryCharge) AS MetricValue
FROM            Reports.CarSalesData
WHERE           ReportingYear = @ReportingYear
                AND ReportingMonth <= @ReportingMonth
GROUP BY        ReportingMonth

-- Code.pr_WD_KeyFiguresForMonthDiscountSparkline

SELECT
ReportingMonth
,SUM(TotalDiscount) AS MetricValue
FROM            Reports.CarSalesData
WHERE           ReportingYear = @ReportingYear
                AND ReportingMonth <= @ReportingMonth
GROUP BY        ReportingMonth

-- Code.pr_WD_KeyFiguresForMonthPartsSparkline

SELECT
ReportingMonth
,SUM(SpareParts) AS MetricValue
FROM            Reports.CarSalesData
WHERE           ReportingYear = @ReportingYear
                AND ReportingMonth <= @ReportingMonth
GROUP BY        ReportingMonth

-- Code.pr_WD_KeyFiguresForMonthSalesSparkline

SELECT
ReportingMonth
,SUM(SalePrice) AS MetricValue
FROM            Reports.CarSalesData
WHERE           ReportingYear = @ReportingYear
                AND ReportingMonth <= @ReportingMonth
GROUP BY        ReportingMonth

-- Code.pr_WD_KeyFiguresForMonth

IF OBJECT_ID('tempdb..#Tmp_Output') IS NOT NULL DROP TABLE tempdb..#Tmp_Output
```

```
CREATE TABLE #Tmp_Output
(
 Sales NUMERIC(18,2)
,SalesTarget NUMERIC(18,2)
,Parts NUMERIC(18,2)
,PartsTarget NUMERIC(18,2)
,Labor NUMERIC(18,2)
,LaborTarget NUMERIC(18,2)
,Delivery NUMERIC(18,2)
,DeliveryTarget NUMERIC(18,2)
,Discount NUMERIC(18,2)
,DiscountTarget NUMERIC(18,2)
,SalesStatus TINYINT
,PartsStatus TINYINT
,LaborStatus TINYINT
,DeliveryStatus TINYINT
,DiscountStatus TINYINT
)

-- Sales

INSERT INTO #Tmp_Output (Sales)

SELECT          SUM(SalePrice)
FROM            Reports.CarSalesData
WHERE           ReportingYear = @ReportingYear
                AND ReportingMonth <= @ReportingMonth

UPDATE      #Tmp_Output
SET     SalesTarget =
               (
               SELECT    SUM(BudgetValue)
               FROM      Reference.Budget
               WHERE     BudgetElement = 'Sales'
                         AND Year = @ReportingYear
                         AND Month <= @ReportingMonth
               )

-- Parts

UPDATE      #Tmp_Output
SET     Parts =
               (
               SELECT         SUM(SpareParts)
               FROM           Reports.CarSalesData
               WHERE          ReportingYear = @ReportingYear
                              AND ReportingMonth <= @ReportingMonth
               )
```

```
UPDATE     #Tmp_Output
SET     PartsTarget =
              (
              SELECT          SUM(BudgetValue)
              FROM            Reference.Budget
              WHERE           BudgetElement = 'Costs'
                              AND BudgetDetail = 'Parts'
                              AND Year = @ReportingYear
                              AND Month <= @ReportingMonth
              )

-- Labor

UPDATE     #Tmp_Output
SET     Labor =
              (
              SELECT          SUM(LaborCost)
              FROM            Reports.CarSalesData
              WHERE           ReportingYear = @ReportingYear
                              AND ReportingMonth <= @ReportingMonth
              )

UPDATE     #Tmp_Output
SET     LaborTarget =
              (
              SELECT            SUM(BudgetValue)
              FROM            Reference.Budget
              WHERE             BudgetElement = 'Costs'
                          AND BudgetDetail = 'Labor'
                          AND Year = @ReportingYear
                          AND Month <= @ReportingMonth
              )

-- Delivery Charge

UPDATE     #Tmp_Output
SET     Delivery =
              (
              SELECT          SUM(DeliveryCharge)
              FROM            Reports.CarSalesData
              WHERE           ReportingYear = @ReportingYear
                              AND ReportingMonth <= @ReportingMonth
              )

UPDATE     #Tmp_Output
SET     DeliveryTarget =
              (
              SELECT          SUM(BudgetValue)
              FROM            Reference.Budget
              WHERE           BudgetElement = 'Costs'
                              AND BudgetDetail = 'DeliveryCharge'
                              AND Year = @ReportingYear
                              AND Month <= @ReportingMonth
              )
```

```
-- Discount

UPDATE          #Tmp_Output
SET         Discount =
              (
              SELECT        SUM(TotalDiscount)
              FROM          Reports.CarSalesData
              WHERE         ReportingYear = @ReportingYear
                            AND ReportingMonth <= @ReportingMonth
              )

UPDATE     #Tmp_Output
SET     DiscountTarget =
              (
              SELECT        SUM(BudgetValue)
              FROM          Reference.Budget
              WHERE         BudgetElement = 'Costs'
                            AND BudgetDetail = 'TotalDiscount'
                            AND Year = @ReportingYear
                            AND Month <= @ReportingMonth
              )

-- StatusIndicators

UPDATE     #Tmp_Output
SET     SalesStatus =
              CASE
              WHEN ((Sales - SalesTarget) / Sales) + 1 <= 0.8 THEN 1
              WHEN ((Sales - SalesTarget) / Sales) + 1  > 1.2 THEN 3
              WHEN ((Sales - SalesTarget) / Sales) + 1  > 0.8
                   AND ((Sales - SalesTarget) / SalesTarget) <= 1.2 THEN 2
              ELSE 0
              END

UPDATE     #Tmp_Output
SET     PartsStatus =
              CASE
              WHEN ((Parts - PartsTarget) / Parts) + 1 > 1.2 THEN 1
              WHEN ((Parts - PartsTarget) / Parts) + 1 <= 0.8  THEN 3
              WHEN ((Parts - PartsTarget) / Parts) + 1  > 0.8
                   AND ((Parts - PartsTarget) / PartsTarget) <= 1.2 THEN 2
              ELSE 0
              END

UPDATE      #Tmp_Output
SET     LaborStatus =
              CASE
              WHEN ((Labor - LaborTarget) / Labor) + 1  > 1.2 THEN 1
              WHEN ((Labor - LaborTarget) / Labor) + 1 <= 0.8 THEN 3
              WHEN ((Labor - LaborTarget) / Labor) + 1  > 0.8
```

```
                    AND ((Labor - LaborTarget) / LaborTarget) <= 1.2 THEN 2
               ELSE 0
               END

UPDATE      #Tmp_Output
SET     DeliveryStatus =
               CASE
               WHEN ((Delivery - DeliveryTarget) / Delivery) + 1 > 1.2 THEN 1
               WHEN ((Delivery - DeliveryTarget) / Delivery) + 1 <= 0.8  THEN 3
               WHEN ((Delivery - DeliveryTarget) / Delivery) + 1  > 0.8
                    AND ((Delivery - DeliveryTarget) / DeliveryTarget) <= 1.2 THEN 2
               ELSE 0
               END

UPDATE      #Tmp_Output
SET     DiscountStatus =
               CASE
               WHEN ((Discount - DiscountTarget) / Discount) + 1 <= 0.8 THEN 1
               WHEN ((Discount - DiscountTarget) / Discount) + 1  > 1.2 THEN 3
               WHEN ((Discount - DiscountTarget) / Discount) + 1  > 0.8
                    AND ((Discount - DiscountTarget) / DiscountTarget) <= 1.2 THEN 2
               ELSE 0
               END

-- Output

SELECT * FROM #Tmp_Output
```

The six outputs are shown in Figure 2-10.

pr_WD_KeyFiguresForMonthDeliverySparkline

	ReportingMonth	MetricValue
1	1	12000.00
2	2	7725.00
3	3	10455.00
4	4	4400.00
5	5	8775.00

pr_WD_KeyFiguresForMonthDiscountSparkline

	ReportingMonth	MetricValue
1	1	6250.00
2	2	4100.00
3	3	9900.00
4	4	12000.00
5	5	11545.01

pr_WD_KeyFiguresForMonthLaborSparkline

	ReportingMonth	MetricValue
1	1	12082.00
2	2	12860.00
3	3	15733.00
4	4	13677.00
5	5	15026.00

pr_WD_KeyFiguresForMonth

	Sales	SalesTarget	Parts	PartsTarget	Labor	LaborTarget	Delivery	DeliveryTarget	Discount	DiscountTarget	SalesStatus	PartsStatus	LaborStatus	DeliveryStatus	DiscountStatus
1	6857300.00	3645000.00	81150.00	25000.00	69378.00	12500.00	43355.00	17500.00	43795.01	17500.00	3	1	1	1	3

pr_WD_KeyFiguresForMonthPartsSparkline

	ReportingMonth	MetricValue
1	1	17600.00
2	2	14100.00
3	3	17470.00
4	4	14990.00
5	5	16990.00

pr_WD_KeyFiguresForMonthSalesSparkline

	ReportingMonth	MetricValue
1	1	1395750.00
2	2	1375500.00
3	3	1461000.00
4	4	1158850.00
5	5	1466200.00

Figure 2-10. *The data used to create a text-based KPI*

How the Code Works

These pieces of T-SQL are quite a bit more complex than most of those that you have seen so far. This reflects the fact that several steps are necessary to collate and extract all the necessary data. This is what happens:

- The five small "sparkline" stored procedures collect the values for the five metrics that are used by the KPI (LaborCost, TotalDiscount, DeliveryCharge, SalePrice, and SpareParts) for each month of the selected year up until the month set in the ReportingMonth parameter.
- The stored procedure pr_WD_KeyFiguresForMonth may be long, but it simply collates the value and target for the five metrics for the selected year up to and including the selected month in a single dataset. Each of the 10 metrics needs a separate process to return the data, so a temporary table is probably the best solution. The status of each metric is then calculated in a way similar to that used in previous examples in this chapter using hard-coded thresholds.

Building a Text-Based KPI

Although this KPI is designed to be read as the sum of its parts, you only really need to learn how to make one of the text blocks. Once an initial KPI is finished, it can be duplicated and then the dataset filtered and any expressions tweaked to display the relevant data.

Here is how to create a text-based KPI. You will build the Sales KPI as the initial model.

1. Create a new SSRS report named KeyFiguresForMonth.rdl, and add the shared data source CarSales_Reports. Name it CarSales_Reports.
2. Add the following four shared datasets (ensuring that you also use the same name for the dataset in the report): CurrentYear, ReportingYear, CurrentMonth, ReportingMonth. Set the parameter properties for ReportingYear and ReportingMonth as defined at the start of Chapter 1.
3. Add six datasets corresponding to the six stored procedures shown previously. Name them KeyFiguresForMonth (using Code.pr_WD_KeyFiguresForMonth) SalesSparkline (using Code.pr_WD_KeyFiguresForMonthSalesSparkline), DiscountSparkline (using Code.pr_WD_KeyFiguresForMonthDiscountSparkline), PartsSparkline (using Code.pr_WD_KeyFiguresForMonthPartsSparkline), LaborSparkline (using Code.pr_WD_KeyFiguresForMonthLaborSparkline) and DeliverySparkline (using Code.pr_WD_KeyFiguresForMonthDeliverySparkline).
4. Add a table to the SSRS report canvas. Set the SalesSparkline as its dataset and remove all borders as described previously. Delete the Detail row by clicking a text box to display the gray grid and clicking the grey square to the left of the row. Then right-click and select Delete Row, and confirm that you want to do this.
5. Add rows and columns until your table is composed of five rows and four columns.
6. The top and bottom rows and leftmost and rightmost columns will be used only as spacers and will not contain any data. So you need to make them as narrow as possible-a couple of millimeters at the most. You also need to select the textboxes in the top and bottom rows and set the font height to 2 points; this prevents the rows growing vertically when displaying the report.

7. Select the two center cells in the third row and merge them. Do the same for the two center cells on the fourth row.

8. Set column 2 to be 2 inches wide, and column 3 to be 1 inch wide.

9. Set row 2 to be 0.385 inches high, row 3 to be 0.5 inches high, and row 4 to be 0.25 inches high.

10. Drag a sparkline to the merged cells in the fourth row. Select Column as the sparkline type (the top left sparkline). Click the sparkline to display the Chart Data pane, then select MetricValue as the ∑ Value. Click the pop-up triangle to the right of (Details) in the Category Groups section and select ReportingMonth.

11. Click twice on any of the data bars. Then right-click and select Series Properties. In the Fill pane, set the following properties:

    a. Fill style: Gradient

    b. Color: Dark Blue

    c. Secondary color: White

    d. Gradient style: Vertical Center

12. Click in the two merged cells in row 3 and add the following expression: `=Sum(Fields!Sales.Value, "KeyFiguresForMonth")`. Set the font to Cambria 22pt black, and center the textbox contents horizontally and vertically. Set the number format to the following custom format: #,#.

13. Set the background color for the two rightmost cells in row 2 to Black.

14. Add the following expression to the third cell in row 2: `=Sum(Fields!Sales.Value, "KeyFiguresForMonth") / Sum(Fields!SalesTarget.Value, "KeyFiguresForMonth")`. Set the font to Arial 14pt, and right-align the text box contents. Set the number format to percentage with two decimals. Set the vertical alignement of the text box to Top.

15. Set the Color property (this will be the text color) for this same cell to the following expression:

```
=Switch(
Sum(Fields!SalesStatus.Value, "KeyFiguresForMonth")=1,"Red",
Sum(Fields!SalesStatus.Value, "KeyFiguresForMonth")=2,"Gray",
Sum(Fields!SalesStatus.Value, "KeyFiguresForMonth")=3,"White"
)
```

16. Add a 1-point black border to the top and left outside border of the table, and a 2-point black border to the bottom and right outside border of the table.

17. In the second cell of the second row, add the text: Sales. Set it to Arial Black 14 point. Leave it left-aligned. Set the vertical alignment of the text box to Top.

You now have your initial KPI. You can now make four copies of the KPI and set the dataset for each to one of the remaining datasets: `Discount Sparkline`, `Parts Sparkline`, `Labor Sparkline`, and `Delivery Sparkline`. As there are no filters applied, you will have to tweak the expressions that are used to show the key metric, the percentage, and the status from the `KeyFiguresForMonth` dataset. In practice, this means replacing the field in step 15 with the field required by the specific KPI. So, for example, in the table that displays Parts KPIs you need to change `SalesStatus.Value` to `PartsStatus.Value`. Similar changes are required in steps 12 and 14.

How It Works

This KPI shows that you can use multiple datasets to populate KPIs. However, it can necessitate some forethought and a little work. At its heart, this KPI is similar to the KPI using a sparkline that was the second KPI example in this chapter. However, each table only shows one set of data; there is no data grouping. The metrics and status indicator are specific to each KPI, and rather than have multiple datasets, they are returned in a single dataset that contains a single row where it is the column that defines the metric. Consequently, when writing the functions to return a value from a dataset that is not linked to the table, you have to specify the column that will be used, as well as the dataset.

Step 10 sets the sparkline properties. You need to set the Category Groups so that the month data will appear on the horizontal axis.

The threshold indicator in this example is simply the text color. While not difficult, this can be effective-if it is not overused. This is why I left the sales figure in black. More generally, in this KPI I have chosen a slightly "harder" color scheme, trading grey for black and primary colors for pastel shades.

The table structure is essentially part of the presentation. However, I think that this part of the KPI development should not be overlooked, as it is the visual quality that will make your KPI memorable. So be prepared to spend some time tweaking row heights (and even the font sizes for blank cells) as well as column widths to add spacing to your table, both outside the central data and also inside the table.

A Complex Text-Based KPI

There are occasions when merely having the value, target, status, and trend for a metric are not enough. You may need to display the value for previous periods, such as the preceding month or the same month last year. All of these figures need to show the percentage difference and a status indicator. It is when faced with this level of complexity that a text-based KPI can be really effective. To give you an idea of what you are looking for, take a glance at Figure 2-11.

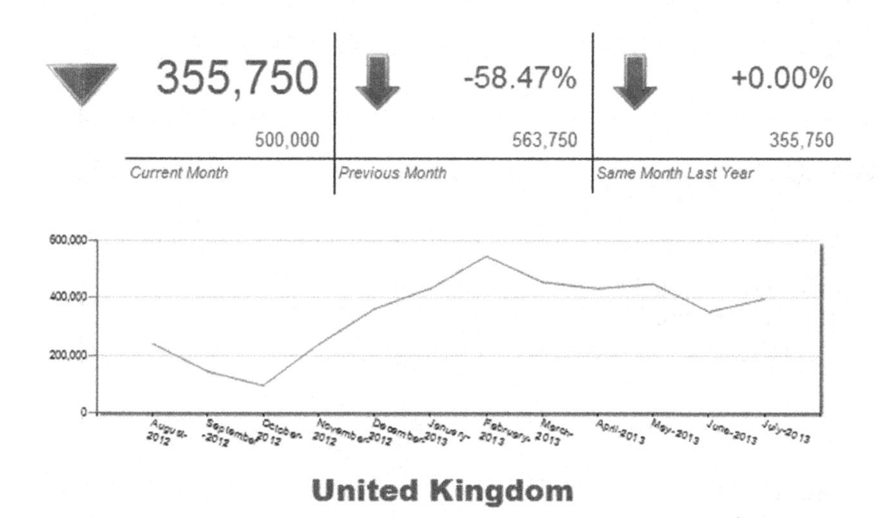

Figure 2-11. *A complex text-based KPI*

The Source Data

This KPI needs two datasets. They are Code.pr_ScorecardTimeCountryAndMake and Code.pr_ScorecardTimeCountryAndMake12MonthSales in the CarSales_Reports database. The code for both is as follows:

```
DECLARE @ReportingYear INT = 2015
DECLARE @ReportingMonth TINYINT = 6

-- Code.pr_ScorecardTimeCountryAndMake

IF OBJECT_ID('Tempdb..#Tmp_KPIOutput') IS NOT NULL DROP TABLE Tempdb..#Tmp_KPIOutput

CREATE TABLE #Tmp_KPIOutput
(
Country NVARCHAR(100) COLLATE DATABASE_DEFAULT
,Sales NUMERIC(18,6)
,SalesTarget NUMERIC(18,6)
,PrevMonthSales NUMERIC(18,6)
,PrevYearSales NUMERIC(18,6)
,Delta NUMERIC(18,6)
,PrevMonthDelta NUMERIC(18,6)
,PrevYearDelta NUMERIC(18,6)
,DeltaPercent NUMERIC(18,6)
,PrevMonthDeltaPercent NUMERIC(18,6)
,PrevYearDeltaPercent NUMERIC(18,6)
,SalesStatus TINYINT
,PrevMonthStatus TINYINT
,PrevYearStatus TINYINT
)

-- Sales

INSERT INTO #Tmp_KPIOutput (Country, Sales)

SELECT
CASE
WHEN CountryName = 'United Kingdom' THEN 'United Kingdom'
WHEN CountryName = 'France' THEN 'France'
WHEN CountryName = 'Switzerland' THEN 'Switzerland'
ELSE 'Other'
END
,SUM(SalePrice)

FROM       Reports.CarSalesData
WHERE      ReportingYear = @ReportingYear
           AND ReportingMonth = @ReportingMonth
GROUP BY
CASE
WHEN CountryName = 'United Kingdom' THEN 'United Kingdom'
WHEN CountryName = 'France' THEN 'France'
```

```
WHEN CountryName = 'Switzerland' THEN 'Switzerland'
ELSE 'Other'
END

-- Targets

-- Previous Months Sales
;
WITH PrevMonthSales_CTE
AS
(
SELECT
CASE
WHEN CountryName = 'United Kingdom' THEN 'United Kingdom'
WHEN CountryName = 'France' THEN 'France'
WHEN CountryName = 'Switzerland' THEN 'Switzerland'
ELSE 'Other'
END AS Country
,SUM(SalePrice) AS MetricValue

FROM     Reports.CarSalesData

WHERE    InvoiceDate BETWEEN DATEADD(mm, -1, DATEADD(dd, 1,
         EOMONTH(CONVERT(DATE, CAST(@ReportingYear AS CHAR(4))
         + RIGHT('0' + CAST(@ReportingMonth AS VARCHAR(2)),2) + '01', 112), -1)))
         AND  DATEADD(mm, -1, EOMONTH(CONVERT(DATE,
         CAST(@ReportingYear AS CHAR(4)) + RIGHT('0'
        + CAST(@ReportingMonth AS VARCHAR(2)),2) + '01', 112)))

GROUP BY
CASE
WHEN CountryName = 'United Kingdom' THEN 'United Kingdom'
WHEN CountryName = 'France' THEN 'France'
WHEN CountryName = 'Switzerland' THEN 'Switzerland'
ELSE 'Other'
END
)

UPDATE       Tmp
SET          Tmp.PrevMonthSales = CTE.MetricValue
FROM         #Tmp_KPIOutput Tmp
             INNER JOIN PrevMonthSales_CTE CTE
             ON Tmp.Country = CTE.Country

-- PrevYearSales
;
WITH PrevYearSales_CTE
AS
(SELECT
CASE
```

```
WHEN CountryName = 'United Kingdom' THEN 'United Kingdom'
WHEN CountryName = 'France' THEN 'France'
WHEN CountryName = 'Switzerland' THEN 'Switzerland'
ELSE 'Other'
END AS Country
,SUM(SalePrice) AS MetricValue

FROM            Reports.CarSalesData
WHERE           ReportingYear = @ReportingYear
                AND ReportingMonth = @ReportingMonth
GROUP BY
CASE
WHEN CountryName = 'United Kingdom' THEN 'United Kingdom'
WHEN CountryName = 'France' THEN 'France'
WHEN CountryName = 'Switzerland' THEN 'Switzerland'
ELSE 'Other'
END
)

UPDATE          Tmp
SET             Tmp.PrevYearSales = CTE.MetricValue
FROM            #Tmp_KPIOutput Tmp
                INNER JOIN PrevYearSales_CTE CTE
                ON Tmp.Country = CTE.Country

-- Targets
;
WITH Targets_CTE
AS
(
SELECT
CASE
WHEN BudgetDetail = 'United Kingdom' THEN 'United Kingdom'
WHEN BudgetDetail = 'France' THEN 'France'
WHEN BudgetDetail = 'Switzerland' THEN 'Switzerland'
ELSE 'Other'
END AS Country
,SUM(BudgetValue) AS BudgetValue

FROM            Reference.Budget
WHERE           BudgetElement = 'Countries'
                AND Year = @ReportingYear
                AND Month = @ReportingMonth
GROUP BY
CASE
WHEN BudgetDetail = 'United Kingdom' THEN 'United Kingdom'
WHEN BudgetDetail = 'France' THEN 'France'
WHEN BudgetDetail = 'Switzerland' THEN 'Switzerland'
ELSE 'Other'
END
)
```

```
UPDATE          Tmp
SET             Tmp.SalesTarget = CTE.BudgetValue
FROM            #Tmp_KPIOutput Tmp
                INNER JOIN Targets_CTE CTE
                ON Tmp.Country = CTE.Country

-- Calculations

UPDATE          #Tmp_KPIOutput

SET              Delta = SalesTarget - Sales
                ,PrevMonthDelta = (Sales - PrevMonthSales)
                ,PrevYearDelta = (Sales - PrevYearSales)
                ,DeltaPercent = (SalesTarget - Sales) / Sales
                ,PrevMonthDeltaPercent = (Sales - PrevMonthSales) / Sales
                ,PrevYearDeltaPercent = (Sales - PrevYearSales) / Sales

-- Status

UPDATE        #Tmp_KPIOutput

SET      SalesStatus =
                  CASE
                  WHEN ((Sales - SalesTarget) / Sales) + 1 <= 0.8 THEN 1
                  WHEN ((Sales - SalesTarget) / Sales) + 1  > 1.2 THEN 3
                  WHEN ((Sales - SalesTarget) / Sales) + 1  > 0.8
                        AND ((Sales - SalesTarget) / SalesTarget) <= 1.2 THEN 2
                  ELSE 0
                  END
      ,PrevMonthStatus =
                     CASE
                     WHEN PrevMonthDeltaPercent <= 0.9 THEN 1
                     WHEN PrevMonthDeltaPercent  > 1.1 THEN 3
                     WHEN PrevMonthDeltaPercent  > 0.9
                          AND PrevMonthDeltaPercent <= 1.1 THEN 2
                     ELSE 0
                     END
      ,PrevYearStatus =
                     CASE
                     WHEN PrevYearDeltaPercent <= 0.7 THEN 1
                     WHEN PrevYearDeltaPercent  > 1.4 THEN 3
                     WHEN PrevYearDeltaPercent  > 0.7
                          AND PrevYearDeltaPercent <= 1.4 THEN 2
                     ELSE 0
                     END
-- Output

SELECT * FROM #Tmp_KPIOutput

-- Code.pr_ScorecardTimeCountryAndMake12MonthSales
```

```
SELECT
CASE
WHEN CountryName = 'United Kingdom' THEN 'United Kingdom'
WHEN CountryName = 'France' THEN 'France'
WHEN CountryName = 'Switzerland' THEN 'Switzerland'
ELSE 'Other'
END AS Country
,SUM(SalePrice) AS Sales
,DATENAME(mm,InvoiceDate) + '-' + CAST(YEAR(InvoiceDate) AS CHAR(4)) AS MonthYear
,CAST(YEAR(InvoiceDate) AS CHAR(4)) + RIGHT('0' + CAST(MONTH(InvoiceDate) AS VARCHAR(2)),2)
AS SortOrder

FROM        Reports.CarSalesData
WHERE       InvoiceDate BETWEEN '2012-08-01' AND '2013-08-01'
GROUP BY
CASE
WHEN CountryName = 'United Kingdom' THEN 'United Kingdom'
WHEN CountryName = 'France' THEN 'France'
WHEN CountryName = 'Switzerland' THEN 'Switzerland'
ELSE 'Other'
END
,DATENAME(mm,InvoiceDate) + '-' + CAST(YEAR(InvoiceDate) AS CHAR(4))
,CAST(YEAR(InvoiceDate) AS CHAR(4)) + RIGHT('0' + CAST(MONTH(InvoiceDate) AS VARCHAR(2)),2)

ORDER BY SortOrder
```

The data returned from these two scripts (although only partially for the second) looks like the output shown in Figure 2-12. This data is for June 2015 in the sample data.

pr_ScorecardTimeCountryAndMake

Country	Sales	SalesTarget	PrevMonthSales	PrevYearSales	Delta	PrevMonthDelta	PrevYearDelta	DeltaPercent	PrevMonthDeltaPercent	PrevYearDeltaPercent	SalesStatus	PrevMonthStatus	PrevYearStatus
France	40440.000000	26000.000000	48550.000000	40440.000000	-14440.000000	-8110.000000	0.000000	-0.357072	-0.200544	0.000000	3	1	1
Other	798190.000000	126000.000000	733100.000000	798190.000000	-672190.000000	65090.000000	0.000000	-0.842143	0.081547	0.000000	3	1	1
Switzerland	46750.000000	26000.000000	120800.000000	46750.000000	-20750.000000	-74050.000000	0.000000	-0.443850	-1.583957	0.000000	3	1	1
United Kingdom	355750.000000	500000.000000	563750.000000	355750.000000	144250.000000	-208000.000000	0.000000	0.405481	-0.584680	0.000000	1	1	1

pr_ScorecardTimeCountryAndMake12MonthSales

	Country	Sales	MonthYear	SortOrder
23	United Kingdom	240000.00	August-2012	201208
24	United Kingdom	143500.00	September-2012	201209
25	United Kingdom	95000.00	October-2012	201210
26	United Kingdom	238500.00	November-2012	201211
27	United Kingdom	361500.00	December-2012	201212
28	United Kingdom	434000.00	January-2013	201301
29	United Kingdom	546750.00	February-2013	201302
30	United Kingdom	456500.00	March-2013	201303
31	United Kingdom	434500.00	April-2013	201304
32	United Kingdom	450750.00	May-2013	201305
33	United Kingdom	352250.00	June-2013	201306
34	United Kingdom	397750.00	July-2013	201307

Figure 2-12. *The data needed to create a complex text-based KPI*

How the Code Works

As was the case for the second example in this chapter (the KPI with a sparkline), it is easier to separate out the data for the sparkline from the data used for all the other metrics in the KPI. This means two queries:

- `pr_ScorecardTimeCountryAndMake12MonthSales`, which returns the sales data for three countries that you want to analyze, as well as an "other" category for any remaining countries. It returns the sales figures for each month (up to and including the selected month) for the selected year.
- `pr_ScorecardTimeCountryAndMake`, which returns the data for sales, sales for the previous month, sales for the previous month, and sales for the same month for the previous year. These figures are added to a temporary table. Then the deltas are calculated so that status indicators for certain sales figures can be added.

While a little long, these procedures are not complicated. If anything, they emphasize that BI depends on the source data and that if you want a visualization to work, you must prepare the source data cogently.

Building a Complex Text-Based KPI

Once again, with the data defined you can start building the KPI itself. Here is how.

1. Create a new SSRS report named `_TimeCountryAndMake.rdl`, and add the shared data source `CarSales_Reports`. Name it (as ever) `CarSales_Reports`.
2. Create a dataset named `CountryAnalysis`. Set it to use the `CarSales_Reports` data source, and the query `Code.pr_ScorecardTimeCountryAndMake`. Add a second dataset named `CountrySalesOver12Months` using the query `Code.pr_ScorecardTimeCountryAndMake12MonthSales`.
3. Add the following four shared datasets (ensuring that you also use the same name for the dataset in the report): `CurrentYear`, `ReportingYear`, `CurrentMonth`, `ReportingMonth`. Set the parameter properties for `ReportingYear` and `ReportingMonth` as defined at the start of Chapter 1.
4. Add a table to the SSRS report canvas. Set `CountryAnalysis` as its dataset. Delete the header row. Ensure that the border style is set to None for all text boxes in the table.
5. Right-click any row or column selector and choose Tablix Properties from the context menu. In the Tablix Properties dialog, select Filters and click Add. Select Country as the expression, and United Kingdom as the value. Click OK. You can, instead, use the Properties window (with the table selected) and click the ellipsis for the Filters property to set the filter if you prefer.
6. Right-click the row selector and select Add Group ➤ Parent Group. Group by the field Country. Confirm with OK.
7. Add two rows (you can insert above or below, it makes no difference) and two columns at the right of the table so that your table is three rows by six columns.
8. Set the row sizes to the following:
    a. Top row: 0.65 inches.
    b. Middle and bottom rows: 0.25 inches.

9. Set the column widths to the following:
   a. Column 1: 0.65 inches.
   b. Columns 2, 4 and 6: 1.55 inches.
   c. Columns 3 and 5: 0.7 inches.
10. Drag an Indicator from the toolbox onto the first column (which is, in effect, a single cell). Select three up/down triangles as the indicator type (top right of the directional indicators). Click OK.
11. Right-click the indicator and set the following indicator properties:

Section	Property	Value
General	Auto fit all gauges in panel	Unchecked
	X position (percent)	4
	Y position (percent)	-15
	Width (percent)	92
	Height (percent)	92
Values and States	Value	`[Sum(SalesStatus)]`
	States Measurement unit	Numeric
	Icon 1	Start: 1, End: 1
	Icon 2	Start: 2, End: 2
	Icon 3	Start: 3, End: 3

12. Add an indicator to the third column of the first row. Set the type as three arrows (colored) and then set the following indicator properties:

Section	Property	Value
General	Auto fit all gauges in panel	Unchecked
	X position (percent)	10
	Y position (percent)	20
	Width (percent)	70
	Height (percent)	70
Values and States	Value	`[Sum(PrevMonthStatus)]`
	States Measurement unit	Numeric
	Icon 1	Start: 1, End: 1
	Icon 2	Start: 2, End: 2
	Icon 3	Start: 3, End: 3

13. Copy this indicator into the fifth column of the first row and alter the Value to `[Sum(PrevYearStatus)]`.

14. Add the following fields:

Cell	Value
Row 1, Column 2	`[Sales]`
Row 1, Column 4	`[PrevMonthDeltaPercent]`
Row 1, Column 6	`[PrevYearDeltaPercent]`
Row 2, Column 2	`[SalesTarget]`
Row 2, Column 4	`[PrevMonthSales]`
Row 2, Column 6	`[PrevYearSales]`

15. On the third row, merge the cells in columns 3 and 4. Do the same for the cells in columns 5 and 6.

16. Format all the cells in the third row as Arial 8pt italic. Format all the cells in the second row as Arial 10pt. On the first row, format the cell in the second column as Arial 24 pt, and in the fourth and sixth columns as 16pt. Set all the font colors to Dim Gray, and the alignment to right aligned.

17. On the first row fourth column, set the Color as the following expression:

```
=Switch(
Fields!PrevMonthStatus.Value=1,"Red",
Fields!PrevMonthStatus.Value=2,"Green",
Fields!PrevMonthStatus.Value=3,"Blue"
)
```

18. Do the same for the cell in the first row, sixth column. Here, however, the field in the switch expression is `PrevYearStatus`.

19. Add a 1-point border to the right of the cells in the second and fourth columns, as well as a 1-point border above all the cells in the third row-except for the first column.

20. Format the numbers in the fourth and sixth cells of the first row as percentage with two decimals. Format the second, fourth, and sixth cells on the second row to the custom format `#,#`.

21. Add a line chart to the report outside and below the table that you have created. Right-click inside the chart and select Chart Properties from the context menu. Set the dataset name as `CountrySalesOver12Months`. Click Filters on the left and then Add. Select Country as the Expression and United Kingdom as the Value, and then Click OK.

22. As the chart is selected, you should see the Chart datapane on the right. Click the plus symbol to the right of ∑ Values and select Sales.

23. In the Chart Data pane, right-click Details in the Category Groups section and select Category Group Properties from the context menu. Select MonthYear as the Label, and click OK.

24. Right-click the legend (outside the chart on the right) and select Delete Legend from the context menu.

25. Right-click (in turn) the horizontal and vertical axis titles and uncheck Show Axis Title.

26. Click inside the chart and display the Properties window. Expand CustomInnerPlotPosition, and set Enabled to True and Width to 95.

27. Right-click the chart title and select Title Properties from the context menu. Set the following properties:

Section	Property	Value
General	Title text, as a function	`=Fields!Country.Value`
	Title position	Bottom center
Font	Font	Arial Black
	Size	16pt
	Color	Black

28. Right-click the horizontal axis and select Horizontal Axis Properties from the context menu. Set the following properties:

Section	Property	Value
Labels	Disable auto-fit	Selected
	Label rotation angle (degrees)	21
Label Font	Font	Arial
	Size	6pt
	Color	Gray
Major Tick Marks	Hide major tick marks	Checked
Minor Tick Marks	Hide minor tick marks	Checked
Line	Line color	Dim Gray

29. Right-click the vertical axis and select Vertical Axis Properties from the context menu. Set the following properties:

Section	Property	Value
Axis Options	Always include zero	Checked
Labels	Disable auto-fit	Selected
	Label rotation angle (degrees)	0
Label Font	Font	Arial
	Size	6pt
	Color	Gray
Number	Category	Number
	Decimal places	0
	Use 1000 separator (,)	Checked
Major Tick Marks	Hide major tick marks	Unchecked
	Position	Outside
	Length	1
	Line color	Gray
Minor Tick Marks	Hide minor tick marks	Checked
Line	Line color	Dim Gray

30. Right-click inside the chart and select Chart Area Properties from the context menu. Select Shadow on the left, and in the Set the Shadow Options pane set the Shadow offset to 1.5pt and the Shadow color to Gray.

31. Ensure that the table and chart are the same width and positioned perfectly one under the other.

Well, that took some time and needed close attention to detail. The result is, however, worth the effort; I am sure that you will agree. This visualization displays not only the data and status for the current and previous month's sales (you will need to choose June 2015 to get the same result as Figure 2-11), but also the sales and status for the same month in the previous year. The trend is a chart, rather than a sparkline, that clearly shows the direction of travel.

You can now duplicate the two elements, table and chart, and set the filters on both to another country, and thus create a complex visualization for data from multiple countries.

How It Works

Once again, the secret to this KPI is in the source data. If the two datasets are accurate, then building the KPI is essentially simple. As the chart is a separate object, the decision to create one stored procedure for the chart data and one for all the data in the table seemed self-evident.

The table is, once again, an "Excel-like" structure where the art is in creating the right structure and merging cells where appropriate. As SSRS does not let you merge cells vertically, you group a parent group to "merge" vertically in the leftmost column. Then you place the appropriate fields and format them, and add three indicators that you customize. As you can see, indicators can be copied just like most other objects, and it can save a lot of time to get the first one right-and then copy and paste it as a basis for other indicators.

To make a change from nested IIF statements, you use a `Switch()` function in step 17. If this is new to you, consider it a bit like a T-SQL CASE statement.

When building complex scorecards like this, you must be careful to apply the relevant filters to all the objects that make up a single visualization. This means applying the filter to the chart *and* to the table for this example.

One interesting thing about the chart is setting the custom inner plot position. Altering this allows you to override SSRS's sense of proportion, and to use the available space more efficiently. Another space-saving trick is to set the horizontal axis titles at an angle.

In this visualization, I chose slightly bolder colors for the indicators (and the status given by text colors where appropriate). You may prefer to use more pastel shades or alter the color scheme completely.

Conclusion

This chapter introduced you to Key Performance Indicators. As you saw, there are many types of KPI that you can build in SSRS. These range from the simple to the complex and from the classic to the adventurous. KPIs can be based on text, image indicators, or gauges-or a combination of some or all of these types. Then you can combine KPIs to make scorecards or dashboards, as you will see later in this book.

CHAPTER 3

Gauges for Business Intelligence

If there is one presentation technique that you really have to master for business intelligence, it is using gauges. A well though-out gauge (or set of gauges) will enable you to do the following:

- Deliver essential information quickly
- Focus your audience's attention on what matters
- Give a clean and efficient-looking impression of your BI team's technical prowess

Fortunately, SSRS has, since the 2008 version, added a wide range of gauges to its available features. There are many types of gauge available, and they are

- Easy to implement
- Effective and interesting to use
- Capable of producing a wide range of effects

All of the gauges supplied with Reporting Services are either linear or radial. Fortunately, this distinction is largely esoteric because you will be using the same techniques to create a gauge whatever its type. What really matters is learning how gauges work and then how you can push and tweak the available properties to deliver business intelligence reporting that really impresses your audience.

Gauge Elements

However varied and whatever the information contained, a gauge will always contain some or all of the following core elements:

- One or more pointers to display the values
- One or more scales to indicate what numeric value the pointer represents
- A range (or ranges) to indicate where the values are relative to a defined threshold

If you are aiming for a more complex representation of your data, you could add any of the following to your gauges:

- Text elements (or labels) to define the gauge content or add further information
- Indicators (to alert the user or to display trends)
- Sub-gauges to display additional information

Of course, any gauge can also be formatted in many, many different ways. You can apply different colors and shading to most gauge elements as well as adding images to the gauge background or selecting a gauge shape and frame from a wide set of built-in options. Using and combining these elements are essential to effective BI delivery.

When you start creating gauges, the sheer wealth of options can seem a little overwhelming. The best advice I can give is to stay calm and persevere. With a little experience you will almost certainly come to appreciate all that can be done once you know your way around the available settings.

Data for Gauges

Feeding the right data into a gauge is the key to successful delivery. Gauges nearly always need only few data elements. Yet preparing the correct data before you leap into designing and building even simple gauges will repay your initial investment in time spent re-tweaking code considerably.

So what do you need for a gauge? Well, you will definitely need

- The metric for each pointer.

You could also need

- The upper limit for each range.
- The minimum and maximum values for each scale; however, in any cases the upper limit for the topmost range is also the maximum value for a scale.
- Threshold indicators if you are using colors in pointers (be they needles, bars, or markers) to indicate where the metric stands relative to a key break-point.

I realize that range limits can be hard-coded, as can scale limits. However, in practice hard-coding these values is only a short-term solution that rapidly becomes unusable in practice. So feel free to hard-code in your initial development, but be prepared to add further data to the dataset (or even to create further datasets) to deliver dynamic range and scale thresholds before you consider moving the gauge into production.

Depending on the intricacy of the report that you are developing and the source data that you are using, you might use multiple data sets for a gauge. My preference once again is to combine all the required data into a single dataset in most cases, as it is easier to apply data to a gauge (and easier to maintain the gauge) when all of the data is in the same set. Consequently, nearly all the examples in this chapter will use various T-SQL techniques to do the following:

- Combine all required data into a single dataset per gauge
- Combine all necessary data into a single dataset for all gauges that form part of a "multi-gauge" visualization

In practice, you are free to use the type and number of datasets and data sourcing techniques that you feel happiest with-or that the business requires for maintenance.

Gauge Development

Creating gauges is frequently a process involving much trial and error until you have settled on the perfect look and feel. In the real world, you will probably have to accept that your users will go through many iterations until they are content. So be prepared to call on vast amounts of patience when developing your BI gauges in SSRS.

Gauges are rarely used singly. Most often they only become truly useful when combined in various ways. This chapter, however, will only concentrate on creating single gauges. The techniques for assembling them and creating composite visualizations are explained in various other chapters, specifically in Chapters 7 and 10.

Classic Gauges

As either a brief introduction to basic gauge use, or as a rapid refresher course if you have already used gauges, I will begin with a fairly standard gauge that contains

- A single pointer
- A single scale
- A range that indicates, in true KPI style, where the figures stand in relation to corporate objectives

This example will explain the basic concepts and elements to give you a solid basis from which you can then proceed to more advanced gauge development.

Figure 3-1 shows what you are trying to create. To give you an idea of how a gauge like this could be used, it is shown alongside a visualization made up of several gauges that are identical in design but based on different data filters. Also, as a break with the "softer" design approach taken so far in the book, I will make this gauge more "edgy." The techniques needed to assemble the six gauges are given in Chapter 7.

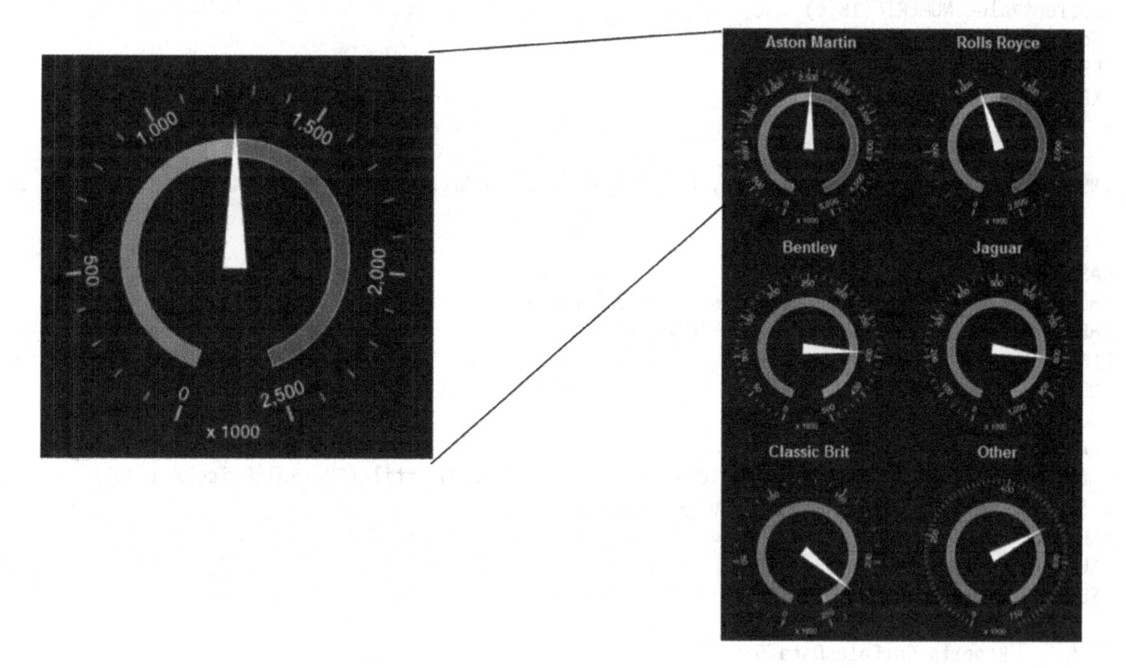

Figure 3-1. *Comparing a metric to a target*

The Source Data

The first thing that you need is your business data. In this example, let's suppose that what you want is a list of the principal makes of car, plus a couple of "catch-all" categories for minor makes, along with the budget figures for a selected year and month. Using the budget figure, you will then calculate the thresholds for a three-part range that you will add to the gauge. This range indicates how well you are doing compared to the target. The following code snippet (available as `Code.pr_SmartPhoneCarSalesGauges` in the `CarSales_Report` database) gives you the data you need for July 2012:

```
DECLARE @ReportingYear INT = 2013
DECLARE @ReportingMonth TINYINT = 6

IF OBJECT_ID('tempdb..#Tmp_Output') IS NOT NULL DROP TABLE tempdb..#Tmp_Output

CREATE TABLE #Tmp_Output
(
KPIMetric VARCHAR(50)
,CurrentValue NUMERIC(18,6)
,ScaleMax INT
,KPI1Threshold NUMERIC(18,6)
,KPI2Threshold NUMERIC(18,6)
)

INSERT INTO #Tmp_Output (KPIMetric, CurrentValue, ScaleMax, KPI1Threshold, KPI2Threshold)

SELECT
CASE
WHEN Make IN ('Aston Martin','Rolls Royce','Bentley','Jaguar') THEN Make
WHEN Make IN ('Triumph','MG') THEN 'Classic Brit'
ELSE 'Other'
END
,SUM(SalePrice) / 1000
,CASE
WHEN SUM(SalePrice) >=  SUM(SalePrice) THEN Code.fn_ScaleQuartile(SUM(SalePrice / 1000))
ELSE Code.fn_ScaleQuartile(SUM(B.BudgetValue / 1000))
END
,SUM(B.BudgetValue) * 0.95 / 1000
,SUM(B.BudgetValue) * 1.05 / 1000

FROM      Reports.CarSalesData S
          INNER JOIN
          (
           SELECT      BudgetValue
                      ,Year
                      ,Month
                      ,BudgetDetail
                      FROM      Reference.Budget
                      WHERE      BudgetElement = 'Sales'
          ) B
```

```
ON
CASE
WHEN Make IN ('Aston Martin','Rolls Royce','Bentley','Jaguar') THEN Make
WHEN Make IN ('Triumph','MG') THEN 'Classic Brit'
ELSE 'Other'
END = B.BudgetDetail
AND ReportingYear = B.Year
AND ReportingMonth = B.Month

WHERE     ReportingYear = @ReportingYear
          AND ReportingMonth = @ReportingMonth

GROUP BY
CASE
WHEN Make IN ('Aston Martin','Rolls Royce','Bentley','Jaguar') THEN Make
WHEN Make IN ('Triumph','MG') THEN 'Classic Brit'
ELSE 'Other'
END

SELECT * FROM #Tmp_Output
```

Running this snippet gives the output shown in Figure 3-2.

	KPIMetric	CurrentValue	ScaleMax	KPI1Threshold	KPI2Threshold
1	Aston Martin	253.000000	500	228.000000	252.000000
2	Bentley	39.500000	50	114.000000	126.000000
3	Classic Brit	22.500000	25	114.000000	126.000000
4	Jaguar	80.750000	100	228.000000	252.000000
5	Other	52.250000	75	228.000000	252.000000
6	Rolls Royce	110.000000	250	114.000000	126.000000

Figure 3-2. *The data used to compare vehicle sales to sales target for a specific month*

How the Code Works

This code block aggregates sales data from the CarSalesData view for the selected year and joins this to the budget table. The output groups the makes of car into a few key makes and two other groupings. All this is done, in this example, in a single query rather than by using a temporary table and multiple steps.

The code then divides the final result by 1,000 to enhance the readability of the scale on the gauge.

This code snippet also uses one of the user-defined functions described in Chapter 1 to round up the largest output to a figure in increments of 25, 250, or 2,500, etc. This also enhances the readability of the scale on the gauge.

Building the Gauge

Now that you have prepared your data, it is time to create the gauge itself. Here is how.

1. Create a new SSRS report named `_CarSalesGauge.rdl`.
2. Add the shared data source `CarSales_Reports`. Name it `CarSales_Reports`.
3. Create a dataset named `SmartPhoneCarSalesGauges`. Have it use the `CarSales_Reports` data source and the query `Code.pr_SmartPhoneCarSalesGauges` that you saw earlier.
4. Add the following four shared datasets (ensuring that you also use the same name for the dataset in the report):
    a. `CurrentYear`
    b. `CurrentMonth`
    c. `ReportingYear`
    d. `ReportingMonth`
5. Set the parameter properties for `ReportingYear` and `ReportingMonth` as defined at the start of Chapter 1.
6. Drag a gauge from the toolbox onto the report surface. Select Radial as the gauge type.
7. Right-click outside the gauge itself, but inside the gauge panel (the square delimiting the gauge) and select Gauge Panel properties. Select `SmartPhoneCarSalesGauges` as the dataset name.
8. Click Filters on the left, followed by Add. Select KPIMetric from the pop-up list as the expression, equals (=) as the operator, and `Rolls Royce` as the value.
9. Click OK.
10. Click the pointer. The Gauge Data panel will appear to the right of the gauge (if it is not already visible). Click the triangle to the right of the pop-up currently containing Unspecified and select CurrentValue. This will make the pointer display the sales metric.
11. Right-click the pointer and select Pointer Properties. Set the following properties:
    a. Pointer Type: Needle
    b. Pointer Style: Triangular
    c. Placement: Inside
    d. Distance from scale (percent): 18
    e. Width (percent): 15
12. Click Pointer Fill on the left and set the fill style to solid and the color to white.
13. Click the Cap Options on the left and check the Hide pointer cap checkbox.
14. Click OK. This will confirm all your changes and close the Pointer Properties dialog.

15. Right-click the gauge scale and select Scale Properties. In the General pane, set the following:

    a. Minimum: 0

    b. Maximum: Click the Expression button (Fx) and add this code:

    ```
    =Fields!ScaleMax.Value
    ```

    c. Set the multiply scale labels by option to 10.

16. Select Layout on the left and set the following properties:

    a. Scale radius (percent): 42

    b. Start angle (degrees): 20

    c. Sweep angle (degrees): 320

    d. Scale bar width (percent): 0

17. Select Labels on the left and set the Placement relative to scale property to Inside.

18. Select Label Font on the left and select Arial 11 point white as the font.

19. Select Major Tick Marks on the left and set the following properties:

    a. Major tick mark shape: Rectangle

    b. Major tick mark placement: Cross

    c. Width (percent): 2

    d. Length (percent): 10

20. Select Number on the left and set the Category to Number, Decimal places to 0, and check the Use 1000 separator check box.

21. Click Ok. This will confirm all your modifications for the gauge.

22. Right-click inside the gauge panel (the square delimiting the gauge) and select Gauge Panel properties. Click Fill on the left and select black as the color.

23. Click Frame on the left and set the following:

    a. Style: Simple.

    b. Shape: Circular2.

24. Click Frame Fill on the left and set the Color to Black.

25. Click OK.

26. Right-click the range and select Range Properties. On the General tab, set the following:

    a. Start Range at scale value: 0

    b. End range at scale value: Click the Expression button (Fx) and add this code:

    ```
    =Fields!KPI1Threshold.Value
    ```

    c. Placement relative to scale: Inside

  d. Distance from scale (percent): 30
  e. Start width (percent): 10
  f. End width (percent): 10

27. Click Fill on the left and set the following:
    a. Fill style: Gradient
    b. Color: Red
    c. Secondary color: Salmon
    d. Gradient style: Start to end
28. Click Border on the left and set the following:
    a. Line style: Solid
    b. Line width: 1 point
    c. Line color: Silver
29. Click OK.
30. Right-click the gauge and select Add range. Do this a second time.
31. Set the properties of the two other ranges as shown below.

Property	Range2	Range3
Start Range at scale value	`=Fields!KPI2Threshold.Value`	`=Fields!KPI1Threshold.Value`
End Range at scale value	`=Fields!ScaleMax.Value`	`=Fields!KPI2Threshold.Value`
Placement relative to scale	Inside	Inside
Distance from scale (percent)	30	30
Start width (percent)	10	10
End width (percent)	10	10
Fill style	Solid	Gradient
Color	Dark Orange	Green
Secondary color	Orange	Dark Green
Gradient style	Start to end	Start to end
Line style	Solid	Solid
Line width	1 point	1 point
Line color	Silver	Silver

32. Click inside the gauge panel and select Add label. Right-click the label and select Label Properties from the context menu. Set the following properties:

Section	Property	Value
General	Text	x 1000
	Top	91
	Left	42
	Width	16
	Height	6
Font	Font	Arial
	Color	White
Fill	Fill style	Solid
	Fill color	No color
Border	Line style	None

33. Click OK.

Your gauge is now complete. Admittedly, there were a considerable number of properties that needed to be set, and it can take a while to familiarize yourself with the structure of a gauge. The tricky part to this visualization is setting the upper and lower thresholds of the ranges correctly. So you may need to pay particular attention to this part of the process, and ensure that you set each of the three ranges correctly. However, the end result is, and I hope you agree, well worth the effort. Once a first gauge is perfect, you can copy it and adjust the filter to have a multi-gauge display. You will be assembling one of these using this gauge in Chapter 7.

How It Works

This gauge simply represents the sales figure as the pointer (the needle). As mentioned earlier, the gauge scale is adjusted to quartile increments as this means more readable graduations on the scale. The gauge has three ranges whose color indicates how the sales figure compares to the budget metric. The range boundaries are set in the code to allow for easier maintenance and more interactive redefinition.

■ **Note** Selecting a range when you have created several ranges can be a little tricky. To select a range, right-click inside the gauge panel and click on the Gauge Panel in the context menu. A sub-menu will appear, listing (among other things) all the existing ranges. All you have to do is click the range that you want to modify.

Now that the initial gauge is defined, you can make multiple copies of the original, and modify the filter on each one so that it shows data for a different make of car, for instance. This composite gauge collection is designed for a smartphone; see Chapter 10 for more information.

Hints and Tips

As this was the first gauge in the chapter, I have quite a few tips for you.

- This gauge uses a fairly plain needle type. However, SSRS offers a wide selection of available needle types, so feel free to experiment and to select the type that best suits the report or dashboard you are creating. Also remember that you can tweak the length or width of a needle to alter the overall effect produced by the gauge. Somewhat counterintuitively, the needle length is set using the Distance from scale property.
- When sizing and positioning gauge elements such as scales, pointers, and ranges, you do not need to know initially the exact figures to enter to set the various properties. To start with, it's best to place and size elements visually. This means moving the dialog away from the gauge, so that both dialog and gauge are visible. Then modify the properties that you want to tweak. Since any modifications that you make in the dialog are reflected interactively in the gauge, you will see the result of your modifications immediately.
- For some elements you will need to reuse the same property several times (height, position, and width are some of these). In these cases, it can be worth noting the values that you want to reuse once you have an initial element that is correct and thus representative of the values that you want to reproduce for other elements.
- Although I defined the size and position of the text box inside the gauge by setting properties, you can position (though not resize) text elements with the mouse.
- It is worth noting that although you set the font for the scale to a specific size, this will only be used as guidance for SSRS. The font will scale if you resize the gauge.
- In step 15, you modified the multiply scale labels by property to 10. This can be set to a positive or negative factor of 10 to act as a multiplier on the axis.

Using Multiple Pointers in Gauges

The truly wonderful thing about gauges is their sheer versatility. One basic technique that can expand the way a gauge is used is to add multiple pointers to a gauge. The following example shows how a gauge can be used to display separate metrics related to different thresholds. This example also tries to vary the presentation effect by using a bar style of pointer. Another extension of the basic elements of a gauge is to change the color of the bar to represent thresholds. This in effect replaces the need to add ranges to the gauge.

Figure 3-3 shows a gauge with two pointers. To give you an idea of how this gauge can be used in a practical application, you can see that it then becomes part of a larger visualization where each gauge displays the results for a month in the selected year.

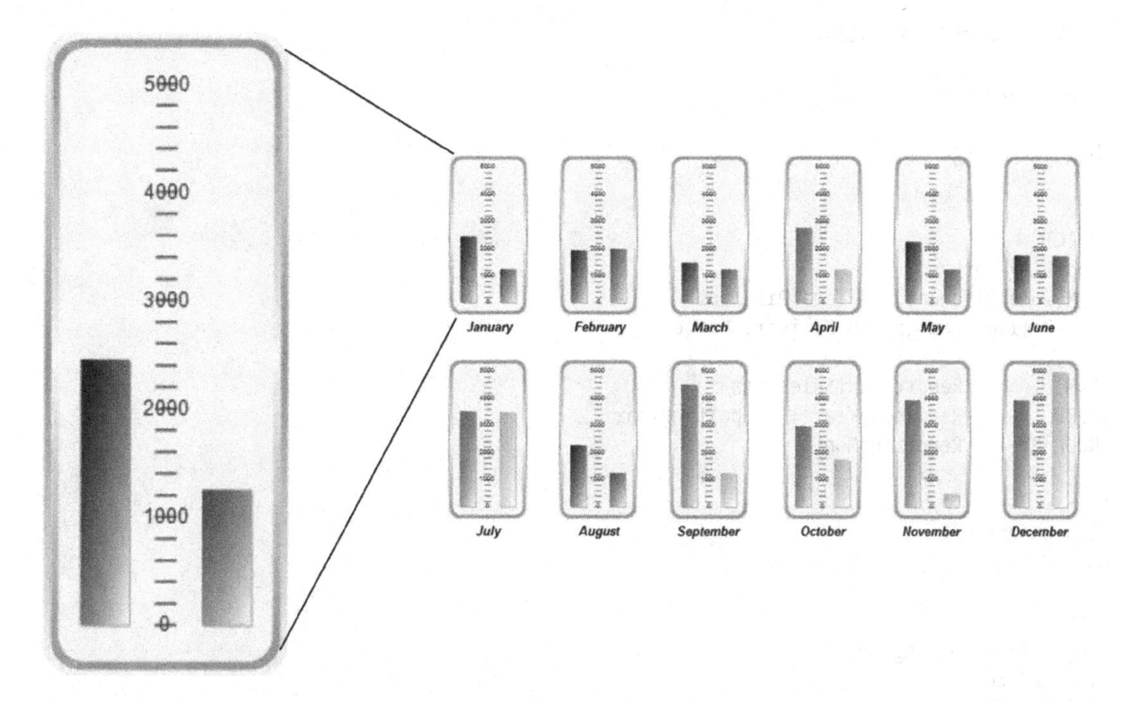

Figure 3-3. *Multiple gauges with multiple pointers*

The Source Data

The data for this series of gauges is not difficult to create. For each month of a selected year you want the delivery charge and the discount given across all cars sold. You then want to compare this with the budget figure and set an alert flag if you have exceeded the budget. The code for this is in the stored procedure `Code.pr_ScorecardCostsGauges` in the `CarSales_Reports` database. You will also in this example need a second dataset to return the maximum value to use for all the gauge pointers. It is in the stored procedure `Code.pr_ScorecardCostsGaugesTopscale`.

The body of the code for these two snippets is as follows:

```
-- Code.pr_ScorecardCostsGauges

DECLARE @ReportingYear INT = 2012
IF OBJECT_ID('tempdb..#Tmp_KPIOutput') IS NOT NULL DROP TABLE tempdb..#Tmp_KPIOutput

CREATE TABLE #Tmp_KPIOutput
(
 ReportingMonth TINYINT
,Discount NUMERIC(18,6)
,DeliveryCharge NUMERIC(18,6)
,DiscountBudget NUMERIC(18,6)
,DeliveryChargeBudget NUMERIC(18,6)
,DiscountAlert TINYINT
,DeliveryChargeAlert TINYINT
)
```

```
INSERT INTO #Tmp_KPIOutput
(
 ReportingMonth
,Discount
,DeliveryCharge
)

SELECT
ReportingMonth
,SUM(TotalDiscount) AS TotalDiscount
,SUM(DeliveryCharge) AS DeliveryCharge

FROM          Reports.CarSalesData
WHERE         ReportingYear = @ReportingYear
GROUP BY      ReportingMonth

;
WITH Budget_CTE
AS
(
SELECT
 Month AS ReportingMonth
,TotalDiscount
,DeliveryCharge

FROM
(
SELECT
 Month
,BudgetDetail
,BudgetValue
FROM      Reference.Budget
WHERE     Year = @ReportingYear
          AND BudgetElement = 'Costs'
) SRC

PIVOT       (
            SUM( BudgetValue)
            FOR  BudgetDetail IN (TotalDiscount,DeliveryCharge)
            ) AS PVT
)

UPDATE        Tmp
SET            Tmp.DiscountBudget = CTE.TotalDiscount
              ,Tmp.DeliveryChargeBudget = CTE.DeliveryCharge
FROM          #Tmp_KPIOutput Tmp
              INNER JOIN Budget_CTE CTE
              ON CTE.ReportingMonth = Tmp.ReportingMonth
```

```
-- Calculations

UPDATE        #Tmp_KPIOutput
SET           DiscountAlert =
                    CASE
                    WHEN DiscountBudget < Discount THEN 1
                    ELSE 0
                    END
             ,DeliveryChargeAlert =
                    CASE
                    WHEN DeliveryChargeBudget < DeliveryCharge THEN 1
                    ELSE 0
                    END
-- Output

SELECT * FROM #Tmp_KPIOutput

--Code.pr_ScorecardCostsGaugesTopscale

SELECT
CASE WHEN MAX(TotalDiscount) > MAX(DeliveryCharge) THEN
Code.fn_ScaleQuartile(MAX(TotalDiscount))
ELSE Code.fn_ScaleQuartile(MAX(DeliveryCharge))
END AS TopScale

FROM          Reports.CarSalesData
WHERE         ReportingYear = @ReportingYear
```

Running these code snippets, or executing the two stored procedures from the `CarSales_Reports` database, should return the two datasets shown in Figure 3-4.

pr_ScorecardCostsGauges

	ReportingMonth	Discount	DeliveryCharge	DiscountBudget	DeliveryChargeBudget	DiscountAlert	DeliveryChargeAlert
1	1	1250.000000	2450.000000	2500.000000	2500.000000	0	0
2	2	2000.000000	1950.000000	2500.000000	2500.000000	0	0
3	3	1250.000000	1500.000000	2500.000000	2500.000000	0	0
4	4	1250.000000	2775.000000	2500.000000	2500.000000	0	1
5	5	1250.000000	2275.000000	2500.000000	2500.000000	0	0
6	6	1750.000000	1775.000000	2500.000000	2500.000000	0	0
7	7	3500.000000	3525.000000	2500.000000	2500.000000	1	1
8	8	1250.000000	2275.000000	2500.000000	2500.000000	0	0
9	9	1250.000000	4525.000000	2500.000000	2500.000000	0	1
10	10	1750.000000	3000.000000	2500.000000	2500.000000	0	1
11	11	500.000000	3950.000000	2500.000000	2500.000000	0	1
12	12	5500.000000	3950.000000	2500.000000	2500.000000	1	1

pr_ScorecardCostsGaugesTopscale

	TopScale
1	5000

Figure 3-4. *Month-on-month data for discount and delivery metrics compared to budget*

How the Code Works

The first procedure (pr_ScorecardCostsGauges) aggregates the total discount and delivery charge for the selected year grouped by month and inserts this into a temporary table. It then updates the temporary table with the budget figures for the same year and metrics. Then it calculates if the results are over or under budget and sets a flag (a 0 or a 1) to indicate the status. You will use this in the gauge to set the pointer color. The second stored procedure (pr_ScorecardCostsGaugesTopscale) calculates the maximum value for the scale based on the largest figure for all months. It uses a user-defined function to round up the scale maximum to a figure that will display figures that are easier to read when placed on the scale.

Creating the Gauge

Here, then, is how to create the gauge. Since I went through the creation of the first gauge in this chapter in some detail, I will not be explaining all the steps in the process quite so exhaustively this time around. Consequently, I suggest that you refer to the previous example if you need a more in-depth explanation of a specific aspect of how to configure a gauge.

1. Create a new SSRS report named Scorecard_Costs.rdl, and add the shared data source CarSales_Reports. Name it CarSales_Reports.
2. Create a dataset named BudgetComparisons. Set it to use the CarSales_Reports data source and the stored procedure Code.pr_ScorecardCostsGauges.
3. Create a dataset named TopScale. Set it to use the CarSales_Reports data source and the stored procedure Code.pr_ScorecardCostsGaugesTopscale.
4. Add the following two shared datasets (ensuring that you also use the same name for the dataset in the report):
    a. CurrentYear
    b. ReportingYear
5. Set the parameter properties for ReportingYear as defined at the start of Chapter 1.
6. Drag a gauge from the toolbox onto the report surface. Select Multiple bar pointers as the gauge type. This is the third of the linear gauges in the Select Gauge Type dialog.
7. Right-click in the gauge itself, and select Gauge Panel ➤ Gauge Panel properties. Select BudgetComparisons as the dataset name. Set the filter to ReportingMonth = 2. When these are set, click OK.
8. Right-click the leftmost vertical bar and select delete to remove it. This is considered the first gauge pointer, and it is named LinearPointer1.
9. Click the leftmost gauge pointer (LinearPointer2) and in the Gauge Data panel set it to use DeliveryCharge as the data field. Set the other gauge pointer (LinearPointer3) to use Discount as its data field.

10. Right-click the scale, select Scale Properties, and set the following properties:

Section	Property	Value
General	Minimum	0
	Maximum (expression)	`=Sum(Fields!TopScale.Value, "TopScale")`
Layout	Position in gauge (percent)	50
	Start margin (percent)	8
	End margin (percent)	8
	Scale bar width (percent)	0
Labels	Placement (relative to scale)	Cross
Font	Font	Arial
	Size	8 point
	Color	Dim Gray
Major Tick Marks	Hide major tick marks	Checked
Minor Tick Marks	Hide minor tick marks	Unchecked
	Minor tick mark shape	Rectangle
	Minor tick mark placement	Cross
	Width (percent)	1
	Length (percent)	9

11. Right-click the leftmost gauge pointer (LinearPointer2) and set the following pointer properties:

Section	Property	Value
Pointer Options	Placement (relative to scale)	Outside
	Distance from scale (percent)	-35
	Width (percent)	20
Pointer Fill	Fill style	Gradient
	Color	`=IIF(Fields!DeliveryChargeAlert.Value=1,"Red","DarkBlue")`
	Secondary color	White
	Gradient style	Diagonal left

12. Right-click the rightmost gauge pointer (LinearPointer1) and set the following pointer properties:

Section	Property	Value
Pointer Options	Placement (relative to scale)	Outside
	Distance from scale (percent)	15
	Width (percent)	20
Pointer Fill	Fill style	Gradient
	Color	`=IIF(Fields!DeliveryChargeAlert.Value=1,"Orange","DarkGreen")`
	Secondary color	White
	Gradient style	Diagonal left

This is all you have to do. Once the "base" gauge is as you want it to be, you can duplicate it for all the months in the year and set the filter on each gauge to the relevant month.

How It Works

What you are doing here is adding a second pointer to a basic gauge, and then setting each pointer to use a specific field from the data source. Once the scale has been set to use the maximum value from the second data source, everything else is pure aesthetics. The pointer color and fill are set to change using a function that uses a field in the source code to switch the color.

Once an initial gauge has been created successfully, it can be copied and pasted once for each month, and the filter can be set to the required month.

Varying the Pointer Types and Assembling Gauges

Gauge pointers do not have to be plain needles. You can instead use markers or bars to represent the data. Indeed, as the following example shows, you can mix pointer types in certain cases. Each pointer will display a different metric. The choice and positioning of the pointer can, as you will see, affect the way in which the information is perceived by the user.

Figure 3-5 shows four gauges using multiple pointer types assembled into a visualization that is designed to separate visually three makes from the rest. However, the essence of this visualization is based on a single gauge, and it is this gauge that you will build. Creating legends for multiple gauges is explained in Chapter 7.

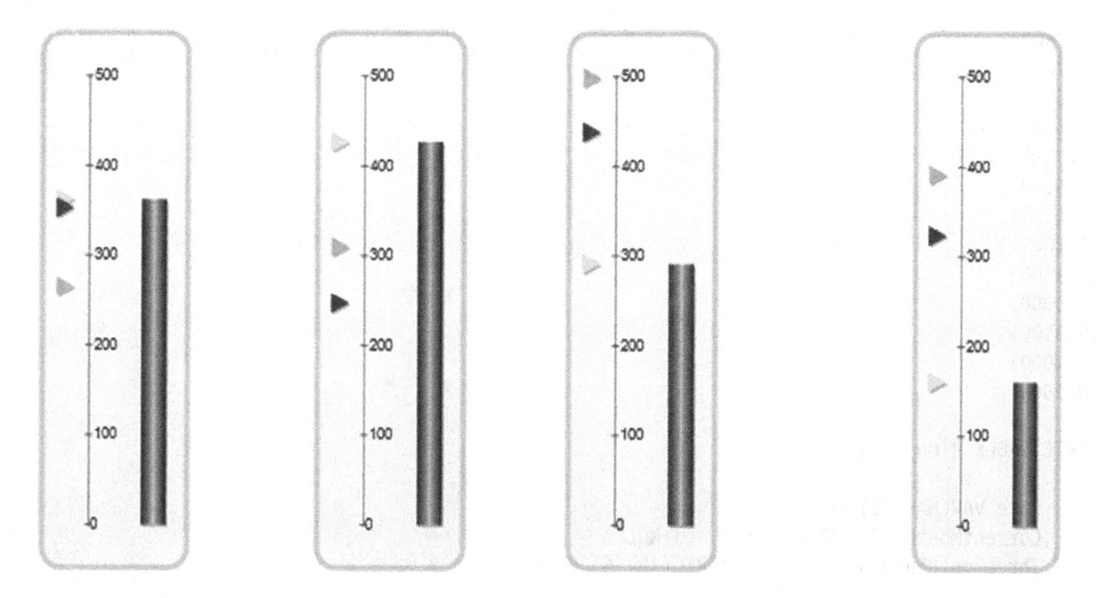

Figure 3-5. *Multiple gauges with multiple pointer types*

The Source Data

The source data for these gauges is a little more complex that the data used in previous gauges. However, there is nothing difficult here. If anything, it proves that if you do the heavy lifting in the source data definition, then creating gauges is generally easier. The only layer of complexity is added by the need to calculate a running average, the previous month's data, and the data for the same month the previous year. I find it easier to do this using a temporary table that can then be updated with the relevant metrics. As a final tweak, you will calculate the maximum value for the gauges and add this to the source dataset, rather than using a separate dataset as you did previously. Also (even if it is a little longer), I will include the code to round up the maximum value for the gauge in the T-SQL snippet itself, rather than use a function.

The code is shown below, and can be found in the stores procedure `Code.pr_WD_MakeSalesComparedOverTime` in the database `CarSales_Reports`. This example uses data for June 2015.

```
DECLARE @ReportingYear INT = 2015
DECLARE @ReportingMonth TINYINT = 6

IF OBJECT_ID('tempdb..#Tmp_Output') IS NOT NULL DROP TABLE tempdb..#Tmp_Output
IF OBJECT_ID('tempdb..#Tmp_MaxOutput') IS NOT NULL DROP TABLE tempdb..#Tmp_MaxOutput
IF OBJECT_ID('Tempdb..#Tmp_ScaleRef') IS NOT NULL DROP TABLE Tempdb..#Tmp_ScaleRef

CREATE TABLE #Tmp_ScaleRef (ScaleThreshold NUMERIC(36,6))

INSERT INTO #Tmp_ScaleRef (ScaleThreshold)
VALUES
 (1.0000)
,(0.9500)
,(0.9000)
,(0.8500)
,(0.8000)
```

```
,(0.7500)
,(0.7000)
,(0.6500)
,(0.6000)
,(0.5500)
,(0.5000)
,(0.4500)
,(0.4000)
,(0.3500)
,(0.3000)
,(0.2500)
,(0.2000)
,(0.1500)

CREATE TABLE #Tmp_Output
(
      Make VARCHAR(30) NULL
     ,CurrentMonthSales NUMERIC(38, 6) NULL
     ,ThreeMonthRunningAverage NUMERIC(38, 6) NULL
     ,PreviousMonthSales NUMERIC(38, 6) NULL
     ,SameMonthLastYearSales NUMERIC(38, 6) NULL
     ,MaxValue NUMERIC(38, 6) NULL
)

CREATE TABLE #Tmp_MaxOutput (MaxValue NUMERIC(38, 6) NULL)

-- Month sales

INSERT INTO #Tmp_Output (CurrentMonthSales, Make)

SELECT
 SUM(SalePrice) AS CurrentMonthSales
,CASE
     WHEN Make = 'Aston Martin' THEN Make
     WHEN Make = 'Rolls Royce' THEN Make
     WHEN Make = 'Jaguar' THEN Make
     ELSE 'Other'
 END AS Make

FROM     Reports.CarSalesData

WHERE    ReportingYear = @ReportingYear
         AND ReportingMonth = @ReportingMonth

GROUP BY
 CASE
     WHEN Make = 'Aston Martin' THEN Make
     WHEN Make = 'Rolls Royce' THEN Make
     WHEN Make = 'Jaguar' THEN Make
     ELSE 'Other'
 END
```

```
 -- Previous 3 month average

 ;
 WITH ThreeMonthSales_CTE
 AS
 (
SELECT
SUM(SalePrice) / 3 AS Sales
,CASE
     WHEN Make = 'Aston Martin' THEN Make
     WHEN Make = 'Rolls Royce' THEN Make
     WHEN Make = 'Jaguar' THEN Make
     ELSE 'Other'
 END AS Make

FROM     Reports.CarSalesData

WHERE    InvoiceDate >= DATEADD(mm, -3 ,CONVERT(DATE, CAST(@ReportingYear AS CHAR(4))
         + RIGHT('0' + CAST(@ReportingMonth AS VARCHAR(2)),2) + '01'))
         AND InvoiceDate <= DATEADD(dd, -1 ,CONVERT(DATE, CAST(@ReportingYear AS CHAR(4))
         + RIGHT('0' + CAST(@ReportingMonth AS VARCHAR(2)),2) + '01'))

GROUP BY
 CASE
     WHEN Make = 'Aston Martin' THEN Make
     WHEN Make = 'Rolls Royce' THEN Make
     WHEN Make = 'Jaguar' THEN Make
     ELSE 'Other'
 END
 )

UPDATE       Tmp
SET          Tmp.ThreeMonthRunningAverage = CTE.Sales
FROM         #Tmp_Output Tmp
             INNER JOIN ThreeMonthSales_CTE CTE
             ON Tmp.Make = CTE.Make

 -- Previous month sales

 ;
 WITH PreviousMonthSales_CTE
 AS
 (
SELECT
SUM(SalePrice) AS Sales
,CASE
     WHEN Make = 'Aston Martin' THEN Make
     WHEN Make = 'Rolls Royce' THEN Make
     WHEN Make = 'Jaguar' THEN Make
     ELSE 'Other'
 END AS Make
```

```
FROM        Reports.CarSalesData

WHERE       InvoiceDate >= DATEADD(mm, -1 ,CONVERT(DATE, CAST(@ReportingYear AS CHAR(4))
            + RIGHT('0' + CAST(@ReportingMonth AS VARCHAR(2)),2) + '01'))
            AND InvoiceDate <= DATEADD(dd, -1 ,CONVERT(DATE, CAST(@ReportingYear AS CHAR(4))
            + RIGHT('0' + CAST(@ReportingMonth AS VARCHAR(2)),2) + '01'))

GROUP BY
 CASE
     WHEN Make = 'Aston Martin' THEN Make
     WHEN Make = 'Rolls Royce' THEN Make
     WHEN Make = 'Jaguar' THEN Make
     ELSE 'Other'
 END
 )

UPDATE      Tmp
SET         Tmp.PreviousMonthSales = CTE.Sales
FROM        #Tmp_Output Tmp
            INNER JOIN PreviousMonthSales_CTE CTE
            ON Tmp.Make = CTE.Make

 -- Same Month Last Year

 ;
 WITH SameMonthLastYear_CTE
 AS
 (
SELECT
SUM(SalePrice) AS Sales
,CASE
     WHEN Make = 'Aston Martin' THEN Make
     WHEN Make = 'Rolls Royce' THEN Make
     WHEN Make = 'Jaguar' THEN Make
     ELSE 'Other'
 END AS Make

FROM        Reports.CarSalesData

WHERE       ReportingYear = @ReportingYear
            AND ReportingMonth = @ReportingMonth

GROUP BY
 CASE
     WHEN Make = 'Aston Martin' THEN Make
     WHEN Make = 'Rolls Royce' THEN Make
     WHEN Make = 'Jaguar' THEN Make
     ELSE 'Other'
 END
 )
```

```
UPDATE        Tmp

SET           Tmp.SameMonthLastYearSales = CTE.Sales

FROM          #Tmp_Output Tmp
              INNER JOIN SameMonthLastYear_CTE CTE
              ON Tmp.Make = CTE.Make

-- MaxValue

INSERT INTO  #Tmp_MaxOutput SELECT CurrentMonthSales FROM #Tmp_Output
INSERT INTO  #Tmp_MaxOutput SELECT ThreeMonthRunningAverage FROM #Tmp_Output
INSERT INTO  #Tmp_MaxOutput SELECT PreviousMonthSales FROM #Tmp_Output
INSERT INTO  #Tmp_MaxOutput SELECT SameMonthLastYearSales FROM #Tmp_Output

-- Maximum value for gauges

INSERT INTO #Tmp_Output (Make, MaxValue)

SELECT  'MaxValue', MIN(ScaleThreshold) * POWER(10, LEN((SELECT FLOOR(MAX(MaxValue))
                                                          FROM #Tmp_MaxOutput)))
FROM    #Tmp_ScaleRef
WHERE   ScaleThreshold >= (SELECT FLOOR(MAX(MaxValue)) FROM #Tmp_MaxOutput) / POWER(10,
                           LEN((SELECT FLOOR(MAX(MaxValue)) FROM #Tmp_MaxOutput)))

-- Output

 SELECT * FROM #Tmp_Output
```

The output from this code should look like Figure 3-6.

	Make	CurrentMonthSales	ThreeMonthRunningAverage	PreviousMonthSales	SameMonthLastYearSales	MaxValue
1	Aston Martin	291380.000000	438266.666600	499200.000000	291380.000000	NULL
2	Jaguar	427000.000000	247166.666600	310000.000000	427000.000000	NULL
3	Other	160250.000000	323583.333300	391500.000000	160250.000000	NULL
4	Rolls Royce	362500.000000	353000.000000	265500.000000	362500.000000	NULL
5	MaxValue	NULL	NULL	NULL	NULL	500000.000000

Figure 3-6. *Data for multiple gauges including the maximum scale value*

How the Code Works

This code block starts by creating a temp table of maximum values. This will be used for rounding up the maximum value for all the scales in the visualization in order to let the gauges compare values across all the filter elements. This is exactly what is done using the helper functions you saw earlier, but this time it is part of the stored procedure.

The sales for the year and month are then aggregated and added to another temporary table. This table is then updated with the running average of the sales figures for the previous three months and then for the previous month and the same month for the previous year. To make things more realistic, the aggregations group car manufacturers into three major makes and an "other" category. The maximum value is then rounded up and added to the table so that it can be used in all gauges, unlike in the previous example where this value was obtained using a second stored procedure.

It is, of course, possible to return all the sales data using correlated subqueries. As with so many things, this is a question of personal preference and deciding on the appropriate method.

Building the Gauge

Here, then, is how you can create a gauge using this data. Let's take the data for the current month as the starting point for the first gauge.

1. Create a new SSRS report named `_MonthSalesByMakeOverTime.rdl`, and add the shared data source `CarSales_Reports`. Name it `CarSales_Reports`.
2. Create a dataset named `MakeSalesComparedOverTime`. Set it to use the `CarSales_Reports` data source and the stored procedure `Code.pr_WD_MakeSalesComparedOverTime`.
3. Add the four shared datasets: `CurrentYear`, `CurrentMonth`, `ReportingYear`, and `ReportingMonth` (ensuring that you also use the same name for the dataset in the report). Set the parameter properties for `ReportingYear` and `ReportingMonth` as defined at the start of Chapter 1.
4. Drag a gauge from the toolbox on to the report surface and select Linear - vertical as the gauge type.
5. Right-click in the gauge itself, and select Gauge Panel ➤ Gauge Panel properties. Select MakeSalesComparedOverTime as the dataset name. Add a filter that you configure so that the `Make = Rolls Royce`. Click OK.
6. Right-click the gauge and select Gauge Properties from the context menu. Set the following properties:

Section	Property	Value
Back Fill	Fill style	Gradient
	Color	White
	Secondary color	Whitesmoke
	Gradient style	Top bottom
Frame	Style	Simple
	Shape	Rounded rectangular
	Width	1 point

(continued)

Section	Property	Value
Frame Fill	Fill style	Gradient
	Color	WhiteSmoke
	Secondary color	DarkGray
	Gradient style	Horizontal center
Frame Border	Line style	Solid
	Line width	2 point
	Line color	Light Gray

7. Right-click the range at the top right of the gauge and select Delete Range from the context menu.

8. Right-click the scale and set the following properties:

Section	Property	Value
General	Minimum	0
	Maximum (expression)	`=Sum(Fields!MaxValue.Value, "MakeSalesComparedOverTime") / 1000`
Layout	Position in gauge (percent)	32
	Start margin (percent)	8
	End margin (percent)	8
	Scale bar width (percent)	1
Labels	Placement (relative to scale)	Outside
	Distance from scale (percent)	-4
Font	Font	Arial Narrow
	Size	11 point
	Color	Indigo
Major Tick Marks	Hide major tick marks	Unchecked
	Major tick mark shape	Diamond
	Major tick mark placement	Cross
	Width (percent)	2
	Length (percent)	7
Minor Tick Marks	Hide minor tick marks	Checked
Fill	Fill style	Solid
	Color	Cornflower Blue

9. Right-click the pointer (the vertical bar) and select Pointer Properties from the context menu. Set the following properties:

Section	Property	Value
Pointer Options	Value (expression)	`=First(Fields!CurrentMonthSales.Value) / 1000`
	Pointer type	Bar
	Placement (relative to scale)	Outside
	Distance from scale (percent)	36
	Width (percent)	17
Pointer Fill	Fill style	Gradient
	Color	Dark Blue
	Secondary color	Light Blue
	Gradient style	Vertical center
Shadow	Shadow Offset	1 Point
	Shadow intensity	25

10. Right-click inside the gauge and select Add Pointer from the context menu. Set the new pointer (LinearPointer2) properties to the following:

Section	Property	Value
Pointer Options	Value	`=First(Fields!PreviousMonthSales.Value) / 1000`
	Pointer type	Marker
	Marker style	Triangle
	Placement (relative to scale)	Inside
	Distance from scale (percent)	10
	Width (percent)	13
	Length	12
Pointer Fill	Fill style	Solid
	Color	Light Steel Blue
Shadow	Shadow Offset	1 point
	Shadow intensity	25

11. Add two more pointers to the gauge. Set the pointer properties as described above for LinearPointer 2, with the following differences:

    a. LinearPointer3:

Section	Property	Value
Pointer Options	Value	`=First(Fields!SameMonthLastYearSales.Value) / 1000`
Pointer Fill	Color	Aqua

    b. LinearPointer4:

Section	Property	Value
Pointer Options	Value	`=First(Fields!SameMonthLastYearSales.Value) / 1000`
Pointer Fill	Color	Blue

Now the gauge is finished. You can then copy the gauge three times and set the filter for each gauge to the other three makes returned by the SQL query (excluding the MaxValue record). After that, all you have to do is position the gauges appropriately, and add text boxes for the description of each make. The way to set vertical text boxes is described in Chapter 7.

How It Works

This gauge extends the principle of gauges with multiple pointers that you used in the previous example. This time, however, you have four pointers and two different pointer types. In most other respects, the approach is the same: add the pointers, attach the correct field from the dataset, and define the aesthetics.

Once built, you place the gauges together in a container and tweak the layout by adding a shared legend to give the impression that the gauges are to be taken as a whole, which is indeed the effect you are trying to create. You then add a "legend" by adding text boxes that contain either the text for the gauge element, or simply a colored background corresponding to the fill of the gauge pointers. You then align these objects to create the effect of a legend for all the gauges. These techniques are explained in greater detail in Chapter 7.

Hints and Tips

- Chapter 2 gave an example of a KPI delivered as a gauge through using two different pointer types in the same gauge. The current example does not attempt to be a "true" KPI because it does not display trend data.

Gauges with Multiple Elements

Sometimes you may want to display several items of data in a single gauge. Most times you will be dealing with a principle metric that will be represented as the main pointer and other metrics that you want to display as secondary gauges or even text. This example uses additional text elements to allow comparison with the previous month, the same month's data for the previous year, and the three month running average.

The type of gauge you are trying to produce in shown in Figure 3-7. In this example (as in the others in this chapter), you will only be creating a single gauge. However, you will craft the code so that it can easily be applied to multiple separate gauges to display similar but comparable data for several data elements.

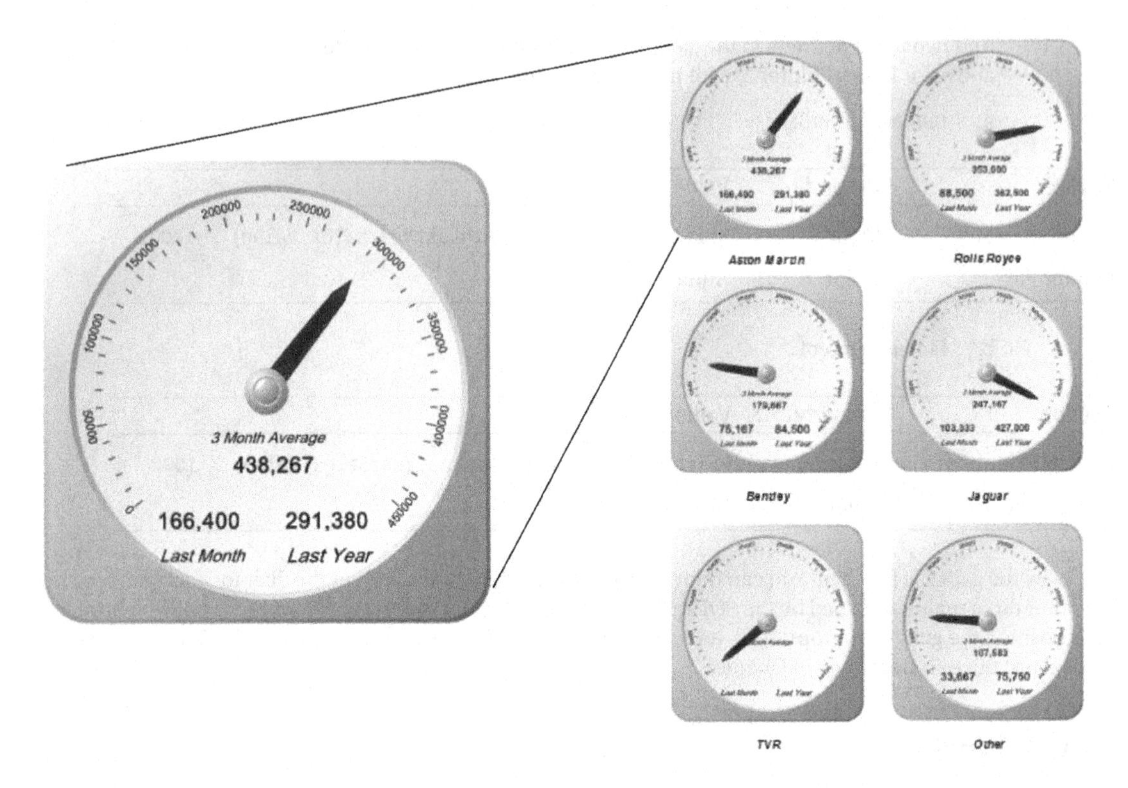

Figure 3-7. *Gauges with multiple data elements added*

The Source Data

The code can be found in the stored procedure `Code.pr_SmartPhoneSalesGauges` in the database `CarSales_Reports`. I am not showing it here, as it is virtually identical to the code used in the previous example, except that it adds one more make of car to the analysis. The output from the SQL is shown in Figure 3-8.

	Make	CurrentMonthSales	ThreeMonthRunningAverage	PreviousMonthSales	SameMonthLastYearSales	MaxValue
1	Aston Martin	291380.000000	438266.666600	166400.000000	291380.000000	NULL
2	Bentley	84500.000000	179666.666600	75166.666600	84500.000000	NULL
3	Jaguar	427000.000000	247166.666600	103333.333300	427000.000000	NULL
4	Other	75750.000000	107583.333300	33666.666600	75750.000000	NULL
5	Rolls Royce	362500.000000	353000.000000	88500.000000	362500.000000	NULL
6	MaxValue	NULL	NULL	NULL	NULL	450000.000000

Figure 3-8. *The output from the stored procedure Code.pr_SmartPhoneSalesGauges*

Creating the Gauge

So, given that you have seen what you want to produce, here is how it is done.

1. Create a new SSRS report named `SmartPhone_SalesOverTime.rdl`, and add the shared data source `CarSales_Reports`. Name the shared data source `CarSales_Reports`.
2. Create a dataset named `SalesOverTime`. Set it to use the `CarSales_Reports` data source and the stored procedure `Code.pr_SmartPhoneSalesGauges`.
3. Add the four shared datasets: `CurrentYear`, `CurrentMonth`, `ReportingYear`, and `ReportingMonth` (ensuring that you also use the same name for the dataset in the report). Set the parameter properties for `ReportingYear` and `ReportingMonth` as defined at the start of Chapter 1.
4. Drag a gauge from the toolbox on to the report surface and select Radial as the gauge type.
5. Delete the range on the gauge, and set the gauge dataset to `SalesOverTime`.
6. Right-click the gauge and set the following properties:

Section	Property	Value
Back Fill	Color	WhiteSmoke
Frame	Style	Edged
	Shape	Circular 15
Frame Fill	Color	WhiteSmoke

7. Right-click the pointer, select Pointer Properties, and set the following properties:

Section	Property	Value
Pointer Options	Value	`[Sum(CurrentMonthSales)]`
	Pointer type	Needle
	Needle style	Tapered
	Placement (relative to scale)	Inside
	Distance from scale (percent)	18
	Width (percent)	13
Pointer Fill	Fill style	Gradient
	Color	Dark Blue
	Secondary color	Blue
	Gradient style	Center
Cap Options	Cap style	Rounded glossy with indentation
	Cap width (percent)	27
Cap Fill	Color	White

8. Right-click the scale, select Scale Properties, and set the following properties:

Section	Property	Value
General	Minimum	0
	Maximum	`=Sum(Fields!MaxValue.Value, "SalesOverTime")`
Layout	Position in gauge (percent)	38
	Start margin (percent)	50
	End margin (percent)	260
	Scale bar width (percent)	0
Labels	Placement (relative to scale)	Outside
	Distance from scale (percent)	0
Font	Font	Arial
	Size	7 point
	Color	Dark Gray
Major Tick Marks	Hide major tick marks	Unchecked
	Major tick mark shape	Rectangle
	Major tick mark placement	Cross
	Width (percent)	1
	Length (percent)	6
Minor Tick Marks	Hide minor tick marks	Unchecked
	Minor tick mark shape	Rectangle
	Minor tick mark placement	Cross
	Width (percent)	1
	Length (percent)	4

9. Right-click the gauge and select Add Label. Right-click the label and set the following properties:

Section	Property	Value
General	Text (function)	`=Microsoft.VisualBasic.Strings.Format(Fields!ThreeMonthRunningAverage.Value, "#,#")`
	Text alignment	Center
	Vertical alignment	Middle
	Top (percent)	61
	Left (percent)	41
	Width (percent)	21
	Height (percent)	10

(continued)

Section	Property	Value
Font	Auto resize text to fit label	Checked
	Font	Arial Light
	Style	Bold
	Color	Dark Blue
Fill	Fill style	Solid
	Color	No color

10. Add a second label using the techniques from the previous step and set its properties identically to the first one except for the following:

Section	Property	Value
General	Text (function)	`=Microsoft.VisualBasic.Strings.Format (Fields!PreviousMonthSales.Value, "#,#")`
	Top (percent)	73
	Left (percent)	24
	Height (percent)	10

11. Add a third label and set its properties identically to the previous two except for the following:

Section	Property	Value
General	Text (function)	`=Microsoft.VisualBasic.Strings.Format (Fields!SameMonthLastYearSales.Value, "#,#")`
	Top (percent)	73
	Left (percent)	53
	Height (percent)	10

12. Add three final labels containing the texts "Three month average", "Last month", and "Last year". Set the font to black and italic, and place them under or over the three labels that you added previously to explain each figure.

Once you have perfected an initial gauge, you can copy it as many times as necessary and tweak the gauge filter so that each gauge displays data for a different make of car.

How It Works

This gauge is fairly similar to the previous example in the chapter except for the fact that you use text boxes to show the comparison figures rather than pointers.

Hints and Tips

Here are some helpful hints and tips for this gauge.

- Formatting text in a text box inside a gauge cannot be done as you would for a standard SSRS text box. This is why you need to use the code given in step 10.
- You can set the position of all the labels inside the gauge manually by dragging them with the mouse. However, if you want perfect alignment, it is probably best to finalize the Top and Left values using the Label Properties dialog and then copy and paste the relevant figures.
- If you format the first label before creating any additional labels, the new labels will inherit the properties of the previous one modified. This makes adding labels much faster (and many thanks to Rodney Landrum, one of the technical reviewers of this book, for this, one of his many excellent suggestions).

Less Traditional Gauges

The classic stand-alone linear and radial gauges in SSRS can probably handle most of your BI reporting requirements. In some cases, however, you may need to create an effect using a less "traditional" approach. This is where a couple of less classic gauge types can be useful. So now is the time to take a look at a couple of less frequently used gauge types, which can nonetheless add real visual value to your BI delivery.

Thermometers

A fairly non-traditional gauge is a thermometer. This is basically just a linear gauge with a bulb at the bottom. However, users will associate a thermometer with progress and warnings, which makes it highly suitable for certain forms of data visualization. The following example shows sales for the year to date, which is ideal for a dashboard that will be used regularly by users who will see the progress over time.

To continue the trend in this book of showing you what you are aiming for, Figure 3-9 shows a thermometer gauge of sales to date. The thermometer will change color to indicate how close it is to the target, shown by the cross bar near the top of the thermometer.

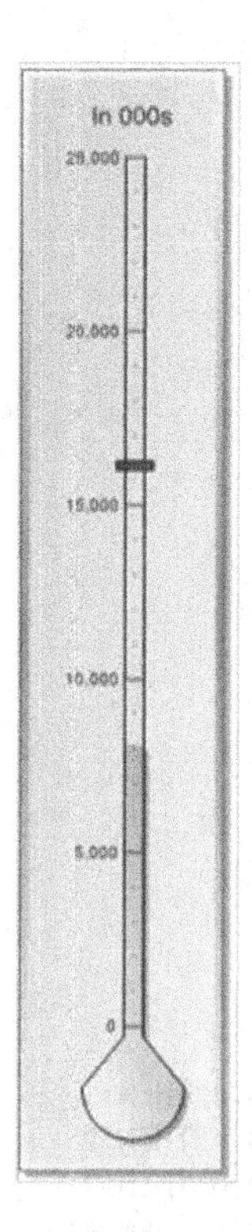

***Figure 3-9.** A thermometer gauge*

The Source Data

The source data for the thermometer is mercifully simple. The only tweak is to include a calculation of a predefined alert level, which will be used to set the color of the gauge fill. The code is shown below, and can be found in the stored procedure `Code.pr_WD_SalesYTD` in the database `CarSales_Reports`. In this example, you will display the data for June 2015.

```
DECLARE @ReportingYear INT = 2015
DECLARE @ReportingMonth TINYINT = 6

IF OBJECT_ID('tempdb..#Tmp_Output') IS NOT NULL DROP TABLE tempdb..#Tmp_Output

CREATE TABLE #Tmp_Output
(Sales NUMERIC(18,2), SalesTarget NUMERIC(18,2), MaxLimit NUMERIC(18,2), AlertLevel TINYINT)

INSERT INTO #Tmp_Output (Sales)

SELECT     SUM(SalePrice)

FROM       Reports.CarSalesData
WHERE      ReportingYear = @ReportingYear
           AND  ReportingMonth <= @ReportingMonth
```

```
-- Target figure

UPDATE      #Tmp_Output
SET         SalesTarget = (SELECT SUM(BudgetValue)
            FROM Reference.Budget
            WHERE BudgetElement = 'Sales' AND Year = @ReportingYear)

UPDATE      #Tmp_Output
SET         MaxLimit = (SELECT Code.fn_ScaleQuartile(
                        (SELECT MAX(MaxValue)
                         FROM (SELECT Sales AS MaxValue FROM #Tmp_Output
                         UNION
                         SELECT SalesTarget FROM #Tmp_Output
                        ) A
                    ) ) )

-- AlertLevel

UPDATE          #Tmp_Output

SET AlertLEvel =
     CASE
     WHEN CAST(Sales AS NUMERIC(18,6)) / CAST(SalesTarget AS NUMERIC(18,6)) <= 0.2 THEN 1
     WHEN CAST(Sales AS NUMERIC(18,6)) / CAST(SalesTarget AS NUMERIC(18,6)) <= 0.4 THEN 2
     WHEN CAST(Sales AS NUMERIC(18,6)) / CAST(SalesTarget AS NUMERIC(18,6)) <= 0.6 THEN 3
     WHEN CAST(Sales AS NUMERIC(18,6)) / CAST(SalesTarget AS NUMERIC(18,6)) <= 0.8 THEN 4
     WHEN CAST(Sales AS NUMERIC(18,6)) / CAST(SalesTarget AS NUMERIC(18,6)) > 0.8 THEN 5
     ELSE 0
     END

-- Output

SELECT * FROM #Tmp_Output
```

The output from this code is shown in Figure 3-10.

	Sales	SalesTarget	MaxLimit	AlertLevel
1	8098430.00	16140500.00	25000000.00	3

Figure 3-10. *The data to display a thermometer gauge*

How the Code Works

If you have been following the code samples given so far in this chapter, I suspect that this code will appear amazingly simple. First, a temporary table is created, and the sales up to and including the selected month are added. Then the target figure is added along with a maximum figure for the scale, suitably rounded up. Finally, a hard-coded set of indicators (on a scale of 1 through 5) are calculated as the basis for the thermometer color.

Building the Thermometer

Here is how the thermometer is put together.

1. Create a new SSRS report named `_ThermometerYTDSales.rdl`, and add the shared data source `CarSales_Reports`. Name the shared data source `CarSales_Reports`.
2. Create a dataset named `SalesYTD`. Set it to use the `CarSales_Reports` data source and the query `Code.pr_WD_SalesYTD`.
3. Add the four shared datasets: `CurrentYear`, `CurrentMonth`, `ReportingYear`, and `ReportingMonth` (ensuring that you also use the same name for the dataset in the report). Set the parameter properties for `ReportingYear` and `ReportingMonth` as defined in Chapter 1.
4. Drag a gauge from the toolbox on to the report surface and select Thermometer as the gauge type. Assign the dataset `SalesYTD` to the gauge.
5. Right-click the gauge and set the following properties:

Section	Property	Value
Back Fill	Fill style	Gradient
	Color	Gainsboro
	Secondary color	WhiteSmoke
	Gradient style	Diagonal left
Frame	Style	Simple
	Shape	Rectangular
	Width	5
Frame Fill	Fill style	Gradient
	Color	WhiteSmoke
	Secondary color	Dark Gray
	Gradient style	Horizontal center
Frame Border	Line style	Solid
	Line width	1 point
	Line color	Dark Gray
Frame Shadow	Shadow offset	3 points

6. Right-click the scale and set the following properties:

Section	Property	Value
General	Minimum	0
	Maximum	`[MaxLimit]`
Layout	Position in gauge (percent)	56
	Start margin (percent)	3
	End margin (percent)	8
	Scale bar width (percent)	0
Labels	Placement (relative to scale)	Inside
	Distance from scale (percent)	2
Font	Font	Arial
	Size	8 point
	Color	Dark Gray
Number	Category	Number
	Decimal places	0
	Use thousands separator	Checked
	Show values in	Thousands
Major Tick Marks	Hide major tick marks	Unchecked
	Major tick mark shape	Rectangle
	Major tick mark placement	Cross
	Width (percent)	1
	Length (percent)	9
Minor Tick Marks	Hide minor tick marks	Unchecked
	Minor tick mark shape	Rectangle
	Minor tick mark placement	Cross
	Width (percent)	1
	Length (percent)	1

7. Right-click the thermometer bulb (the first pointer) and set the following pointer options:

Section	Property	Value
Pointer Options	Value	`[Sum(Sales)]`
	Pointer type	Thermometer
	Thermometer style	Flask
	Placement (relative to scale)	Cross
	Distance from scale (percent)	0
	Width (percent)	9.5
	Offset (from zero position)	5
	Size (percent)	50
Pointer Fill	Fill style	Gradient

	Color	`=Switch`
		`(`
		`Fields!AlertLevel.Value = 1, "Red"`
		`,Fields!AlertLevel.Value = 2, "Yellow"`
		`,Fields!AlertLevel.Value = 3, "Orange"`
		`,Fields!AlertLevel.Value = 4, "Green"`
		`,Fields!AlertLevel.Value = 5, "Blue"`
		`)`

	Secondary color	White
	Gradient style	Diagonal right

8. Click inside the gauge and select Add Pointer from the context menu. Right-click the new pointer (it is already selected) and set the following properties:

Section	Property	Value
Pointer Options	Value	`[Sum(SalesTarget)]`
	Pointer type	Marker
	Marker style	Rectangle
	Placement (relative to scale)	Cross
	Distance from scale (percent)	0
	Width (percent)	4
	Length (percent)	18

(continued)

Section	Property	Value
Pointer Fill	Fill style	Solid
	Color	Dim Gray
Pointer Border	Line style	Solid
	Line width	1 point
	Line color	Gray

9. Click inside the thermometer and select Add Label. Right-click the label, select Label Properties, and set the following:

Section	Property	Value
General	Text	In 000s
	Text alignment	Center
	Vertical alignment	Default
	Top (percent)	3
	Left (percent)	20
	Width (percent)	69
	Height (percent)	3
Font	Auto resize text to fit label	Checked
	Font	Arial
	Style	Bold
	Color	Dark Gray
Fill	Fill style	Solid
	Color	No color
Border	Line style	None

This is all you need to do. You now have a thermometer that displays the target for the year and the sales for the year to date.

How It Works

Since this type of gauge is quite simple, there is little that has not already been covered previously in this chapter. You simply define the gauge type and then apply the correct data to the single pointer. The only slightly quirky tweak is setting the fill color using an expression based on the AlertLevel field.

Hints and Tips

Follow these tips for this gauge.

- You can set the color of the fill in the T-SQL itself if you prefer and refer to the field in the expression. There is an example of this in the next chapter.
- SSRS thermometers allow for a couple of different bulb shapes, so feel free to experiment.

Sub-Gauges

In certain cases, you may want your gauge to look like a wristwatch that contains one or more minor gauges that represent other metrics. In reality, I find that sub-gauges are not very readable, and that too many of them in a gauge can make the whole gauge difficult to read. Nonetheless, if used sparingly, this can be an interesting technique.

In this example, you will display the sales for a country in the main gauge and show the profit in the minor gauge. The profit metric is given more as a comparison element when multiple gauges are used, and is nonetheless an interesting addition that conveys useful information. As sub-gauges are difficult to read in detail, you will make the figure into a tooltip, so the reader can verify any profit ratios that seem suspect. An example of this is shown in Figure 3-11.

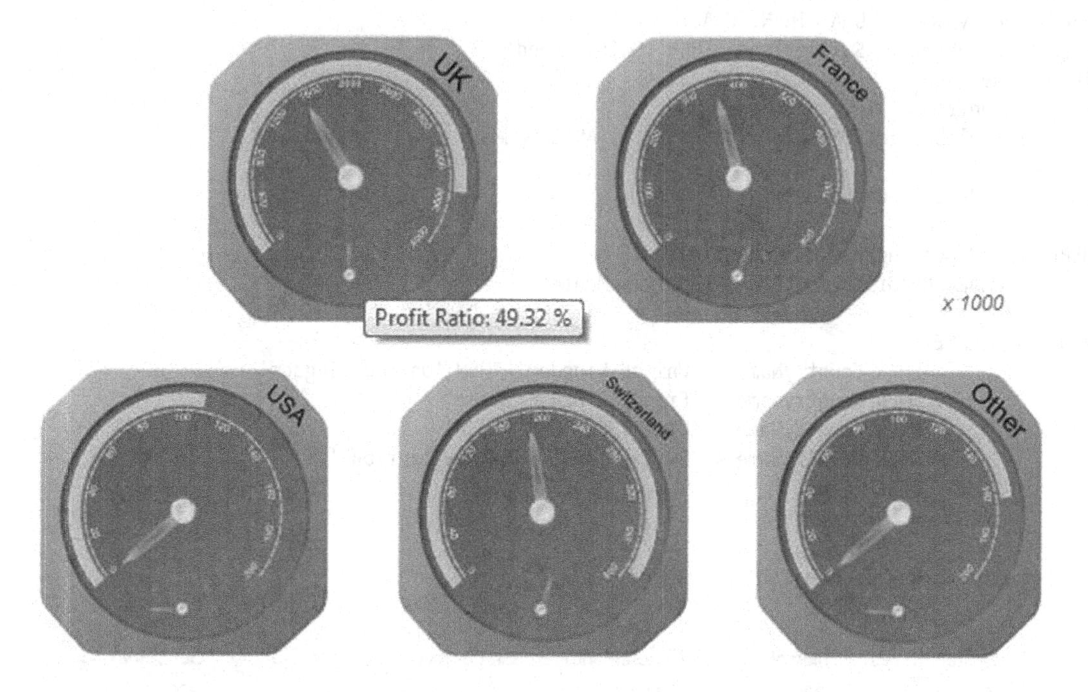

Figure 3-11. *Gauges with sub-gauges*

Once again you will only create a single gauge, but define source data that can be used for multiple gauges together.

The Source Data

The code follows, and can be found in the stored procedure Code.pr_WD_MonthlyProfitByMake in the database CarSales_Reports.

```
DECLARE @ReportingYear INT = 2013
DECLARE @ReportingMonth TINYINT = 8

SELECT
SUM(SalePrice) / 1000  AS CurrentMonthSales
,(SUM(SalePrice) -  SUM(CostPrice) - SUM(TotalDiscount) - SUM(DeliveryCharge) -
SUM(SpareParts) - SUM(LaborCost)) / 1000 AS MonthlyProfit
,(SUM(SalePrice) - (SUM(CostPrice) - SUM(TotalDiscount) - SUM(DeliveryCharge) -
SUM(SpareParts) - SUM(LaborCost))) / SUM(SalePrice) AS CurrentMonthPercentageProfit
,CASE
 WHEN CountryName = 'United Kingdom' THEN 'United Kingdom'
 WHEN CountryName = 'France' THEN 'France'
 WHEN CountryName = 'USA' THEN 'USA'
 WHEN CountryName = 'Switzerland' THEN 'Switzerland'
 ELSE 'Other'
 END AS Country
,Code.fn_ScaleDecile(SUM(SalePrice) / 1000) AS GaugeMax

FROM        Reports.CarSalesData

WHERE       ReportingYear = @ReportingYear
            AND Reportingmonth <= @ReportingMonth

GROUP BY    CASE
                 WHEN CountryName = 'United Kingdom' THEN 'United Kingdom'
                 WHEN CountryName = 'France' THEN 'France'
                 WHEN CountryName = 'USA' THEN 'USA'
                 WHEN CountryName = 'Switzerland' THEN 'Switzerland'
                 ELSE 'Other'
            END
```

The output from this code is shown in Figure 3-12.

	CurrentMonthSales	MonthlyProfit	CurrentMonthPercentageProfit	Country	GaugeMax
1	787.59	336.639000	0.561067	France	800
2	162.75	-8.886000	0.002371	Other	200
3	428.49	203.793000	0.597299	Switzerland	500
4	3915.75	1667.584990	0.482838	United Kingdom	4000
5	108.75	-23.761000	-0.071163	USA	200

Figure 3-12. *The data to display a sub-gauge inside a gauge*

How the Code Works

This code snippet calculates sales and profits (up to and including a selected month for a selected year) and also calculates the profit as a percentage of sales. This time, each gauge will have its own maximum value set for each individual scale.

Creating Gauges with Sub-Gauges

Now that you have your source data, you can configure the gauge itself.

1. Create a new SSRS report named _GaugeSalesUK.rdl, and add the shared data source CarSales_Reports. Name the shared data source CarSales_Reports.
2. Create a dataset named MonthlyProfitByMake. Set it to use the CarSales_Reports data source and the query Code.pr_WD_MonthlyProfitByMake.
3. Add the four shared datasets: CurrentYear, CurrentMonth, ReportingYear, and ReportingMonth (ensuring that you also use the same name for the dataset in the report). Set the parameter properties for ReportingYear and ReportingMonth as defined at the start of Chapter 1.
4. Drag a gauge from the toolbox on to the report surface and select radial with mini gauge as the (radial) gauge type.
5. Right-click the gauge and set the following gauge properties:

Section	Property	Value
Back Fill	Color	Dim Gray
Frame	Style	Edged
	Shape	Circular 13
Frame Fill	Color	Dark Gray
Frame Shadow	Shadow offset	0 pt.

6. In either the Gauge Panel properties dialog or in the Gauge data pane, set the dataset name to MonthlyProfitByMake. In the Filters pane, add a filter and set the Country to United Kingdom.

7. Right-click the pointer (RadialPointer1) and set the following pointer properties:

Section	Property	Value
Pointer Options	Value	`[Sum(MonthlyProfit)]`
	Pointer type	Needle
	Needle Type	Tapered
	Placement (relative to scale)	Inside
	Distance from scale (percent)	18
	Width (percent)	13
Pointer Fill	Fill style	Gradient
	Color	Gray
	Secondary color	Light Gray
	Gradient style	Center
Cap options	Cap style	Flattened with wide indentation
	Cap width (percent)	27
Cap fill	Color	White

8. Add a second pointer, and set the following properties:

Section	Property	Value
Pointer Options	Value	`[Sum(CurrentMonthSales)]`
	Pointer type	Bar
	Bar start	ScaleStart
	Placement (relative to scale)	Inside
	Distance from scale (percent)	-18
	Width (percent)	13
Pointer Fill	Fill style	Gradient
	Color	Gray
	Secondary color	Light Gray
	Gradient style	Center

9. Right-click the scale and set the following properties (having selected Scale Properties from the context menu):

Section	Property	Value
General	Minimum	0
	Maximum (expression)	`=Fields!GaugeMax.Value`
Layout	Scale radius (percent)	35
	Start angle (degrees)	50
	Sweep angle (degrees)	260
	Scale bar width (percent)	2
Labels	Placement (relative to scale)	Inside
	Distance from scale (percent)	0
Label Font	Font	Arial
	Size	10 point
	Color	White
Major Tick Marks	Hide major tick marks	Unchecked
	Major tick mark shape	Rectangle
	Major tick mark placement	Cross
	Width (percent)	1
	Length (percent)	6
Minor Tick Marks	Hide minor tick marks	Unchecked
	Minor tick mark shape	Rectangle
	Minor tick mark placement	Cross
	Width (percent)	1
	Length (percent)	4
Fill	Fill style	Solid
	Color	Light Gray

10. Right-click the sub-gauge and select Gauge Properties from the context menu. Set the following properties:

Section	Property	Value
General	Tooltip (Expression)	`="Profit Ratio: " & Microsoft.VisualBasic.Strings.Format(Fields!CurrentMonthPercentageProfit.Value, "0.00 %")`
	X Position (percent)	26
	Y Position (percent)	64
	Width (percent)	48
	Height (percent)	26
	Pivot Point X position	50
	Pivot Point Y position	75
Frame	Style	None

11. Right-click the sub-gauge and select Gauge Panel ➤ Pointer Properties from the context menu. Set the following properties:

Section	Property	Value
Pointer options	Value	`[Sum(CurrentMonthPercentageProfit)]`
	Pointer type	Needle
	Needle style	Tapered
	Placement (relative to scale)	Inside
	Distance from scale (percent)	17
	Width (percent)	15
Pointer Fill	Fill style	Gradient
	Color	Dim Gray
	Secondary color	Light Gray
	Gradient style	Diagonal left
Cap Options	Cap style	Rounded with wide indentation
	Cap width (percent)	33
Cap Fill	Color	White

12. Right-click the scale for the sub-gauge and set the following properties:

Section	Property	Value
General	Minimum	0
	Maximum	1
Layout	Scale radius (percent)	57
	Start angle (degrees)	90
	Sweep angle (degrees)	180
	Scale bar width (percent)	0
Labels	Placement (relative to scale)	Inside
	Distance from scale (percent)	0
Label Font	Font	Arial
	Size	14 point
	Color	Orange
Major Tick Marks	Hide major tick marks	Unchecked
	Major tick mark shape	Rectangle
	Major tick mark placement	Cross
	Width (percent)	2
	Length (percent)	18
Minor Tick Marks	Hide minor tick marks	Unchecked
	Minor tick mark shape	Rectangle
	Minor tick mark placement	Cross
	Width (percent)	1
	Length (percent)	8

13. Add a label and set the following properties:

Section	Property	Value
General	Text	UK
	Anchor label to	`RadialGauges.RadialGauge1`
	Top (percent)	9
	Left (percent)	75
	Width (percent)	20
	Height (percent)	13
	Angle (degrees)	42

(continued)

Section	Property	Value
Font	Auto resize to fit label	Checked
	Font	Arial
	Color	Black
Fill	Solid	No Color

And that is all that you have to do. You can then copy the gauge and apply a different filter to each copy to create a composite visualization. You will be doing this in Chapter 7.

How It Works

By this stage of this chapter you are probably getting used to creating BI gauges. So I will not explain all the details of the techniques shown previously. The minor gauge is a gauge like any other, only smaller. This is why I suggest adding a tooltip to make the figure for the percentage profit readable. One thing to note is that you need to format tooltips as an expression (just as you did for gauge text boxes).

■ **Note** If you are using multiple gauges, you could well want to use a table to control how they are laid out. Using tables to assemble objects is described in Chapter 7.

Hints and Tips

Follow these hints and tricks for this example.

- You can create multi-line tooltips by adding a VB `CrLf` function to your tooltip expression.
- You can add multiple sub-gauges if you really want to create the visual aura of certain wristwatches. However, be warned that such a visualization can become unreadable, and that sub-gauges are easily hidden by a major gauge pointer.
- Setting the second bar pointer outside the scale (on the main gauge) separates the two pointers visually.
- Placing descriptive text inside the gauge frame is a useful space-saving technique. However, it can mean a lot of fiddling with custom font sizes for each individual gauge.

Interesting Tricks with Gauges

So far in this chapter you have seen a good few techniques that allow you to present information in a clear, concise, and even stylish way using the built-in support for gauges in SSRS. To conclude this overview, I want to show you one idea that you might find useful when developing your own dashboards and reports.

Often you need multiple gauges to convey all the information that is required. In cases like this, you can consider that all the gauges taken together are really one large gauge. When creating multi-gauge displays like this, the difficulty can be in making the whole as interesting as the parts. One idea can be to use color effectively, as the following example shows (Figure 3-13). Or, rather, as the sample file shows, because unlike in the book, it is in color.

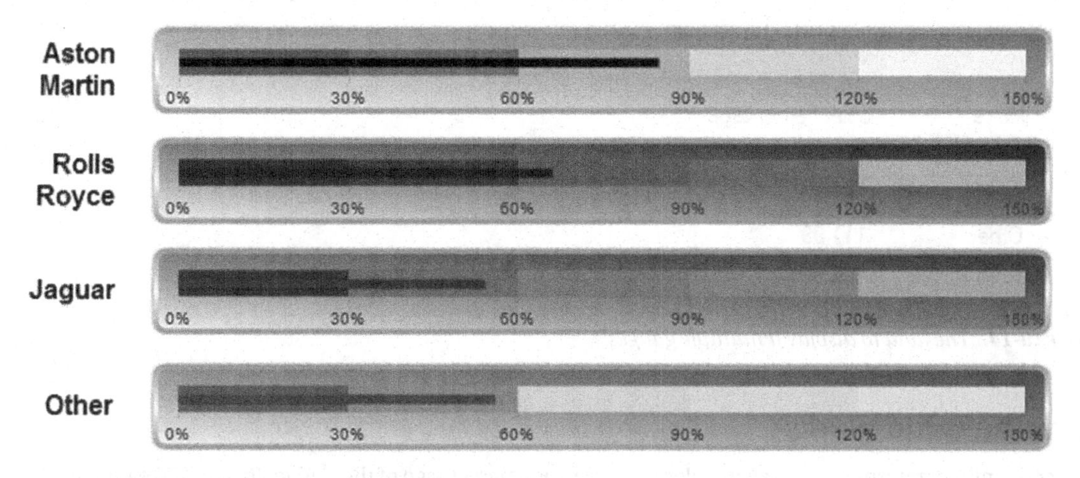

Figure 3-13. *Multiple gauges with multiplecolors*

This gauge is essentially a composite visualization, so you will create the entire collection of four gauges at once in this example.

The Source Data

The code follows, and can be found in the stored procedure `Code.pr_WD_SalesPercentageByMakein` the database `CarSales_Reports`.

```
DECLARE @ReportingYear INT = 2013
DECLARE @ReportingMonth TINYINT = 8
SELECT CASE
       WHEN Make IN ('Aston Martin','Jaguar','Rolls Royce') THEN Make
       ELSE 'Other'
       END AS Make
       ,(SUM(SalePrice)  /
                           (
                           SELECT   SUM(BudgetValue)
                           FROM     Reference.Budget
                           WHERE    BudgetElement = 'Sales'
                                    AND  BudgetDetail = 'Aston Martin'
                                    AND  Year = @ReportingYear
                                    AND  Month <= @ReportingMonth
                            ) * 100) AS SalesPercentage

FROM       Reports.CarSalesData
WHERE      ReportingYear = @ReportingYear
           AND  ReportingMonth <= @ReportingMonth
GROUP BY   CASE
                WHEN Make IN ('Aston Martin','Jaguar','Rolls Royce') THEN Make
                ELSE 'Other'
            END
```

The output from this code is shown in Figure 3-14.

	Make	SalesPercentage
1	Aston Martin	170.79
2	Jaguar	98.95
3	Other	111.09
4	Rolls Royce	126.30

Figure 3-14. *The data to display a mutliple gauges*

How the Code Works

This code snippet returns the aggregate sales expressed as a percentage of the budget for the make for four selected makes.

Creating and Assembling the Gauges

This process is rather long, I warn you, but it is not difficult. You begin by making one gauge that you will then copy and rework (slightly) three times. Here is how to go about it.

1. Create a new SSRS report named `_SalesPercentageByMake.rdl`, and add the shared data source `CarSales_Reports`. Name the shared data source `CarSales_Reports`.
2. Create a dataset named `SalesPercentageByMake`. Set it to use the `CarSales_Reports` data source and the stored procedure `Code.pr_WD_SalesPercentageByMake`.
3. Add the four shared datasets: `CurrentYear`, `CurrentMonth`, `ReportingYear`, and `ReportingMonth` (ensuring that you also use the same name for the dataset in the report). Set the parameter properties for `ReportingYear` and `ReportingMonth` as defined at the start of Chapter 1.
4. Drag a gauge from the toolbox on to the report surface and select Horizontal as the (linear) gauge type.

5. Right-click the gauge and set the following gauge properties (most of these could be the default on your system but I prefer to specify them anyway):

Section	Property	Value
Back Fill	Fill style	Gradient
	Color	Gray
	Secondary color	WhiteSmoke
	Gradient style	Diagonal right
Frame	Style	Edged
	Shape	Rounded Rectangular
	Width	8
Frame Fill	Fill style	Gradient
	Color	WhiteSmoke
	Secondary color	Dark Gray
	Gradient style	Horizontal center
Frame Border	Line style	None
Frame Shadow	Shadow offset	0 pt.

6. In either the Gauge Panel properties dialog or in the Gauge data pane, set the dataset name to `SalesPercentageByMake`. In the Filters pane, add a filter and set the Make to Aston Martin.

7. Right-click the pointer (or use the Gauge Data pane), select Pointer Properties, and set the following properties:

Section	Property	Value
Pointer Options	Value	`[Sum(SalesPercentage)]`
	Pointer type	Bar
	Bar start	Scale start
	Placement (relative to scale)	Cross
	Distance from scale (percent)	0
	Width (percent)	10
Pointer Fill	Fill style	Solid
	Color	Black

8. Right-click the scale and set the following properties:

Section	Property	Value
General	Minimum	0
	Maximum	150
Layout	Position in gauge (percent)	41
	Start margin (percent)	3
	End margin (percent)	3
	Scale bar width (percent)	18
Labels	Placement (relative to scale)	Outside
	Distance from scale (percent)	20
Font	Font	Arial
	Size	18 point
	Color	Dark Gray
Number	Custom	0\%
Major Tick Marks	Hide major tick marks	Unchecked
	Major tick mark shape	Rectangle
	Major tick mark placement	Cross
	Width (percent)	0.5
	Length (percent)	5
Minor Tick Marks	Hide minor tick marks	Checked

9. Add four new ranges to the existing one. Set the following common properties for all five ranges:

Section	Property	Value
General	Start width (percent)	30
	End width (percent)	30
	Placement relative to scale	Cross
Fill	Solid	
Border	Line style	None
Shadow	Shadow offset	0pt

10. Now you need to differentiate the ranges by setting their start and end values and color. These are as follows:

Range	Section	Property	Value
1	General	Start range at scale value	0
		End range at scale value	30
	Fill	Color	Black
2	General	Start range at scale value	30
		End range at scale value	60
	Fill	Color	Gray
3	General	Start range at scale value	60
		End range at scale value	90
	Fill	Color	Silver
4	General	Start range at scale value	90
		End range at scale value	120
	Fill	Color	Gainsboro
5	General	Start range at scale value	120
		End range at scale value	150
	Fill	Color	Whitesmoke

11. Copy the gauge three times, and set the filter for each gauge to a different make of car from the dataset (Rolls Royce, Jaguar, and Other in this example).

12. Set the color properties of the gauge back fill, pointer, and five ranges so that each gauge uses shades of a base color. I will not go into all of the details of how to do each one because you saw the principles in previous steps. Figure 3-13 uses the following colors:

Gauge	Property	Value
Rolls Royce	Back fill	Midnight Blue
	Pointer fill	Dark Blue
	Range 1 Fill	Medium Blue
	Range 2 Fill	Dark Slate Blue
	Range 3 Fill	Royal Blue
	Range 4 Fill	Steel Blue
	Range 5 Fill	Sky Blue

(*continued*)

Gauge	Property	Value
Jaguar	Back fill	Dark Red
	Pointer fill	Dark Red
	Range 1 Fill	Maroon
	Range 2 Fill	Red
	Range 3 Fill	Orange Red
	Range 4 Fill	Tomato
	Range 5 Fill	Light Salmon
Other	Back fill	Dark Green
	Pointer fill	Dark Green
	Range 1 Fill	Green
	Range 2 Fill	Medium Sea Green
	Range 3 Fill	Lime
	Range 4 Fill	Lawn Green
	Range 5 Fill	Pale Green

How It Works

The main task here is to define the five ranges that serve as a background for the pointer bar. Each must be of the same height and width, and have start and end ranges that do not leave any gaps. Apart from this, it is up to you to define colors for each gauge that differentiate the gauges while remaining readable.

Hints and Tips

These tips will help with this example.

- You will probably need to use custom formatting for the scale. Specifically, the number of commas defines the multiple of thousands, millions, etc. that will be applied.
- The fewer ranges that you use, the easier it will be to see a clearly delineated set of colors.
- You can set the gauge pointer color so that it changes, too. This way you can set a contrasting color to the underlying range. In this example, for instance, you would set the pointer to be a light shade if the figure is lower and dark if the figure is higher.

Conclusion

This was a fairly long chapter, but there is so much that can be done with gauges to deliver compelling business intelligence that I thought it worth the effort to explain as many useful gauge types and tricks as possible. Given the vast potential that gauges can bring to business intelligence when using SSRS, we have not finished with gauges in this book. You will see a few more techniques using gauges in Chapter 10. So if you want to see how gauges can be adapted to BI delivery on mobile devices, this is the chapter for you.

Hopefully, the ideas that you have seen here will inspire you to create your own BI reports using gauges where necessary (and within the bounds of clarity and good taste) so that your users will be wowed, rather than bored-and who knows, perhaps the boss will be secretly impressed too.

CHAPTER 4

Charts for Business Intelligence

Charts may be a staple of spreadsheets and PowerPoint presentations, but that does not mean that you cannot use them in dashboards (and in BI generally) to great effect. The main thing to remember if you want your dashboards to be striking and memorable is that a BI chart needs to convey key information succinctly. So you may well find that you end up avoiding the more traditional (not to say overused) chart types in your BI reports in favor of less tired ways of visualizing your data.

Charts have been around in SSRS for nearly 15 years, so I won't provide a chart primer here. What I want to do is to outline some of the ways in which a well-designed chart can be an intrinsic part of an effective BI dashboard. Charts are, however, infinitely varied and a matter of taste. So I will not be prescribing which chart type to use in which scenarios; I can only advise you to consult some of the excellent books on this subject. Nonetheless, one fundamental point to bear in mind when creating your reports is that charts used for BI will nearly always benefit considerably from being *simplified* rather than overloaded. It is nearly always better to create several charts rather than to create a single chart containing many data series.

What I will do is show you some of the ways in which charts can be used (almost) as KPIs, and how using some of the rarer chart types can add real visual effect to a dashboard. This chapter aims to be largely a collection of tips and tricks which I hope you will find useful. I am not attempting to provide complete coverage of how to use charts in Reporting Services for business intelligence because there are simply too many ways that charts can be used to convey information succinctly and coherently in dashboards and other BI output to be handled in a short book. However, this chapter should encourage you to experiment with charts for business intelligence and to deliver some really cool visualizations to your users.

Some Chart Presentation Ideas

All the charts that you will be creating in this chapter are designed to be used in a business intelligence context. For me, this means that they *must* be designed for clarity and simplicity, so that the key message and metrics are conveyed succinctly and transparently.

We all have our ideas about what constitutes taste and effective delivery, especially where design and color are concerned. Consequently, I expect many people to disagree with my ideas, and I can only encourage you to do your own thing. In this chapter, however, you will generally be adopting the following design principles for all of the charts that you will develop:

- Chart title: Arial, 10 point, gray
- Axis titles: Arial narrow, 8 point, italic, gray
- Legend title: Arial Narrow, 9 point, gray
- Legend: Shadow, dark gray also for lines
- Lines, gridlines, tick marks: Light gray 0.5 or 1 point

- Right and bottom border of the chart: 1.75 point
- Top & left border of the chart: 1 point all in dark gray
- All backgrounds white or transparent

The logic behind this design is to make text and boundaries less obtrusive so that the data can stand out. Equally, I will be using mostly lighter colors except in a few cases. However, this is my approach, and you are, of course, free to apply your own design ideas and strictures.

As was the case in previous chapters, I will be fairly expansive when explaining the first example, and a little more concise for the remaining examples in this chapter.

Charts to Compare Metrics with Targets

Charts can be ideal when you need to show how a series of metrics compares to target values. They have the advantage of being largely intuitive, as can be seen in the number of charts in BI reports and dashboards the world over.

Admittedly, comparing targets to results is largely what most KPIs do. Where charts are concerned, though, I will not be adding any form of trend indicator. As this would really be necessary for a true KPI, I do not consider target-metric charts as KPIs. This does not in any way, however, mitigate their usefulness and value in a BI context.

Basic Target Comparison Charts

In its most simple form, a target comparison chart could have only two series of data:

- The required metric
- The target value

The only trick here is to differentiate the two series sufficiently. Mixing a standard chart type that represents one series with custom markers for another series can be an efficient way of doing this, as Figure 4-1 shows for June 2015 in the sample data.

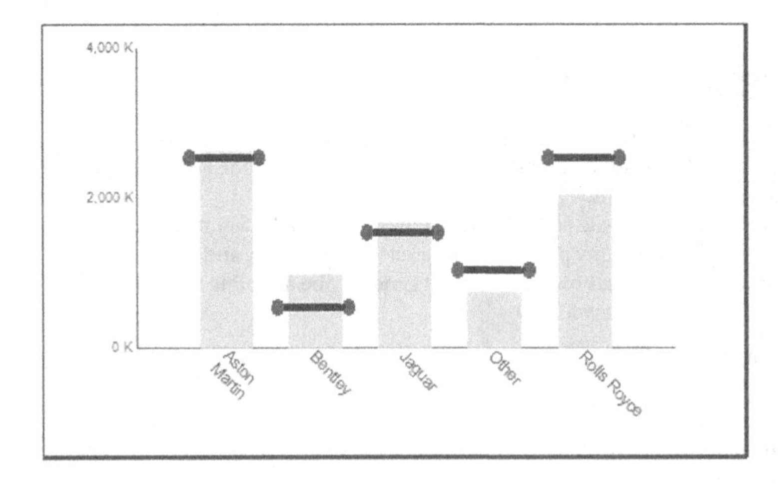

Figure 4-1. *Comparing a metric to a target*

The Source Data

The first thing that you need is your business data. In this example, let's suppose that all you want is a list of the principal makes of car, plus a catch-all category for minor makes, along with the budget figures for a selected year and month. The following code snippet (available as `Code.pr_MonthlyCarSalesWithTarget` in the `CarSales_Report` database) provides the data:

```
DECLARE @ReportingYear INT = 2015
DECLARE @ReportingMonth TINYINT = 6

IF OBJECT_ID('Tempdb..#Tmp_Output') IS NOT NULL DROP TABLE Tempdb..#Tmp_Output

CREATE TABLE #Tmp_Output
(
Make NVARCHAR(80) COLLATE DATABASE_DEFAULT
,Sales NUMERIC(18,6)
,SalesBudget NUMERIC(18,6)
)

INSERT INTO #Tmp_Output
(
Make
,Sales
)

SELECT     CASE
           WHEN Make IN ('Aston Martin','Bentley','Jaguar','Rolls Royce') THEN Make
           ELSE 'Other'
           END AS Make
           ,SUM(SalePrice)
FROM       Reports.CarSalesData
WHERE      ReportingYear = @ReportingYear
           AND ReportingMonth <= @ReportingMonth
GROUP BY   CASE
           WHEN Make IN ('Aston Martin','Bentley','Jaguar','Rolls Royce') THEN Make
           ELSE 'Other'
           END
;
WITH Budget_CTE
AS
(
SELECT     SUM(BudgetValue) AS BudgetValue
           ,BudgetDetail
FROM       Reference.Budget
WHERE      BudgetElement = 'Sales'
           AND Year = @ReportingYear
           AND Month = @ReportingMonth
GROUP BY   BudgetDetail
)
```

```
UPDATE      Tmp
SET         Tmp.SalesBudget = CTE.BudgetValue
FROM        #Tmp_Output Tmp
            INNER JOIN Budget_CTE CTE
            ON Tmp.Make = CTE.BudgetDetail

-- Output

SELECT      *
FROM        #Tmp_Output
```

Running this snippet gives the output shown in Figure 4-2.

	Make	Sales	SalesBudget
1	Aston Martin	2629680.000000	2500000.000000
2	Bentley	989250.000000	500000.000000
3	Jaguar	1681000.000000	1500000.000000
4	Other	749000.000000	1000000.000000
5	Rolls Royce	2049500.000000	2500000.000000

Figure 4-2. *The data used to compare vehicle sales to sales target for a specific month*

How the Code Works

This short code snippet calculates the sales for the year to date (as defined by the selected month) and places the result into a temporary table. It then adds the budgetary figure for the same period to the table. This data only looks at four major makes of car plus an "Other" category.

Building the Chart

Time, then, to build the chart. As this is the first chart you're creating, I will go into some detail of the base techniques that are necessary to create it.

1. Create a new SSRS report named `_CountryChartPercentageToTarget.rdl`. Resize the report so that it is sufficiently large to hold a chart.
2. Add the shared data source `CarSales_Reports`. Name it `CarSales_Reports`.
3. Create a dataset named `CountryChartPercentageToTarget`. Have it use the `CarSales_Reports` data source and the stored procedure `Code.pr_MonthlyCarSalesWithTarget`, whose code is shown above.
4. Add the following four shared datasets (ensuring that you also use the same name for the dataset in the report):
    a. `CurrentYear`
    b. `CurrentMonth`
    c. `ReportingYear`
    d. `ReportingMonth`

5. Set the parameter properties for ReportingYear and ReportingMonth as defined at the start of Chapter 1.

6. Right-click Images in the Report Data window, select Add Image from the context menu, and navigate to the file C:\BIWithSSRS\Images\DumbbellTarget.png to embed this image in the report.

7. Click the Chart item in the SSRS toolbox, and either draw a sufficiently large area for the chart or drag a chart onto the report surface that you will resize once the chart is created. Select 3D Column as the chart type, and click OK.

8. Right-click in a blank part of the chart and select Chart Properties from the context menu. Select CountryChartPercentageToTarget as the dataset name and click OK.

9. In the Chart Data pane (at the right of the chart; it will have appeared when you right-clicked in the previous step) add the Sales and SalesBudget fields as the ∑ values (in this order), and the Make field as the Category Group by clicking on Details in the Category Groups and selecting Make from the pop-up.

10. Right-click inside the chart area, but not on a column, and choose Chart Area Properties from the context menu. In the 3D Options pane set the Rotation, Inclination, and Wall Thickness to 0.

11. Click Fill on the left, leave the fill style as Solid, and set the fill color to No Color. Click OK.

12. Click one of the columns for SalesBudget (or the SalesBudget field in the Chart Data pane) and display the Properties window (pressing F4 should do it).

13. In the Chart Series properties, expand Marker, then Image, and set the following properties to apply the image that you imported earlier:

    a. Mime type: image/png

    b. Source: Embedded

    c. Value: DumbbellTarget

    d. TransparentColor: White

14. While in the Properties window, set the Color (of this chart series) to No Color.

15. Leave the Properties window visible and click the Sales series (or the Sales field in the Chart Data pane). Set the following properties:

    a. Color: Khaki (or a suitably quiet and non-virulent color)

    b. Border Color: No Color

    c. Custom Attributes ➤ Point Width: 0.6

16. Right-click the legend and select Delete Legend. Do the same for the title.
17. Right-click each axis title and uncheck Show Axis Title.
18. Set the number format of the vertical axis to the following custom format: `#,0," K";(#,0," K")`
19. While you are setting vertical axis properties, disable auto fit for the labels and set the font size to 7 point.
20. Right-click the horizontal axis and select Horizontal Axis Properties from the context menu. Set the following properties:

Section	Property	Value
Labels	Disable auto-fit	Selected
	Label rotation angle (degrees)	45
Label Font	Font	Arial
	Size	8 point
	Color	Gray
Major Tick Marks	Hide major tick marks	Checked
Minor Tick Marks	Hide minor tick marks	Checked
Line	Line color	Dim gray

21. Right-click the vertical axis and select Vertical Axis Properties from the context menu. Set the following properties:

Section	Property	Value
Axis Options	Always include zero	Checked
Labels	Disable auto-fit	Selected
	Label rotation angle (degrees)	0
Label Font	Font	Arial
	Size	6 point
Number	Category	Custom
	Custom format	`#,0," K";(#,0," K")`
Major Tick Marks	Hide major tick marks	Checked
Minor Tick Marks	Hide minor tick marks	Checked
Line	Line color	Dim gray

That is it. You have delivered a chart that avoids the more "classic" approach by using an image rather than a column to display the target value. Moreover, you have superposed the two data series by using a 3D chart that you have "flattened" to remove the perspective.

■ **Note** You may prefer to leave the axes and fonts on the axes in black, or another color of your choice. I am merely trying to show that you can avoid distracting the user's gaze if you tweak axes and number formatting (and remove any superfluous decoration such as titles) in this way.

Of course, when it comes to setting properties in SSRS charts, you have several alternative ways of achieving your objectives. I am merely suggesting one way of doing this here. If you have another method and prefer it, the choice is yours.

How It Works

This chart required quite a few steps to complete, and some of the techniques you applied may not be immediately comprehensible. Some may even seem superfluous. Here are some of the reasons for doing what you did.

- The principal technique was to replace one of the columns with an image. In reality, the column is still there, but as its color is the same as the chart area, it is invisible. The upper edge of the column is where the image will appear.
- Creating and applying a personalized image could seem like overkill. If it does, then you can use a standard marker in the place of the image. Applying a marker is described later in this chapter. The reason for using an image is that using a bar or a line as the point of comparison is far too traditional. So I chose to customize the visualization to avoid provoking indifference on the part of the reader.
- Formatting the vertical axis is a great way to save space and to avoid distracting the user with multiple sets of 000,000,000, which add nothing to the visualization. SSRS number formats are a little abstruse, but it is well worth playing with them so that you can enhance your charts with unobtrusive axes.
- Equally, reducing the amount of numbers displayed (by tweaking the vertical axis interval) can also enhance the chart by reducing the quantity of numbers on the axis.
- Tweaking the column width to make the chart seem less standard is another way of preventing the user from feeling a sense of déjà vu. Remember that delivering information also means avoiding repetition, even if the sense of repetition has been created by other people in other visualizations.

An Advanced KPI Chart

Another way of showing how a metric compares to a target value is to use the same chart type for both data series in a stacked bar or column chart. One of these charts is shown in Figure 4-3. If you are only opening the sample file, set the year to 2015 and the month to 6 to get the same result.

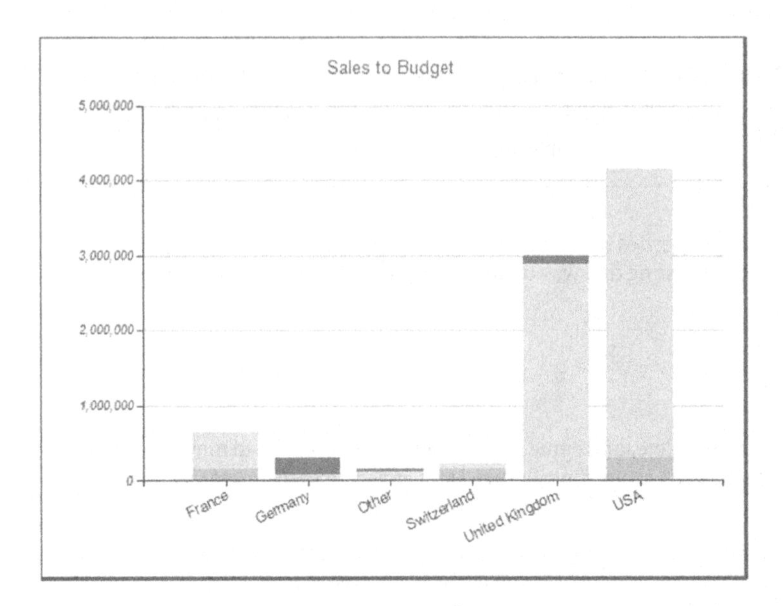

Figure 4-3. *Comparing metrics to target values in a chart*

If the black and white example from a printed book does not make the point (and you are not reading this in a color e-book) then you can always display the file `_SalesToBudget.rdl` from the example solution. This way you should get the effect of displaying the sales over budget in one color and the sales under budget in another color.

This approach requires a little work on the source data, as a simple calculation will be required to do the following:

- See if the metric is less than the target, and if so, set the metric to one series, and the shortfall to the second series. This way the two together equal the target figure.
- In the case where the metric is greater than the target, set the first series to equal the target value, and the second series to represent the surplus between the metric and the target.

Moreover, to make it clearer when there is a shortfall and when the target has been exceeded, the source data will include a flag that will indicate the state of the second data series (shortfall or excess) that can then be used to set the color of the category for each data point.

The Source Data

The source SQL not only return the sales and budget figures, but will also apply the business logic that is required to set the chart column colors. You can find this code (wrapped in a stored procedure) in the `CarSales_Reports` database as `Code.pr_WD_CountrySalesToBudgetRatio`.

```
DECLARE @ReportingYear INT = 2015
DECLARE @ReportingMonth TINYINT = 6

IF OBJECT_ID('tempdb..#Tmp_MyData') IS NOT NULL DROP TABLE tempdb..#Tmp_MyData
```

```
CREATE TABLE #Tmp_MyData
(
CountryName VARCHAR(50) COLLATE DATABASE_DEFAULT
,Sales MONEY
,Budget MONEY
,SalesChartValue MONEY  -- The first data series
,BudgetChartValue MONEY -- The second data series
,Indicator TINYINT
)

-- Sales Data

INSERT INTO #Tmp_MyData (CountryName, Sales)

SELECT
CASE
WHEN CountryName IN ('France','Germany','Switzerland','United Kingdom','USA') THEN
CountryName
ELSE 'Other'
END AS CountryName
,SUM(SalePrice)
FROM      Reports.CarSalesData
WHERE     ReportingYear = @ReportingYear
          AND ReportingMonth <= @ReportingMonth
GROUP BY
CASE
WHEN CountryName IN ('France','Germany','Switzerland','United Kingdom','USA') THEN
CountryName
ELSE 'Other'
END

-- Budget Data
;
WITH Budget_CTE
AS
(
SELECT
BudgetDetail AS CountryName
,SUM(BudgetValue) AS BudgetValue

FROM      Reference.Budget

WHERE     Year = @ReportingYear
          AND  Month <= @ReportingMonth
          AND BudgetElement = 'Countries'

GROUP BY  BudgetDetail
)
```

```
UPDATE      Tmp
SET         Tmp.Budget = CTE.BudgetValue
FROM        Budget_CTE CTE
            INNER JOIN #Tmp_MyData Tmp
            ON CTE.CountryName = Tmp.CountryName

-- Calculations

UPDATE      #Tmp_MyData

SET         SalesChartValue =
                        CASE
                        WHEN Sales < Budget THEN Sales
                        ELSE Budget
                        END
            ,BudgetChartValue =
                        CASE
                        WHEN Sales < Budget THEN Budget - Sales
                        ELSE Sales - Budget
                        END
            ,Indicator =
                        CASE
                        WHEN Sales < Budget THEN 1
                        ELSE 2
                        END
SELECT CountryName, SalesChartValue, BudgetChartValue, Indicator from #Tmp_MyData
```

Running this T-SQL gives the output shown in Figure 4-4.

	CountryName	SalesChartValue	BudgetChartValue	Indicator
1	France	156000.00	486790.00	2
2	Germany	74750.00	225250.00	1
3	Other	116000.00	40000.00	1
4	Switzerland	156000.00	76800.00	2
5	United Kingdom	2888500.00	111500.00	1
6	USA	300000.00	3843590.00	2

Figure 4-4. *Comparing metrics to target values and adding an indicator*

How the Code Works

Here you are using a single stored procedure to output all the data that you need for the chart. As the two data series that are used in the chart are calculated, and not taken directly from the source data, you use a temporary table to store the sales and budget data first. Then a simple calculation sets the values of the two data series. You use the indicator value to set the column color: 1 means that you are under budget, 2 means that you are over budget. Note that the `SalesChartValue` and `BudgetChartValue` figures are *not* the raw data for sales and budget.

Building the Chart

Let's see how to apply this data to your chart to get the result you desire. As you just saw many of the essential techniques in some detail in the previous chart, I will be a little more succinct when describing any repeated elements.

1. Create a new SSRS report named `_SalesToBudget.rdl`. Resize the report so that it is sufficiently large to hold a chart. Add the shared data source CarSales_Reports. Name it `CarSales_Reports`.
2. Create a dataset named `CountrySalesToBudget`. Have it use the `CarSales_Reports` data source and the stored procedure `Code.pr_WD_CountrySalesToBudgetRatio`, which you saw earlier.
3. Add the following four shared datasets (ensuring that you also use the same name for the dataset in the report): `CurrentYear`, `CurrentMonth`, `ReportingYear`, and `ReportingMonth`. Set the parameter properties for `ReportingYear` and `ReportingMonth` as defined at the start of Chapter 1.
4. Add a stacked column chart from the SSRS toolbox and resize it to suit your needs. Attach the dataset `CountrySalesToBudget`.
5. In the Chart Data pane (at the right of the chart), add the `SalesChartValue` and `BudgetChartValue` fields as the ∑ values (in this order). Set the `CountryName` field as the Category Group.
6. Right-click the upper segment of any of the columns (or `BudgetChartValue` in the Chart Data pane) and select Series Properties. Click Fill on the left, and add the following code as the expression for the color:

```
=Iif(Fields!Indicator.Value=1,"Tomato","Khaki")
```

7. Right-click the lower segment of any of the columns (or `SalesChartValue` in the Chart Data pane) and select Series Properties. Click Fill on the left, and add the following code as the expression for the color:

```
=Iif(Fields!Indicator.Value=1,"Khaki","LightBlue")
```

8. Right-click the horizontal axis and select Horizontal Axis properties. Click Labels on the left, click the Disable Auto Fit radio button, and set the Label angle rotation (degrees) to -25. While in this dialog, set the text and line to dim gray.

9. Right-click the vertical axis and select Vertical Axis properties. Set the following properties:

Section	Property	Value
Axis Options	Always include zero	Checked
Labels	Disable auto-fit	Selected
	Label rotation angle (degrees)	0
Label Font	Font	Arial
	Size	9 point
	Color	Dim gray
Number	Category	Number
	Decimal places	0
	Use 1000 separator	Checked
Major Tick Marks	Hide major tick marks	Unchecked
	Position	Outside
	Length	1
	Line color	Dim gray
Minor Tick Marks	Hide minor tick marks	Checked
Line	Line color	Dim gray

10. Add the title text Sales to Budget. Format any other axes and title properties as you think fit.

How It Works

The key to this chart is in the work done by the source SQL. This code essentially sets the second series to display the over- or underachievement of the sales targets. Then SSRS applies colors to the columns to indicate much more visually whether the target has been attained or undershot.

- The expression simply sets the fill color of the second series depending on whether the target has been exceeded or not.
- The choice of colors for the second series can have an immediate visual effect. Traditionally, red is a negative color in finance, whereas blue is positive. You can, of course, define your own color scheme as long as you explain it to your users.

■ **Note** Some graphic artists object strongly to presenting axis elements at an angle, as you have done in the last two examples. I am not saying who is right or wrong, merely showing how it can be done and the effect that it produces. The choice is yours.

Ordering Chart Elements

One trick that can help your audience appreciate the essence of a simple data set (and assuming that you are presenting it as a bar, column or even pie chart) is to order the data in the series. This is particularly useful if you are displaying the top or bottom elements in a set, as it makes the hierarchy of data (such as sales, for instance) really stand out. An example of such a chart is shown in Figure 4-5. The data shown is for June 2015 again.

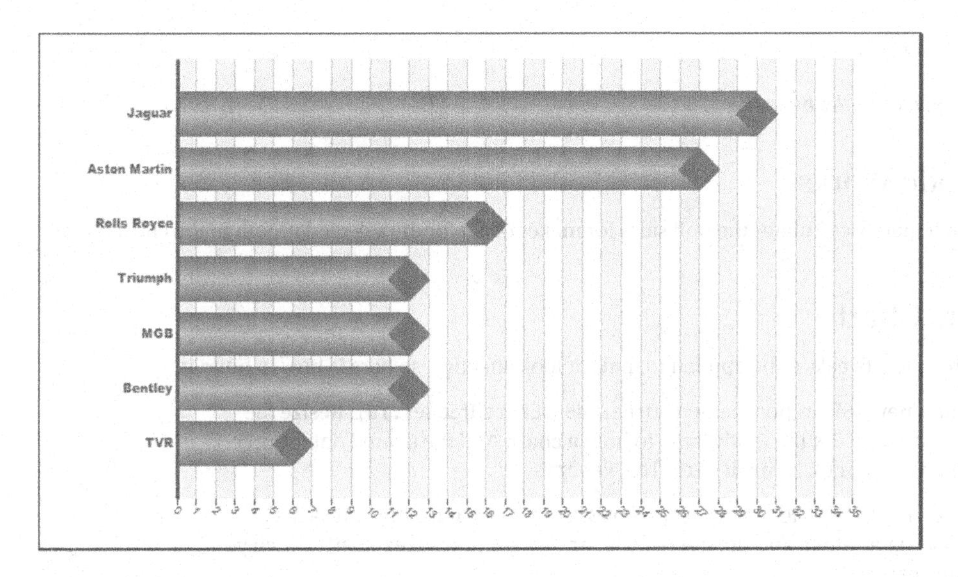

Figure 4-5. *Ordering the series in a chart*

The Source Data

As this is a simple example, the data is also extremely simple to produce. It uses the following code snippet, which is in the stored procedure `Code.pr_MonthlyCarSalesCount`, to return the car sales for the year to date:

```
DECLARE @ReportingYear INT = 2015
DECLARE @ReportingMonth TINYINT = 6

SELECT     Make
           ,COUNT(Make) AS NbSales
FROM       Reports.CarSalesData
WHERE      ReportingYear = @ReportingYear
           AND ReportingMonth <= @ReportingMonth
GROUP BY   Make
```

This T-SQL returns the output shown in Figure 4-6.

	Make	NbSales
1	Aston Martin	30
2	Bentley	13
3	Jaguar	32
4	MGB	12
5	Rolls Royce	16
6	Triumph	10
7	TVR	3

Figure 4-6. *Monthly unit sales by make*

How the Code Works

This simple snippet merely calculates the unit sales for makes of car for the selected month in the chosen year.

Building the Chart

This chart requires a few tweaks to be applied to make it look effective, so here is how to build it.

1. Create a new SSRS report named `_OrderedBarChartOfSales.rdl`. Resize the report so that it is sufficiently large to hold a chart. Add the shared data source `CarSales_Reports`. Name it `CarSales_Reports`.
2. Create a dataset named `MonthlyCarSalesCount`. Have it use the `CarSales_Reports` data source and the query `Code.pr_MonthlyCarSalesCount`, which you created earlier.
3. Add the following four shared datasets (ensuring that you also use the same name for the dataset in the report): `CurrentYear`, `CurrentMonth`, `ReportingYear`, and `ReportingMonth`. Set the parameter properties for `ReportingYear` and `ReportingMonth` as defined at the start of Chapter 1.
4. Add a bar chart from the SSRS toolbox and resize it to suit your needs. Attach the dataset `MonthlyCarSalesCount`.
5. In the Chart Data pane (at the right of the chart), add the `NbSales` fields as the ∑ value, and the `Make` field as the Category Group.
6. Right-click Make in the category groups section of the Chart Data pane and select Category group properties in the context menu. Select Sorting on the left, and in the Sort By pop-up, select `NbSales` instead of `Make` (which is set by default) as the column to order on, set the Order to A to Z, and click OK.

7. Right-click any of the bars and select Series Properties. Click Markers on the left, and set the following attributes:
    a. Marker type: Diamond
    b. Maker size: 24 point (or a suitable size relative to the width of the bars and the total dimensions of the chart)
    c. Marker color: Cornflower blue
    d. Marker border width: 0.5 point
    e. Marker border color: Automatic
8. Click Fill the left, and set the following attributes:
    a. Fill style: Gradient
    b. Color: Cornflower blue
    c. Secondary color: Light blue
    d. Gradient style: Horizontal center
9. Confirm your settings by clicking OK.
10. Click the horizontal axis. Press F4 to display the Properties window (unless it is already visible). Set the following properties:
    a. Interlaced: True
    b. InterlacedColor: Whitesmoke
    c. LineStyle: None
    d. LabelsColor: Gray
    e. MajorGridLines ➤ Enabled: True
    f. MajorGridLines ➤ LineColor: Gainsboro
    g. MajorTickMarks ➤ Enabled: True
    h. MajorTickMarks ➤ LineColor: Gray
    i. MajorTickMarks ➤ Type: Outside
    j. LabelsFont ➤ FontSize: 6 point
11. As a decorative tweak, right-click the vertical axis and select Axis Properties from the context menu. Click Line on the left and set the Line width to 2.5 point and the Line color to gray. Ensure that neither the major nor the minor tick marks are visible.
12. Format the axes, add title text (or remove titles as you see fit), and add a chart border. I have broken my own rules here, and placed the vertical axis text in the font Arial Black (and the font color to gray) to draw attention to the vehicle makes.

That is your chart finished. It should look like Figure 4-5.

How It Works

I have only the following few points to make here:

- Ordering the bars in a chart was as simple as setting the sort order for the Category Group, but it makes the chart considerably easier to read, and makes it immediately clearer which makes sell best. This is, after all, what business intelligence is all about.
- Adding striplines, if they are suitably discreet, can be more readable than gridlines. Be careful, though, to ensure that they enhance the visualization rather than just distract the user. In this example, the striplines are enhanced by the presence of the major gridlines. You can achieve a "softer" effect by hiding the major gridlines.
- Setting a gradient style for the bar can be overkill, as can adding a marker. If you prefer a more sober presentation, go for it. As much as anything I am trying to show that charts need to be varied (and sufficiently different from classic Excel charts as possible) to gain attention.
- I removed the horizontal axis line to emphasize the "horizontal" aspect of the chart. Also, as there are striplines I found a horizontal line on the axis to be a little heavy. You may prefer to keep the line.

Superposed Bar Charts

Bar charts have been around since charts first appeared on PCs (some readers may even remember them gaining popularity with Lotus 1-2-3). Their sheer ubiquity can have a detrimental effect, because readers are so used to bar charts that they do not make any conscious or unconscious attempt to interpret them.

So you, the BI developer, may need to assist the user. One way to do this is to add a little pizazz to a bar chart so that the differences in the data series stand out more. Superimposing bars and tweaking their respective widths is one way to do this.

As this chart is, as the case with so many BI charts, a comparison of two metrics, you will reuse an existing dataset from earlier in the chapter. This dataset is `Code.pr_MonthlyCarSalesWithTarget`. You will use it to present another way of using charts to compare sales with targets. An example is given in Figure 4-7. This example also uses the data for June 2015.

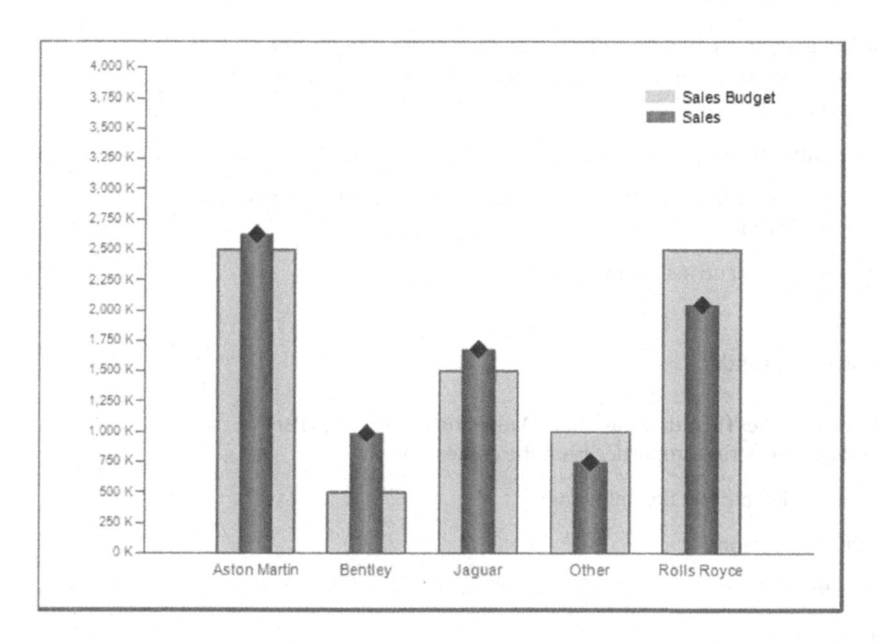

Figure 4-7. *Superposed bar charts*

The Source Data

As the source data for this chart is virtually identical to that used in the previous example, I will not show it here. Feel free to look at it in the `CarSales_Reports` database if you really want a closer look.

Creating the Chart

Now that you have seen what you are aiming for, here is how to create it.

1. Create a new SSRS report named `_MetricAndTargetSuperposed.rdl`. Resize the report so that it is sufficiently large to hold a chart. Add the shared data source `CarSales_Reports`. Name it `CarSales_Reports`.
2. Create a dataset named `MonthlyCarSalesWithTarget`. Have it use the `CarSales_Reports` data source and the query `Code.pr_MonthlyCarSalesWithTarget`, which you created earlier.
3. Add the following four shared datasets (ensuring that you also use the same name for the dataset in the report): `CurrentYear`, `CurrentMonth`, `ReportingYear`, and `ReportingMonth`. Set the parameter properties for `ReportingYear` and `ReportingMonth` as defined at the start of Chapter 1.
4. Add a 3D Column chart from the SSRS toolbox and resize it to suit your needs. Attach the dataset `MonthlyCarSalesWithTarget`.
5. In the Chart Data pane (at the right of the chart), add the `SalesBudget` and `Sales` fields as the ∑ values (in this order), and the `Make` field as the Category Group.
6. Right-click inside the chart area, but not on a column, and choose Chart Area Properties from the context menu. In the 3D Options pane, set the Rotation, Inclination, and Wall Thickness to 0, and the Fill to No Color.

7. Click the SalesBudget series in the Chart Data pane. Press F4 to display the Properties window (unless it is already visible). Expand CustomAttributes and set thePointWidth to 0.7.

8. In the Properties window, set the Color to Gold.

9. Click the Sales series in the Chart Data pane. Set the Color to cornflower blue. Set the BorderColor to No Color.

10. Expand CustomAttributes and set the following:

    a. PointWidth to 0.3

    b. DrawingStyle to Cylinder

11. Right-click the Sales series (in either the Chart Data pane or the series itself in the chart area), and choose Series Properties from the context menu.

12. Click Markers on the left and set the following:

    a. MarkerType to Diamond

    b. MarkerSize to 10 point

    c. MarkerColor to Blue

13. Click OK.

14. Hide or delete the Axis and Chart titles, and configure the chart border and axis attributes as described at the start of this chapter.

15. Right-click the vertical axis, select Vertical Axis Properties from the context menu, and enter 250000 as the Interval. Set the number format of the vertical axis to the following custom format: `#,0," K";(#,0," K")`. While you are setting vertical axis properties, disable auto fit for the labels and set the font size to 7 point.

16. Right-click the legend and select Legend Properties from the context menu. In the General tab, uncheck Show legend outside chart area.

17. Right-click any of the horizontal gridlines and select Major Gridline Properties from the context menu. Set the Line style to None.

18. Right-click the horizontal axis, select Horizontal Axis Properties from the context menu, and click Major tick marks on the left. Check the Hide major tick marks box.

That is all that you need to do to create the chart.

How It Works

You may have a couple of questions as to why you carried out some of the steps in this process. I am not suggesting that all of the aesthetic choices are necessary, but I took the opportunity to add some further effects to your SSRS BI armory. Anyway, the points I want to make are the following:

- Placing the legend inside the chart gains valuable space on the right of the chart. Of course, depending on your actual data, this may or may not prove possible, and you could have to place the legend elsewhere inside the chart.
- Removing the horizontal gridlines is purely to declutter the chart and remove a visual distraction.
- Tweaking the vertical axis interval is simply to obtain a balanced interval on the axis, and to prevent SSRS from creating too few intervals.
- You cannot use gradients in 3D charts, and this chart is technically a 3D chart. This is why you used CustomAttributes/DrawingStyle: Cylinder instead.
- Adding a marker is purely decorative because in this case I wanted to draw attention to the sales series. You may prefer not to add a marker.

You are free to disagree with these choices, and to reinstate or alter any elements that you choose.

Hints and Tips

This chart could potentially be extended in a couple of ways.

- You may want to sort the horizontal axis to leave the Other category at the right. This can be done by adding a `SortOrder` column to the source data that you use to sort the axis.
- Creating "bucket" categories for charts (as you did frequently with gauges in the previous chapter) can be an excellent way of reducing visual clutter by grouping les important metrics in a single element. I find it easiest to hard-code this in the T-SQL.

Radar Charts

Much business intelligence consists of comparing values and being alerted to anomalies. One chart type that lends itself easily to this kind of analysis is a radar chart. This kind of chart is great when you have a few (say between four and eight) data points that are on the same scale. It is instantly evident if any data point varies compared to the others.

As it is easier to understand the techniques if you can visualize the result, an example of a radar chart is shown in Figure 4-8. What is important to remember is that each data point is essentially a percentage (so that disparate figures have been reduced to a similar scale). In this case, the percentage is the over- or under-shoot compared to the budget.

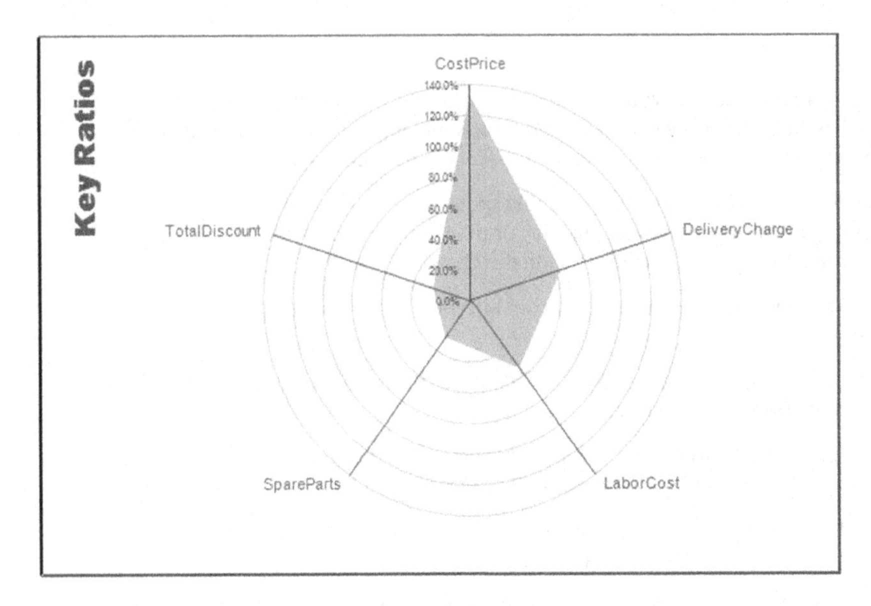

Figure 4-8. *A radar chart*

The Source Data

For this example, the data is also extremely simple to produce. It uses the following code snippet, which is in the stored procedure `Code.pr_WD_RatioAnalysis`, to return the car sales for the year to date (June 2015 in this example):

```
DECLARE @ReportingYear INT = 2015
DECLARE @ReportingMonth TINYINT = 6

SELECT
((1/(SELECT BudgetValue FROM Reference.Budget WHERE [Year] = @ReportingYear AND
BudgetElement = 'KeyRatio' AND BudgetDetail = 'CostPrice')) * SUM(CostPrice)) /
SUM(SalePrice) AS Ratio, 'CostPrice' AS RatioName
FROM   Reports.CarSalesData
WHERE  ReportingYear = @ReportingYear
       AND ReportingMonth <= @ReportingMonth
UNION
SELECT
((1/(SELECT BudgetValue FROM Reference.Budget WHERE [Year] = @ReportingYear AND
BudgetElement = 'KeyRatio' AND BudgetDetail = 'TotalDiscount')) * SUM(TotalDiscount)) /
SUM(SalePrice), 'TotalDiscount'
FROM   Reports.CarSalesData
WHERE  ReportingYear = @ReportingYear
       AND ReportingMonth <= @ReportingMonth
UNION
SELECT
((1/(SELECT BudgetValue FROM Reference.Budget WHERE [Year] = @ReportingYear AND
BudgetElement = 'KeyRatio' AND BudgetDetail = 'DeliveryCharge')) * SUM(DeliveryCharge)) /
SUM(SalePrice), 'DeliveryCharge'
```

```
FROM    Reports.CarSalesData
WHERE   ReportingYear = @ReportingYear
        AND ReportingMonth <= @ReportingMonth
UNION
SELECT
((1/(SELECT BudgetValue FROM Reference.Budget WHERE [Year] = @ReportingYear AND
BudgetElement = 'KeyRatio' AND BudgetDetail = 'LabourCost')) * SUM(LaborCost)) /
SUM(SalePrice), 'LaborCost'
FROM    Reports.CarSalesData
WHERE   ReportingYear = @ReportingYear
        AND ReportingMonth <= @ReportingMonth
UNION
SELECT
((1/(SELECT BudgetValue FROM Reference.Budget WHERE [Year] = @ReportingYear AND
BudgetElement = 'KeyRatio' AND BudgetDetail = 'SpareParts')) * SUM(SpareParts)) /
SUM(SalePrice), 'SpareParts'
FROM    Reports.CarSalesData
WHERE   ReportingYear = @ReportingYear
        AND ReportingMonth <= @ReportingMonth
```

This T-SQL returns the data shown in Figure 4-9.

	Ratio	RatioName
1	0.267681	TotalDiscount
2	0.295551	SpareParts
3	0.543472	LaborCost
4	0.630400	DeliveryCharge
5	1.332744	CostPrice

Figure 4-9. *Key ratios expressed as a percentage of target*

How the Code Works

This chart requires five different sales metrics to be returned from the CarSalesData view and then expressed as a percentage of the corresponding budgetary figure. In this case, I preferred (for no real reason other than to show that it is possible) to use UNION statements to assemble the data set.

Building the Chart

Now that you have your source data, you can build the chart to display it.

1. Create a new SSRS report named _RatioAnalysis.rdl. Resize the report so that it is sufficiently large to hold a radar chart. Add the shared data source CarSales_Reports. Ensure that its name in the report is CarSales_Reports.
2. Create a dataset named RatioAnalysis. Have it use the CarSales_Reports data source and the stored procedure Code.pr_WD_RatioAnalysis, which you saw earlier.

3. Add the following four shared datasets (ensuring that you also use the same name for the dataset in the report): CurrentYear, CurrentMonth, ReportingYear, and ReportingMonth. Set the parameter properties for ReportingYear and ReportingMonth as defined at the start of Chapter 1.
4. Add a radar chart from the SSRS toolbox (it is at the bottom with the polar charts) and resize it to suit your needs. Attach the dataset RatioAnalysis.
5. In the Chart Data pane (at the right of the chart), add the Ratio field as the ∑ values, and the RatioName field as the Category Group.
6. Delete the legend and hide the two axis titles.
7. Right-click the radial axis (the vertical one with the figures) and select Radial Axis Properties from the context menu. Select Disable auto-fit in the Labels pane. Set the Label Font to Arial Narrow, 7 point in gray. Set the Number Format to Percentage with one decimal place. Set the Line Color for both the Line and the Major Tick Marks (with Position: Outside and Length: 1) to light gray. Ensure that the minor tick marks are hidden.
8. Right-click one of the concentric circles and select Major Gridline Properties from the context menu. Set the Line Color to light gray.
9. Right-click the data series (the colored shape in the chart) and select Series Properties from the context menu. Select Fill on the left and set the Color to light blue.
10. Right-click the chart title and select Title Properties from the context menu. Set the Title Text to Key Properties, the font to Arial Black 14 point gray, and (in the General pane) click the top left hand vertical button for the Title Position.
11. Select the chart area and display the Properties window. Click the ellipses to the right of the Category Axes property. The ChartAxisCollection editor dialog will appear. Select Primary on the left and set the color to gray. Expand the LabelsFont property and the font size to 8 point. Confirm the warning about labels autofit being disabled, then click OK.

How It Works

The main point to note for this chart is that this is a classic example of a chart where most of the work is done when preparing the data. If the data is right, then creating the chart is a piece of cake. As you require an output table of five elements, the easiest way to deliver this is using a UNION query where each record is customized to return the required ratio.

Hints and Tips

Here are two hints for this example.

- Striplines can also be extremely effective in radar charts.
- Vertical titles can help you save a lot of space that can be used efficiently to make the chart clearer. As ever, however, they should be used discreetly.

Bubble Charts

Bubble charts seem to be in vogue at the moment. This is probably because, if properly constructed, they can display four data series and how they relate to each other. An example of such a chart is shown in Figure 4-10. This chart displays sales and profit by make of car by age range. One data series displays the make on the X (horizontal) axis. The second data series shows how (on the Y or vertical axis) the sales relate to the age of the car. The third data series shows the bubble size for each intersection of sales and age range. Finally, the fourth data series (the color) indicates the percentage profit.

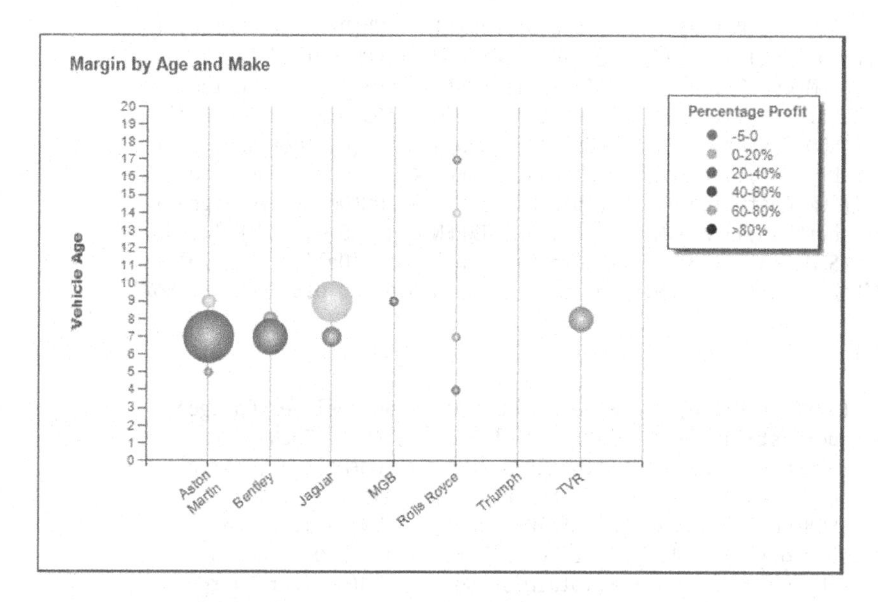

Figure 4-10. *A bubble chart*

The Source Data

The data for this chart is produced by the following code snippet, which is in the stored procedure `Code.pr_WD_MarginByAgeAndMake`. The values returned are used as follows:

- Make, Category Group: Horizontal axis
- VehicleAge, Y Value: Vertical axis
- NbSales, Size: Bubble size
- Percentage Profit, Series Groups: Legend

The code for this chart (for June 2015) is

```
DECLARE @ReportingYear INT = 2015
DECLARE @ReportingMonth TINYINT = 6

SELECT
 Make
,COUNT(SalePrice) AS NbSales
,CASE
WHEN (SUM(SalePrice) - (SUM(CostPrice) + SUM(TotalDiscount) + SUM(DeliveryCharge) +
SUM(SpareParts) + SUM(LaborCost))) / SUM(SalePrice) BETWEEN -5 AND 0 THEN '-5-0'
WHEN (SUM(SalePrice) - (SUM(CostPrice) + SUM(TotalDiscount) + SUM(DeliveryCharge) +
SUM(SpareParts) + SUM(LaborCost))) / SUM(SalePrice) BETWEEN 0 AND 0.2 THEN '0-20%'
WHEN (SUM(SalePrice) - (SUM(CostPrice) + SUM(TotalDiscount) + SUM(DeliveryCharge) +
SUM(SpareParts) + SUM(LaborCost))) / SUM(SalePrice) BETWEEN 0.2 AND 0.4 THEN '20-40%'
WHEN (SUM(SalePrice) - (SUM(CostPrice) + SUM(TotalDiscount) + SUM(DeliveryCharge) +
SUM(SpareParts) + SUM(LaborCost))) / SUM(SalePrice) BETWEEN 0.4 AND 0.6 THEN '40-60%'
WHEN (SUM(SalePrice) - (SUM(CostPrice) + SUM(TotalDiscount) + SUM(DeliveryCharge) +
SUM(SpareParts) + SUM(LaborCost))) / SUM(SalePrice) BETWEEN 0.6 AND 0.8 THEN '60-80%'
ELSE '>80%'
END AS PercentageProfit
,CASE
WHEN (SUM(SalePrice) - (SUM(CostPrice) + SUM(TotalDiscount) + SUM(DeliveryCharge) +
SUM(SpareParts) + SUM(LaborCost))) / SUM(SalePrice) BETWEEN -5 AND 0 THEN 1
WHEN (SUM(SalePrice) - (SUM(CostPrice) + SUM(TotalDiscount) + SUM(DeliveryCharge) +
SUM(SpareParts) + SUM(LaborCost))) / SUM(SalePrice) BETWEEN 0 AND 0.2 THEN 2
WHEN (SUM(SalePrice) - (SUM(CostPrice) + SUM(TotalDiscount) + SUM(DeliveryCharge) +
SUM(SpareParts) + SUM(LaborCost))) / SUM(SalePrice) BETWEEN 0.2 AND 0.4 THEN 3
WHEN (SUM(SalePrice) - (SUM(CostPrice) + SUM(TotalDiscount) + SUM(DeliveryCharge) +
SUM(SpareParts) + SUM(LaborCost))) / SUM(SalePrice) BETWEEN 0.4 AND 0.6 THEN 4
WHEN (SUM(SalePrice) - (SUM(CostPrice) + SUM(TotalDiscount) + SUM(DeliveryCharge) +
SUM(SpareParts) + SUM(LaborCost))) / SUM(SalePrice) BETWEEN 0.6 AND 0.8 THEN 5
ELSE 6
END AS ProfitOrder
,DATEDIFF(yy, Registration_Date, GETDATE()) AS VehicleAge

FROM       Reports.CarSalesData
WHERE      ReportingYear = @ReportingYear
           AND ReportingMonth <= @ReportingMonth
GROUP BY   Make, DATEDIFF(yy, Registration_Date, GETDATE())
ORDER BY   Make
```

This T-SQL returns several rows of data, a few of which are displayed in Figure 4-11.

	Make	NbSales	PercentageProfit	ProfitOrder	VehicleAge
1	Aston Martin	2	0-20%	2	5
2	Aston Martin	12	20-40%	3	7
3	Aston Martin	6	20-40%	3	8
4	Aston Martin	4	40-60%	4	9
5	Aston Martin	2	20-40%	3	15
6	Aston Martin	2	40-60%	4	17
7	Aston Martin	2	20-40%	3	20
8	Bentley	12	40-60%	4	7

Figure 4-11. *The data for a bubble chart*

How the Code Works

Here the data is the count of the number of sales for the selected year up to and including the selected month. However, the code groups the number of sales per make according the whole number of years representing the age of the car. It then calculates the percentage profit of each age range using a hard-coded setting to produce five profit ranges.

The data may seem more complex than it really is. All that the CASE statements do is to calculate profit in discreet ranges to avoid a plethora of different values, and then add a sort order from smallest to largest, which is used by the legend to provide visual coherence by ordering the ranges from smallest to largest, rather than in alphabetical order.

Building the Chart

Now that you have your source data you can build the chart to display it.

1. Create a new SSRS report named `_BubbleChartMarginByAgeAndMake.rdl`. Resize the report so that it is sufficiently large to hold a largish chart. Add the shared data source `CarSales_Reports`. Ensure that its name in the report is `CarSales_Reports`.

2. Create a dataset named `MarginByAgeAndMake`. Have it use the `CarSales_Reports` data source and the query `MarginByAgeAndMake`, whose code is shown above.

3. Add the following four shared datasets (ensuring that you also use the same name for the dataset in the report): `CurrentYear`, `CurrentMonth`, `ReportingYear`, and `ReportingMonth`. Set the parameter properties for `ReportingYear` and `ReportingMonth` as defined at the start of Chapter 1.

4. Add a 3D bubble chart from the SSRS toolbox (it is at the bottom with the scatter charts) and resize it to suit your needs. Attach the dataset `MarginByAgeAndMake` as the data source.

5. Right-click inside the chart area, but not on a column, and choose Chart Area Properties from the context menu. In the 3D Options pane, set the Rotation, Inclination, and Wall Thickness to 0. Select Fill and set the Fill style to Solid and the Color to No Color.

6. In the Chart Data pane (at the right of the chart), add the `VehicleAge` field as the ∑ values. Do not worry about for the moment about the “extra” fields Size and X Value that appear.
7. Add the `Make` field as the Category Group, and PercentageProfit as the Series Group.
8. Right-click VehicleAge in the Chart Data pane and select Series Properties from the context menu. Set the properties shown below. This will alter the fields that are shown in the Chart Data pane.

Section	Property	Value
Series Data	Bubble Size	`[Sum(NbSales)]`
	Category Field	`[Make]`
Markers	Marker Type	Circle
	Marker Size	30 point
	Marker Color	Automatic

9. Hide the horizontal axis title.
10. Right-click the vertical axis and select Vertical Axis Properties from the context menu. Set the Label font to Arial Bold 9 point gray.
11. Right-click the title and select Title Properties from the context menu. Set the font to Arial Bold 10 point gray. Set the Title Text to Margin by Age and Make, and in the General pane, the Title Position to top left (the left-hand button of the top row).
12. Right-click any of the bubbles (or on either of the ∑ values in the Chart Data pane) and choose Series Properties from the context menu. Select Markers to the left, and set the Marker Size to 30 point.
13. Right-click the legend and select Legend Properties from the context menu. Add a border (in light gray) and set the font to Arial Narrow 9 point gray. Ensure (in the General pane) that the legend is on the upper right and is set to Tall Table.
14. Right-click the Series Group (PercentageProfit) in the Chart Data and select Series Group properties from the context menu. Select Sorting on the left, and select ProfitOrder from the Sort By pop-up list instead of PercentageProfit, and click OK.

You can now preview your bubble chart. If all has gone well, it will look like Figure 4-10, above.

How It Works

I have only a three short points to make about what you have done here.

- Setting the marker size only affects the size of the points in the legend. Indeed, the actual size of the marker is largely irrelevant. However, if you leave it at the default, you will produce circles in the legend that are too small to identify clearly.
- One again, you do not have to adopt all of the stylistic modifications that I am suggesting here. They are outlined so that you know how to find them should you want to create your own style or add a house style.
- You applied a custom sort order to the series group so that the legend would be easier to read. Without this the legend, elements would start with >80% and show -5-0 next to last.

Waterfall Charts

A waterfall chart is perfect when you need to isolate the various data elements that make up a whole, such as the determination of profit for a sale, for instance. A waterfall chart will only display one data series, essentially, but you can display multiple waterfall charts in a multiple-chart structure if you need to compare and contrast several elements.

An example of a set of waterfall charts is shown in Figure 4-12, using the data for June 2015. In the example, you will only be creating a single chart, but the source data allows you subsequently to alter the filter on the chart to use it as part of a composite visualization.

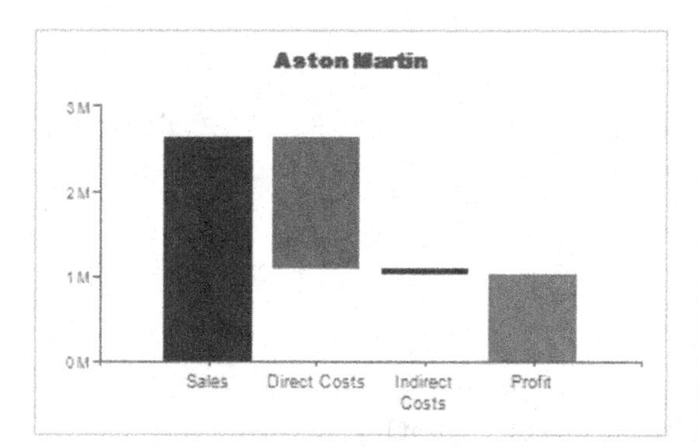

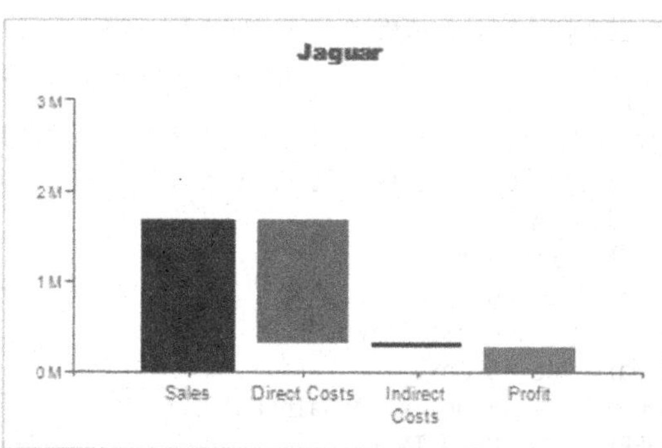

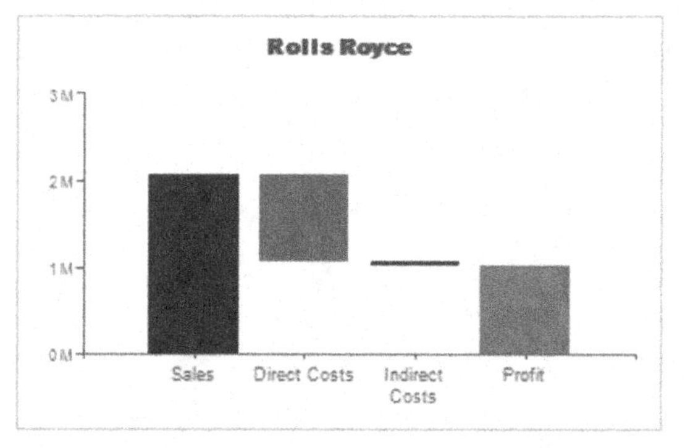

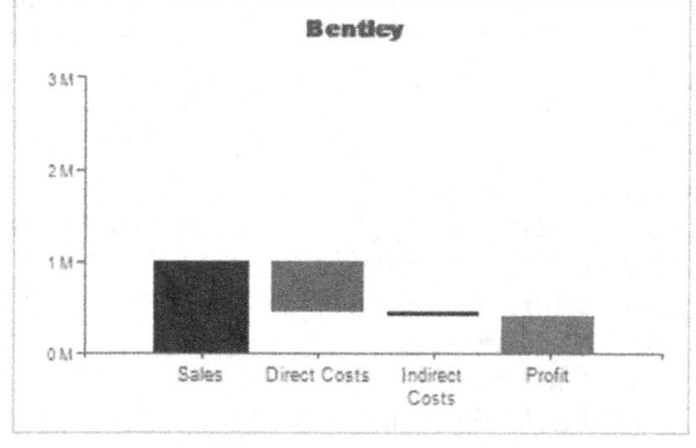

Figure 4-12. *A waterfall chart*

The Source Data

The code for this dataset can be found in the stored procedure Code.pr_WaterfallChartOfProfitByMake in the CarSales_Reports database. It is as follows (for June 2015 again; a pattern is emerging here):

```
DECLARE @ReportingYear INT = 2015
DECLARE @ReportingMonth TINYINT = 6
DECLARE @MaxValUE BIGINT

;
WITH Mx_CTE
AS
(
     SELECT
     Code.fn_ScaleSparse(MAX(SalePrice)) AS MaxValue
     FROM
          (
          SELECT    SUM(SalePrice) AS SalePrice, Make
          FROM      Reports.CarSalesData
          WHERE     Make IN ('Rolls Royce', 'Jaguar', 'Aston Martin', 'Bentley')
                    AND ReportingYear = @ReportingYear
                    AND ReportingMonth <= @ReportingMonth
          GROUP BY  Make
          ) Mx
)

SELECT @MaxValue = MaxValue FROM Mx_CTE

SELECT
Make
,Sum(SalePrice) AS Result
,'Sales' AS Element
,1 AS SortOrder
,'Blue' AS ColumnColor
,'Lower' AS ChartElement
,@MaxValue AS MaxValue
FROM        Reports.CarSalesData
WHERE       Make IN ('Rolls Royce', 'Jaguar', 'Aston Martin', 'Bentley')
            AND ReportingYear = @ReportingYear
            AND ReportingMonth <= @ReportingMonth
GROUP BY    Make
UNION
SELECT
Make
,Sum(CostPrice) + SUM(TotalDiscount) AS Result
,'Direct Costs' AS Element
,2 AS SortOrder
,'Red' AS ColumnColor
,'Upper' AS ChartElement
,@MaxValue AS MaxValue
FROM        Reports.CarSalesData
```

```
WHERE       Make IN ('Rolls Royce', 'Jaguar', 'Aston Martin', 'Bentley')
            AND ReportingYear = @ReportingYear
            AND ReportingMonth <= @ReportingMonth
GROUP BY    Make
UNION
SELECT
Make
,Sum(SalePrice) - (Sum(CostPrice) + SUM(TotalDiscount)) AS Result
,'Direct Costs' AS Element
,2 AS SortOrder
,'white' AS ColumnColor
,'Lower' AS ChartElement
,@MaxValue AS MaxValue
FROM        Reports.CarSalesData
WHERE       Make IN ('Rolls Royce', 'Jaguar', 'Aston Martin', 'Bentley')
            AND ReportingYear = @ReportingYear
            AND ReportingMonth <= @ReportingMonth
GROUP BY    Make
UNION
SELECT
Make
,Sum(DeliveryCharge) + SUM(SpareParts) + SUM(LaborCost) AS Result
,'Indirect Costs' AS Element
,3 AS SortOrder
,'Purple' AS ColumnColor
,'Upper' AS ChartElement
,@MaxValue AS MaxValue
FROM        Reports.CarSalesData
WHERE       Make IN ('Rolls Royce', 'Jaguar', 'Aston Martin', 'Bentley')
            AND ReportingYear = @ReportingYear
            AND ReportingMonth <= @ReportingMonth
GROUP BY    Make
UNION
SELECT
Make
,(Sum(SalePrice) - (Sum(CostPrice) + SUM(TotalDiscount))) - (Sum(DeliveryCharge) +
SUM(SpareParts) + SUM(LaborCost)) AS Result
,'Indirect Costs' AS Element
,3 AS SortOrder
,'White' AS ColumnColor
,'Lower' AS ChartElement
,@MaxValue AS MaxValue
FROM        Reports.CarSalesData
WHERE       Make IN ('Rolls Royce', 'Jaguar', 'Aston Martin', 'Bentley')
            AND ReportingYear = @ReportingYear
            AND ReportingMonth <= @ReportingMonth
GROUP BY    Make
UNION
SELECT
Make
,Sum(SalePrice) - (Sum(CostPrice) + SUM(TotalDiscount) + Sum(DeliveryCharge) +
SUM(SpareParts) + SUM(LaborCost)) AS Result
```

```
,'Profit' AS Element
,4 AS SortOrder
,'CornflowerBlue' AS ColumnColor
,'Lower' AS ChartElement
,@MaxValue AS MaxValue
FROM        Reports.CarSalesData
WHERE       Make IN ('Rolls Royce', 'Jaguar', 'Aston Martin', 'Bentley')
            AND ReportingYear = @ReportingYear
            AND ReportingMonth <= @ReportingMonth
GROUP BY    Make
```

This T-SQL returns many rows of data. Figure 4-13 shows the data for one make of vehicle.

	Make	Result	Element	SortOrder	ColumnColor	ChartElement	MaxValue
1	Aston Martin	69092.00	Indirect Costs	3	Purple	Upper	3000000
2	Aston Martin	1014442.99	Indirect Costs	3	White	Lower	3000000
3	Aston Martin	1014442.99	Profit	4	CornflowerBlue	Lower	3000000
4	Aston Martin	1083534.99	Direct Costs	2	white	Lower	3000000
5	Aston Martin	1546145.01	Direct Costs	2	Red	Upper	3000000
6	Aston Martin	2629680.00	Sales	1	Blue	Lower	3000000

Figure 4-13. *The data for a waterfall chart*

How the Code Works

This type of chart does not take its data directly from a data source, but needs to apply a little logic first. The arithmetic is not difficult (it is essentially a series of accounting operations) but it requires two records for each column:

- One series for the lower section of the column
- One series for the upper section of the column

The two series are defined in the `ChartElement` column and are hard-coded. Equally, and to show that it can be done, the color for the series is hard-coded in the T-SQL. To finish, a manually-defined sort order is added. This will be used to present the columns in the correct sequence on the horizontal axis.

This code could be extended, if you wish, to set the colors as variables or as data sourced from another table.

Creating a Waterfall Chart

Now that you have your source data, you can build the waterfall chart to display it.

1. Create a new SSRS report named `_WaterfallChartOfProfitByMake.rdl`. Resize the report so that it is sufficiently large to hold a largish chart. Add the shared data source `CarSales_Reports`. Ensure that its name in the report is `CarSales_Reports`.
2. Create a dataset named `WaterfallChartOfProfitByMake`. Have it use the `CarSales_Reports` data source and the query `Code.pr_WaterfallChartOfProfitByMake`, which you created earlier.
3. Add the following four shared datasets (ensuring that you also use the same name for the dataset in the report): `CurrentYear`, `CurrentMonth`, `ReportingYear`, and `ReportingMonth`. Set the parameter properties for `ReportingYear` and `ReportingMonth` as defined at the start of Chapter 1.
4. Add a Stacked Column chart from the SSRS and resize it a little, making it smaller than the charts that you have created so far in this chapter. Attach the dataset `WaterfallChartOfProfitByMake` as the data source.
5. Right-click the chart (not on any elements of the chart) and select Chart Properties from the context menu. Click Filters on the left and then click Add to add a filter. Set the filter to `Make = Aston Martin`.
6. In the Chart Data pane (at the right of the chart), add the Result field as the Σ values. Add the Element field as the Category Group, and ChartElement as the Series Group.
7. Right-click the Element field in the Category Groups area of the Chart Data pane and select Category Group Properties from the context menu. Click Sorting on the left and then select SortOrder from the Sort By pop-up (instead of Element) and click OK.
8. Right-click one of the columns and select Series Properties from the context menu. Click Fill on the left, and then the expression (Fx) button for the Color. Enter the following expression:

   ```
   =Fields!ColumnColor.Value
   ```

9. Click OK to return to the Series Properties dialog and then click OK again.
10. Hide the two axis titles and delete the legend.
11. Set the chart title to 12 point Arial Black in gray. Place the title at the top center of the chart. In the General pane of the Title Properties dialog, click the expression (Fx) button for the title. Enter the following expression:

    ```
    =Fields!Make.Value
    ```

12. Confirm your modifications to the title with OK.
13. Right-click one of the horizontal gridlines and uncheck Show Gridlines.

14. Format the vertical and horizontal axes as described for previous charts, or indeed in any way that you prefer.

15. Apply the following number format to the vertical axis: `#,0.0,," M";(#,0.0,," M")`. In the Axis Options pane of the vertical axis properties dialog, set the Interval to 1000000. Click the expression (Fx) button for the Maximum and enter the following expression:

```
=Fields!MaxValue.Value
```

16. Make three copies of the chart and organize the charts into a 2 x 2 matrix.

17. Set the filters for the copied charts respectively to Jaguar, Rolls Royce, and Bentley.

You should now have an initial waterfall chart that shows how the final profit figure for a make of car is computed.

How It Works

This chart uses the two data series that were calculated in the T-SQL to produce the waterfall effect. The trick is to set the lower series to appear transparent for the final three columns. This is done by applying the same color as the chart background.

This group of charts really needs to be used as a set. This is why the maximum value is calculated using a CTE and not left to SSRS to decide. This way the same maximum figure is applied to each of the charts. Doing this allows sales for different makes to be compared without multiple vertical scales distorting the picture and making comparison difficult.

The final trick is to force the order of the horizontal axis, again using a hard-coded value set in the source SQL.

Hints and Tips

Here are several tips for this example.

- You can use a separate dataset to return maximum value for vertical axis if you prefer, as was the case in previous examples.
- The rounding function (fn_ScaleSparse) is one of the three helper functions explained in Chapter 1.
- You can, when setting the reference to database fields, click Fields on the left of the Expression dialog and then double-click the field that you want to add to the expression.
- In this example, you set the vertical axis maximum value to be the same for all the charts. This allows the user to compare the metrics across the different makes.

Trellis Charts

Sometimes a chart itself is extremely simple. Its value, however, is drawn from the fact that it allows multiple elements to be compared. This is where trellis charts (also known as lattice charts) can be useful.

A trellis chart in SSRS is simply a table that holds a matrix of charts. They are usually between two and four rows and two and five columns. Each chart in the matrix is a copy of the same chart, filtered to display a different record from the source data.

For this type of chart, creating the source data is the easy part (though, as ever, this is important). Most of the attention is spent setting up the table that will hold the multiple charts. It is worth noting that the charts that you use really must be simple, or their density will make the information unreadable, and consequently invalidate the whole purpose of the chart. An example of a trellis chart is shown in Figure 4-14.

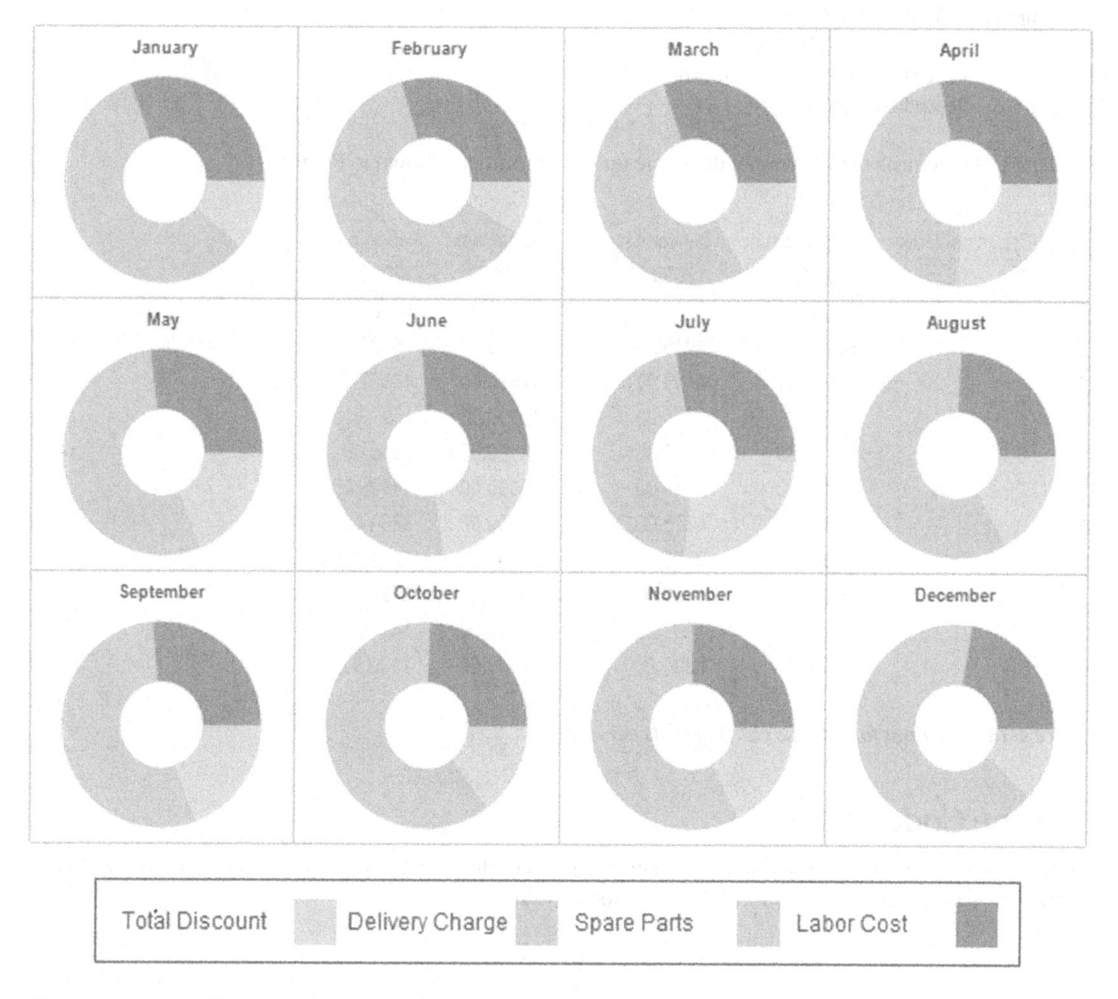

***Figure 4-14.** A trellis chart*

The Source Data

The data for this chart is produced by the following code snippet, which is in the stored procedure Code.pr_Trellis:

```
DECLARE @ReportingYear INT = 2015

SELECT
ReportingMonth
,SUM(TotalDiscount) AS TotalDiscount
,SUM(DeliveryCharge) AS DeliveryCharge
,SUM(SpareParts) AS SpareParts
,SUM(LaborCost) AS LaborCost
FROM        Reports.CarSalesData
WHERE       ReportingYear = @ReportingYear
GROUP BY    ReportingMonth
```

This T-SQL returns twelve rows of data, one for each month, as shown in Figure 4-15.

	ReportingMonth	TotalDiscount	DeliveryCharge	SpareParts	LaborCost
1	1	6250.00	12000.00	17600.00	12082.00
2	2	4100.00	7725.00	14100.00	12860.00
3	3	9900.00	10455.00	17470.00	15733.00
4	4	12000.00	4400.00	14990.00	13677.00
5	5	11545.01	8775.00	16990.00	15026.00
6	6	10400.00	7700.00	14590.00	9845.00
7	7	13750.00	5450.00	15190.00	13606.00
8	8	6650.00	4875.00	13200.00	8864.00
9	9	8100.00	2325.00	14940.00	10712.00
10	10	6600.00	6250.00	16040.00	9702.00
11	11	11950.00	11875.00	19520.00	14446.00
12	12	5650.00	8300.00	15310.00	7880.00

Figure 4-15. *The data for a trellis chart of costs per month*

How the Code Works

You have, for once, a code snippet that is mercifully easy to understand. It aggregates a set of metrics from a source view and groups them by month for the selected year.

Building a Trellis Chart

So, armed with your source data, you can create the trellis chart.

1. Create a new SSRS report named `_TrellisChartCostsPerMonth.rdl`. Resize the report so that it is sufficiently large to hold a large trellis chart. Add the shared data source `CarSales_Reports`. Name it `CarSales_Reports`.
2. Create a dataset named `Trellis`. Have it use the `CarSales_Reports` data source and the stored procedure `Code.pr_Trellis` (which you saw earlier).
3. Add the following two shared datasets (ensuring that you also use the same name for the dataset in the report): `CurrentYear` and `ReportingYear`. Set the parameter properties for `ReportingYear` as defined at the start of Chapter 1.
4. Add a second, embedded dataset named `Dummy` that contains the following SQL code: `SELECT ''`.
5. Add a table to the report. Right-click to the left of the data row on the row selector (the grey box), and select Insert Row ➤ Outside Group - Below. Do this a total of three times.
6. Delete the data row and header row. Your table should only have the three rows that you inserted in the previous step.
7. Add a fourth column to the table.
8. Set the table's dataset to `Trellis`.
9. Set the width of a textbox in each of the three rows to 1.5 inches. Set the height of a textbox in each of the columns to 1.5 inches. This will set all the textboxes to 1.5 inches square.
10. Drag a chart from the toolbox into any of the textboxes in the table and choose the Donut chart type.
11. Right-click in the chart on the legend and select Delete Legend from the context menu.
12. In the Chart Data pane (which should have appeared; if not, click to display it) add the values of TotalDiscount, DeliveryCharge, SpareParts, and LaborCost. Remove the Category Group.
13. In the Chart Properties window, set the Palette to Custom.
14. In the Chart Properties window, click the ellipses to the right of the `CustomPaletteColors` property. In the ChartColor Collection Editor dialog, add four members and set their colors to Aqua, Gold, Lime, and Tan.
15. Set the chart title font to Arial Narrow 7 point gray.
16. Right-click the chart and select Chart Properties. Select Filter on the left, and click Add. Set the Expression to ReportingMonth = 1.
17. Copy the chart to the 11 other text boxes in the table.
18. Type in the month name for each chart as shown previously in Figure 4-14.
19. Set the filter for each chart to correspond to the month number of the month that the chart represents.

This creates the trellis chart. However, it is missing a legend. This is something that you will learn in Chapter 7.

How It Works

The key to a successful trellis chart is a clean table structure from the outset. Once you have created a table with the required number of columns and rows, everything else is a copy-and-paste operation once the initial chart has been created.

Once again, as so often with multiple charts and gauges, getting the initial chart right is key. Otherwise, you will be deleting and re-copying and re-pasting over and over again as you make and reapply changes multiple times.

Hints and Tips

Keep these hints and tips in mind for this example.

- The reason that you defined a custom palette is that the order of palette colors corresponds to the order of the series in the Chart Data pane. This makes it easy to know which color corresponds to which series. Consequently, if you alter the series order, you will have to adjust the colors in the custom legend. This is explained in Chapter 7.
- This chart compares the months in a year, so the source data does not need the month as a parameter.
- You may find it useful to create a set of empty preconfigured tables that you can use to hold trellis charts (and other composite visualizations) in your BI presentations.

A Pyramid Chart

As mentioned, one problem that you as a BI developer will face is avoiding the repetition of constantly used chart types. To avoid "chart fatigue" on the part of your users, you should vary chart types if only to waken your public from their slumbers.

Given the general consensus against using pie charts in BI, one alternative is to use pyramid charts. An example of a pyramid chart is shown in Figure 4-16.

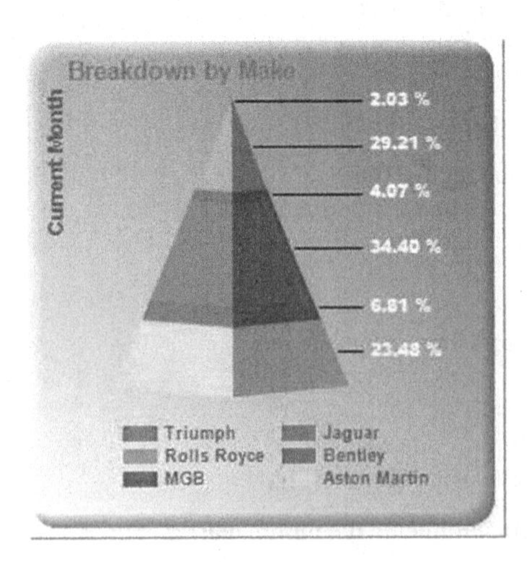

Figure 4-16. *A pyramid chart*

This chart simply shows sales for a selected month (June 2015) by make. It has the added advantage of sorting the makes by sales amount and making the relative percentages clear.

The Source Data

The data for this chart is produced by the following code snippet, which is in the stored procedure Code.pr_WD_PyramidSales_Simple:

```
DECLARE @ReportingYear INT = 2015
DECLARE @ReportingMonth TINYINT = 6
IF OBJECT_ID('tempdb..#Tmp_Output') IS NOT NULL DROP TABLE tempdb..#Tmp_Output

CREATE TABLE #Tmp_Output (ChartType VARCHAR(25) COLLATE DATABASE_DEFAULT, Make VARCHAR(25)
COLLATE DATABASE_DEFAULT, Sales NUMERIC(18,6))

INSERT INTO #Tmp_Output (ChartType, Make, Sales)

SELECT
'MonthlySales'
,Make
,SUM(SalePrice) AS Sales
FROM        Reports.CarSalesData
WHERE       ReportingYear = @ReportingYear
            AND ReportingMonth = @ReportingMonth
GROUP BY    Make

SELECT      * FROM #Tmp_Output
```

This T-SQL returns six rows of data, one for each make of car, as shown in Figure 4-17. The ChartType field is not used yet, but you will see it in action in Chapter 7.

	ChartType	Make	Sales
1	MonthlySales	Aston Martin	291380.000000
2	MonthlySales	Bentley	84500.000000
3	MonthlySales	Jaguar	427000.000000
4	MonthlySales	MGB	50500.000000
5	MonthlySales	Rolls Royce	362500.000000
6	MonthlySales	Triumph	25250.000000
7	YearlySales	Aston Martin	10686040.000000
8	YearlySales	Bentley	4951250.000000
9	YearlySales	Jaguar	6289250.000000
10	YearlySales	MGB	1011000.000000
11	YearlySales	Rolls Royce	7356900.000000
12	YearlySales	Triumph	875000.000000
13	YearlySales	TVR	528500.000000
14	QuarterlySales	Aston Martin	2217470.000000
15	QuarterlySales	Bentley	1461500.000000
16	QuarterlySales	Jaguar	1825000.000000
17	QuarterlySales	MGB	224500.000000
18	QuarterlySales	Rolls Royce	1940250.000000
19	QuarterlySales	Triumph	168500.000000

Figure 4-17. *The data for a trellis chart of costs per month*

How the Code Works

You have, once again, a code snippet that is mercifully easy to understand. It returns the total sales from a source view for a selected month and year.

Building a Pyramid Chart

Now that you have the data, you can create the pyramid chart.

1. Create a new SSRS report named `_PyramidSales_Single.rdl`. Resize the report so that it is sufficiently large to hold a large trellis chart. Add the shared data source `CarSales_Reports`. Name it `CarSales_Reports`.
2. Create a dataset named `PyramidSales`. Have it use the `CarSales_Reports` data source and the stored procedure `Code.pr_WD_PyramidSales_Simple` (which you saw earlier).
3. Add the following two shared datasets (ensuring that you also use the same name for the dataset in the report): `CurrentYear` and `ReportingYear`. Set the parameter properties for `ReportingYear` as defined at the start of Chapter 1.
4. Drag a chart from the toolbox on to the report area. Select 3-D Pyramid as the chart type.
5. Set `PyramidSales` as the dataset, Sales as the ∑ Values, and Make as the Category Groups.

6. Right-click the chart area and select Chart Properties. Set the following properties:

Section	Property	Value
Fill	Fill style	Gradient
	Color	Dim gray
	Secondary color	White
	Gradient style	Diagonal right
Border	Border type	Raised

7. Select the legend, and in the Properties pane set the following properties:

Property	Value
Position	Bottom center
Font	Arial 7 point bold
Color	Dim gray
Layout	Wide table

8. Select the chart title and set the following properties in the Properties pane:

Property	Value
Caption	Breakdown by Make
Color	Gray
Custom Position	Enabled
	Height: 7.97
	Left: 7.07
	Top: 2
	Width: 80
Font	Arial 8 point bold

9. Right-click inside the chart and select Add New Title from the context menu. Select the new title and set the following properties in the Properties pane:

Property	Value
Caption	Current Month
Color	Dim gray
DockOffset	-4
Position	LeftTop
Font	Arial 7 point bold

10. Right-click the pyramid chart and select Add Data Labels from the context menu.
11. Right-click one of the data labels and select Series Label Properties. Set the following properties:

Section	Property	Value
General	Label data	#PERCENT
Font	Font	Arial Black
	Size	7 point
	Color	White
Number	Category	Percentage
	Decimal places	2

That is it. You now have a stylish pyramid chart to display sales per make for a specific month.

How It Works

Setting up a pyramid chart is really very easy, so this one was spiced up with a few stylistic points. Firstly, the titles were tweaked to be positioned closer to the chart area edges than is the default. This automatically lets the chart grow and take up a relatively larger amount of the available space. Then you set the legend so that it does not occupy too much space either, which lets the chart itself grow even more.

A Few Ideas on Using Charts for Business Intelligence

In this chapter, you saw only a very few of the ways that charts can be used to deliver your business intelligence with SSRS. There are many other possibilities, some traditional, some more adventurous, that you can also apply. The trick is to use charts effectively, without trying to impress by using exaggerated stylistic effects, and also without underwhelming the audience through delivering chart types and styles that they have seen too many times before.

So instead of detailing all the potential application of charts, I want to present a few tips on effective BI chart use. This is not a definitive guideline, nor a list of rules to be followed blindly. They are just some ideas that I have applied over time, and that you may find useful.

Keep Charts Simple

BI dashboards rarely have the space, and their readers even more rarely the attention span, for complex or even large charts. So a first principle is to reduce the number of elements in a chart so that it is readable. Another idea is to accentuate the information that really matters, so consider using greys for text unless you need to stress something. Then you can put in black the text that needs the audience's attention. This is better than using large or bold text to attract the reader's attention: you gain space and avoid multiple confusing and conflicting text styles in a tiny visualization. You could use pastel colors for the less important chart elements, and a primary color for a bar, line, or column that you need your audience to focus on.

I always advise simple chart design in BI reports. This can mean avoiding the temptations of 3-D charts. These are essentially a matter of taste, but they can distract from the information being displayed, and they take up more space to deliver the same information.

Use Multiple Charts of the Same Type

If you have a dense amount of information to deliver, consider creating multiple charts with few data series rather than a single chart with multiple data series. As you will see in Chapter 8, there are ways to add interactivity to an SSRS BI report. Consequently, you can switch between multiple charts on the same report if you do not have the space for several charts to be displayed at once.

Minimize Titles

Titles take up space. If space is at a premium, see if you can avoid them. This is not always possible, but if you can reduce the size of titles, or even remove them, then do so. As you saw earlier, you can always rotate titles on axes to save space, and you can also set titles to be vertical or pivoted, all of which can save space. These tweaks can add to development time, but the result is nearly always worth it.

Avoid Extraneous Elements

If your chart does not need a legend, remove it. The same logic applies to most chart elements. Think about the chart as a whole, and remember that in most cases "less is more."

Vary Chart Types

If you are afraid of your readers not reacting to the information on a dashboard because the chart type is too classic (once again, possibly because it is one of the overused Excel standard chart types), then consider varying a basic chart type. A few possibilities are given in Table 4-1, though the final choice will inevitably depend on the data and the effect that you want to create.

Table 4-1. *Chart Variations*

Classic Chart	Alternative
Pie	Donut
	Pyramid
Bar	Horizontal Cylinder
Column	Cylinder
Line	Radar

Conclusion

In this chapter, you saw a few ideas on ways to produce charts that suit the more specific requirements of business intelligence reports. These ideas are not the only ones that you can apply, but hopefully they will give you some ideas for your own dashboards and reports. Take these examples and extend and adapt them. Even better, perhaps you now have ideas that were not even touched upon here.

There are also ways to use charts to deliver greater levels of interactivity in a dashboard. Some techniques to enhance the user experience using charts to highlight and slice data are given in Chapter 10. In any case, charts are a fundamental part of BI delivery. I hope that you will enjoy using them and pushing the envelope to deliver some really eye-catching BI.

CHAPTER 5

Maps in Business Intelligence

Few visual elements are capable of conveying quite as much information as a map. It follows that in certain circumstances a well thought-out map can be a tremendous enhancement to a dashboard or mobile BI report suite.

The advantages of adding a map to a dashboard are the following:

- Maps are an immediate and visual way to display information for countries and regions.
- Countries, regions, and states (or counties) are, for the most part, instantly recognizable your users. They do not need added text in the visualization. This frees up space on screen.
- Maps allow drill-through, which lets you hierarchize your information.

There can be a couple of drawbacks to using maps in BI, which include the following:

- Obtaining, and occasionally adjusting, the geographical data can take time and money.
- Maps are slow to render, which can affect the user experience or require caching (as described in Chapter 12).
- Ensuring that the geographical data set can join to the business data often requires extending or modifying ETL processes.

In most cases, however, the positives far outweigh the negatives when it comes to using maps to convey business information in a comprehensible format that is both intuitive and easy to understand. Moreover, most of the downsides to using maps can be overcome in practice. This chapter will show you some examples of using geographical visualization to deliver your BI insights.

Geographical Data

Before you can even contemplate adding a map to a dashboard you need to find geographical data (also known as spatial data) that will allow SSRS to draw the map. If you are reporting on US data, then you are in luck, as SSRS comes with basic North American data built in. Unfortunately, if you need to report on Europe, Asia, or the rest of the world, you need to find your own geographical data. Fortunately, there are many suppliers of mapping data, so you have a fairly wide choice, if you are prepared to pay for it. If you do not want to invest sometimes considerable sums of money to obtain the data that you need, there are certain publicly available data sources, both freeware and shareware. You will, however, have to be prepared to spend some time searching for them and then then testing them.

In any case, you should probably be aware that there is a de facto file format for geographical data. It is the `.Shp` file, or shapefile. If you can obtain data in this format, then you have every chance of reading it successfully into SSRS and generating maps. Many of the publicly available geographical data sets are in shapefile format, as are much commercial data.

Loading Geographical Data into SQL Server from Shapefiles

Although SSRS can read shapefiles directly, there are several reasons why you may want to add an extra step of loading the geographical data into an SQL Server table, which you can then use as the source for SSRS maps. These reasons include the following:

- It is easy to reuse the same source data from a database for multiple reports.
- You can read the geography table directly in SSMS and verify the data that it contains.
- SSRS files based on shapefiles can be huge because they contain the geographical data. SSRS files that link to geographical data in a database are much smaller.
- You can remove any unwanted data from the data table in SQL Server, and consequently reduce the database size.
- Data tables can be filtered just like any other data source to subset map output. This makes the final report smaller than a similar report based on a shape file containing all the map data.
- Using a data table allows you to add columns that contain the data you will use to join business data to the geographical data. For instance, your business data may not be using standard country or region names. If you load a shapefile into SQL Server, you can subsequently add or modify the data so that your two data sets can be joined.

In any case, there is one freeware program that you need to know about if you want to load shapefiles into SQL Server tables. It is the remarkable and superbly easy-to-use application called Shape2SQL. Download this from `www.sharpgis.net/page/shape2sql` and add it to your armory. You will see how to use this superb application later in this chapter.

■ **Note** You can also obtain geographical data that is not a shapefile but that can be loaded into SQL Server as geospatial data. The intricacies and complexities of this technology place it outside the scope of this book. Should you need more information, I suggest that you consult *Pro Spatial with SQL Server 2012* (Apress 2012) by Alastair Aitchison.

In my experience, even advanced SSRS users have not used maps often, if at all. Consequently, in this chapter I will go into slightly greater detail of the process of map creation for all the examples in this chapter.

A Simple Map of US States with Sales

As a simple first example, let's suppose that you want to add a map of some of the American states where Brilliant British Cars sells its vehicles. This will give you an idea of the geographical penetration that you have achieved in the United States.

Fortunately, SSRS comes with the requisite US mapping data out of the box, so you will be able to create this initial map really easily. This final map will look like Figure 5-1. Clearly the company's North American sales leave room for improvement.

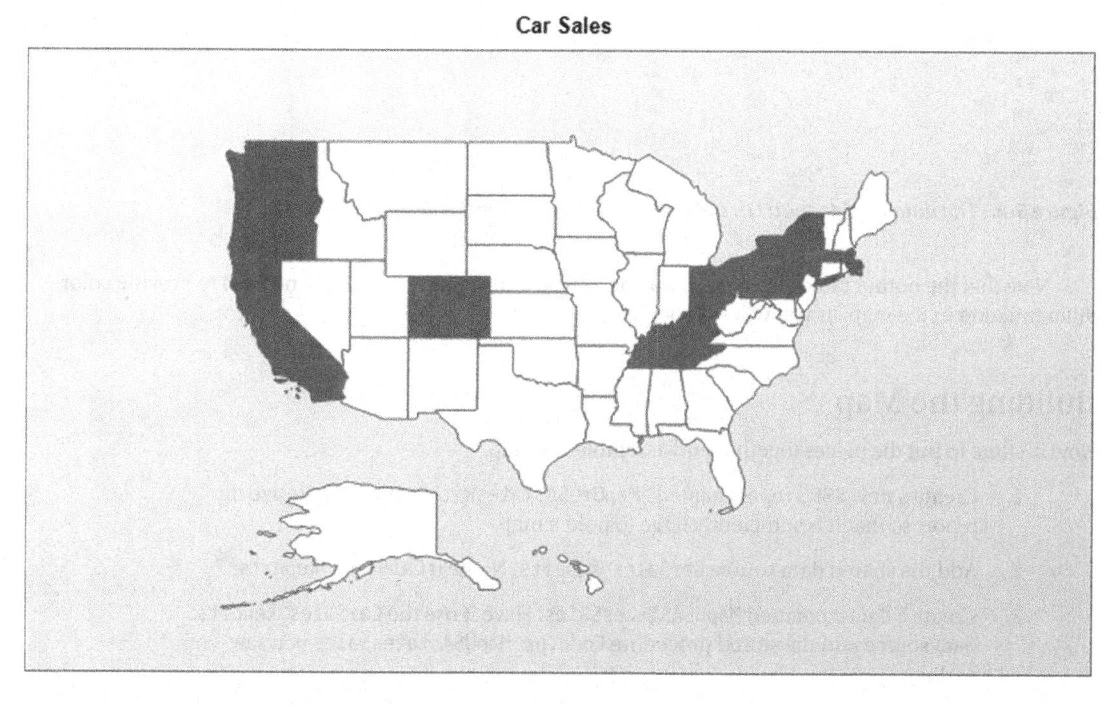

Figure 5-1. *A map of US states where you have made sales to date*

The Source Data

The first thing that you need is your business data. In this example, let's suppose that all you want is a list of North American states where you have made sales to date. You can obtain this using the following SQL (`Code.pr_MapUSAStatesSales`). Fortunately, your business data is sufficiently standardized and contains the state codes used by SSRS to draw the US map.

```
SELECT
OuterPostode
,1 AS ActiveFlag
FROM      Reports.CarSalesData
WHERE     CountryISOCode = 'USA'
GROUP BY  OuterPostode
```

Running this snippet gives the output shown in Figure 5-2.

	OuterPostode	ActiveFlag
1	CA	1
2	CO	1
3	KY	1
4	MA	1
5	NY	1
6	OH	1
7	OR	1
8	PA	1
9	TN	1
10	WA	1

Figure 5-2. *The data used to map US states where you have made sales*

Note that the output table also contains a flag as well as the state code. It will be used to provide color differentiation in the map, as you will see soon.

Building the Map

Now it's time to put the pieces together and assemble the map.

1. Create a new SSRS report named `_MapOfUSAStatesWithSales.rdl`. Resize the report so that it is sufficiently large to hold a map.
2. Add the shared data source `CarSales_Reports`. Name it `CarSales_Reports`.
3. Create a dataset named `MapUSAStatesSales`. Have it use the `CarSales_Reports` data source and the stored procedure `Code.pr_MapUSAStatesSales` you saw earlier.
4. Click the Map item in the SSRS toolbox and drag the mouse in the report area to define the area that the map will cover in the report. Ensure that it covers most of the available report. The New Map Layer dialog will appear showing the Choose a source of spatial data pane.
5. Leave Map Gallery selected.
6. Select USA by State inset. The dialog will look like Figure 5-3.

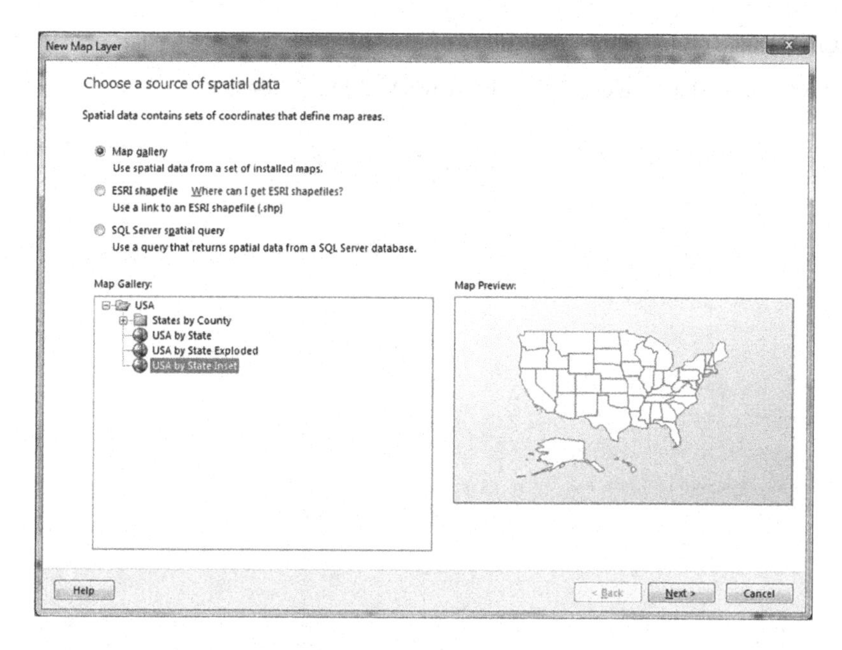

Figure 5-3. *The Map Layer dialog*

7. Click Next. The Map dialog will display the Choose spatial data and map view options pane.
8. Adjust the size and placement of the map using the arrows and the plus and minus buttons on the left of the dialog (or using the mouse if you wish). The dialog will look like Figure 5-4.

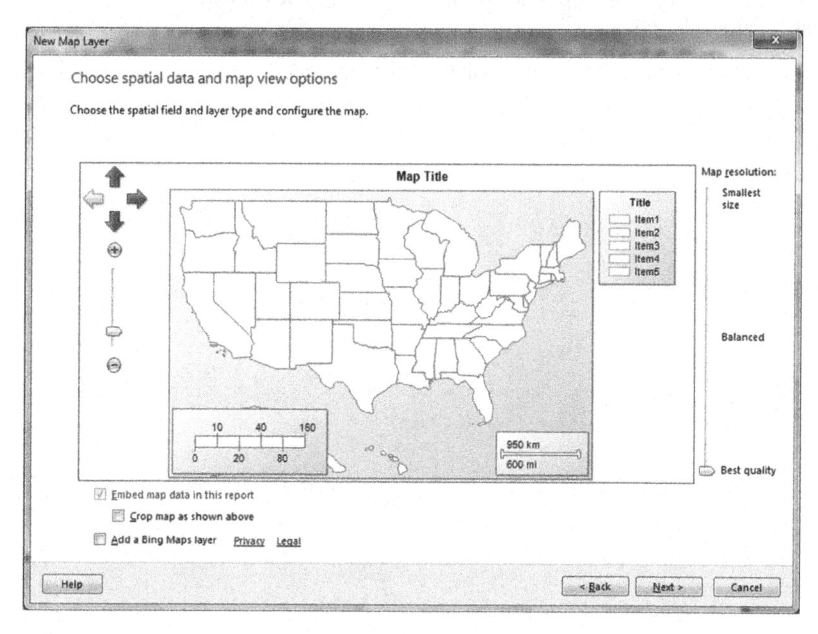

Figure 5-4. *The Choose spatial data and map view options pane*

9. Click Next. The Choose map visualization pane will be displayed.
10. Ensure that Color Analytical Map is selected. The dialog will look like Figure 5-5.

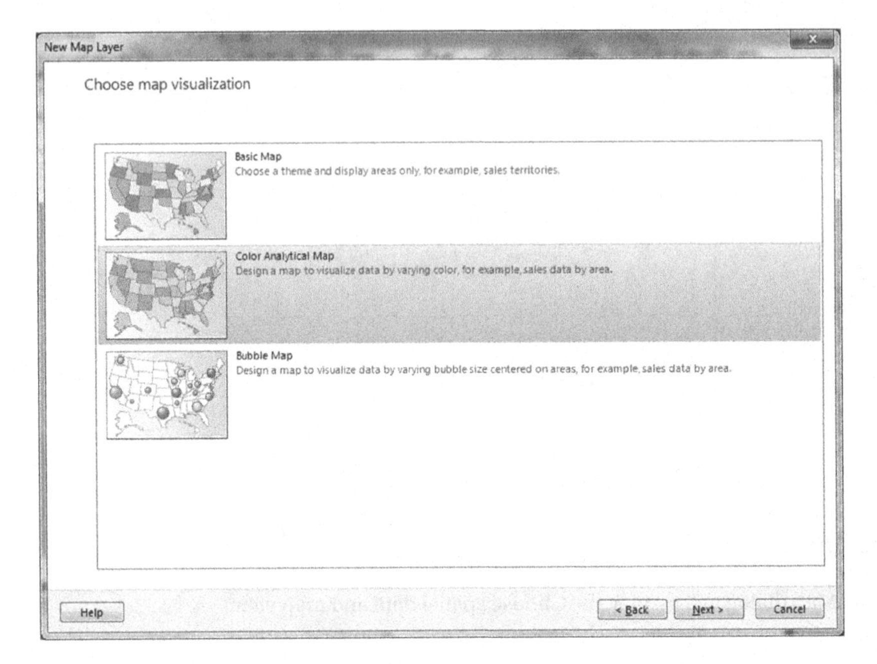

Figure 5-5. *The Color Analytical Map pane*

11. Click Next (or double-click the type of map that you want to select). The Choose the analytical dataset pane will appear.
12. Select the `MapUSAStateSales` dataset. The dialog will look like Figure 5-6.

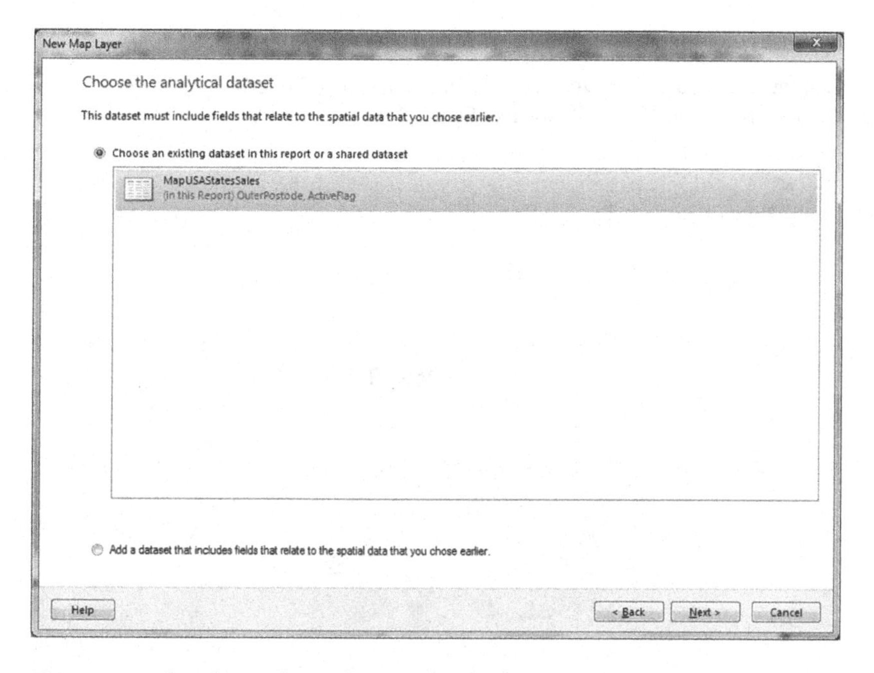

Figure 5-6. *The Choose the analytical dataset pane*

13. Click Next (or double-click the relevant dataset). The Specify the match fields for spatial and analytical data will be displayed.

14. Check the box next to the STUSPS field in the upper section of the dialog, and then select OuterPostcode from the pop-up menu. The dialog will look like Figure 5-7.

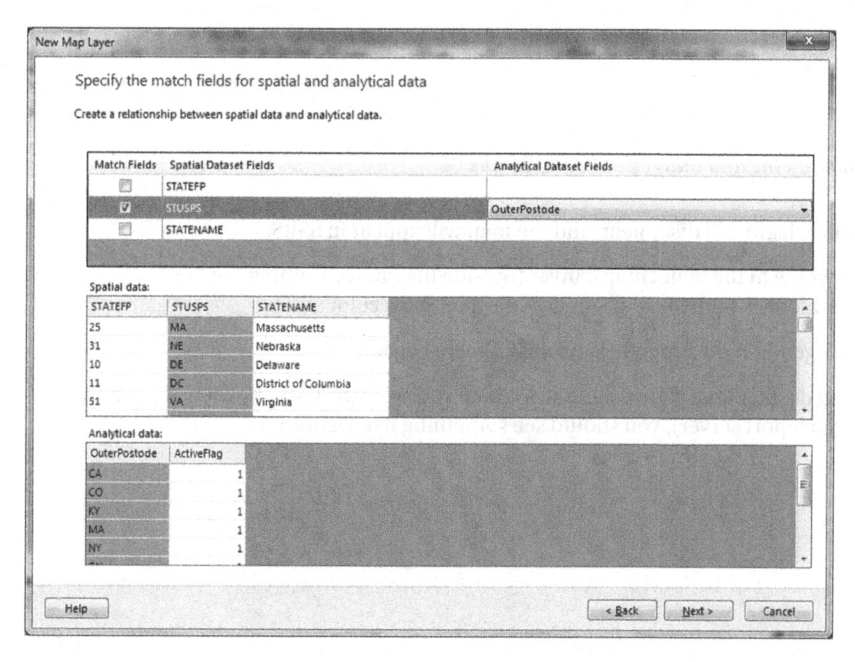

Figure 5-7. *The Specify the match fields for spatial and analytical data pane*

15. Click Next. The Choose color theme and data visualization pane will appear.

16. Select Slate as the theme and [Sum(ActiveFlag)] as the field to visualize and Dark-Light as the Color rule. The pane will look like Figure 5-8.

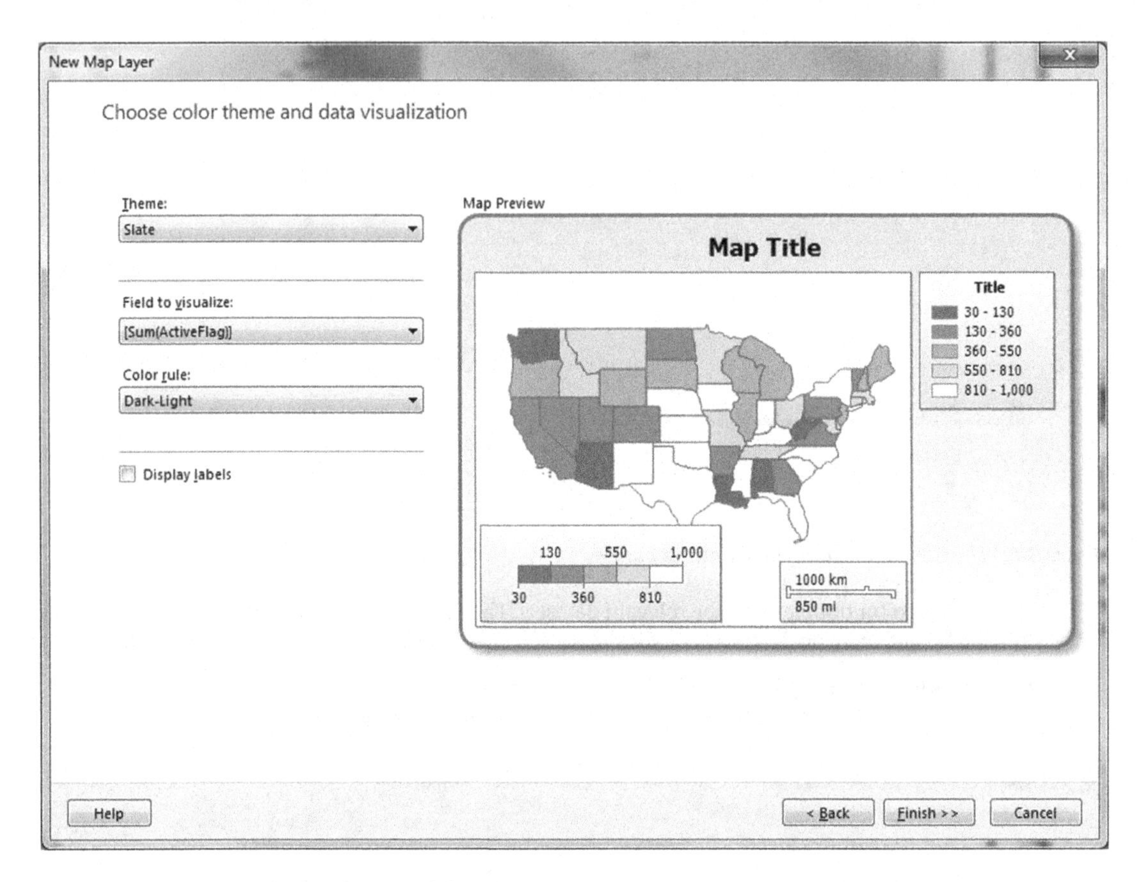

Figure 5-8. The Choose color theme and data visualization pane

17. Click Finish. The wizard will disappear, and the map will appear in SSRS.

18. Right-click anywhere in the blank map canvas (outside the part containing the map itself) and uncheck both Show distance scale and Show color scale.

19. Right-click the legend (at the top right) and click Delete Legend.

That's it! You have created a map that shows the states where you sell cars. If you preview the report (or better still, deploy it to the report server), you should see something like Figure 5-1.

■ **Note** The Map Preview in Figure 5-8 gives a pretty good idea of what the final map will look like. If the map object that you have added to your report is too small, the preview will also make this evident.

How It Works

Matching fields for spatial and analytic data is the key to effective maps in BI. It boils down to having one field in each of the datasets (spatial and business data) that can be used to join the two datasets. In this example, there is a field that contains the state abbreviation in the `Clients` table, and fortunately it matches the data in the Microsoft-supplied map data for the US. To belabor the point somewhat, you *need* to ensure that the spatial and the analytical data can always be joined in this way.

A second important point is having the geographical data that SSRS can use. Fortunately, when displaying data relative to US geography you have the luxury of being able to use the built-in North American geospatial reference data.

Hints and Tips

There is one interesting thing to note when creating color analytical maps in SSRS.

- In Figure 5-5 (Step 10), you also had the option to create a Basic Map. This is simply a map used as pure decoration, with no link to business data. Consequently, it is relatively rare to use this type of map in business intelligence.

A Heat Map of European Sales by Country

As a slightly more advanced example of using maps in business intelligence, let's suppose that you need to see the European countries where you make sales. Not only that, but ideally you would like each country to be displayed in a color that indicates the extent of the sales for that country. Fortunately, SSRS can deliver just what you are looking for. It's called a heat map, and it requires two things:

- Geographical data for the countries of Europe. In this case, I suggest using the World Borders data from `http://thematicmapping.org/downloads/world_borders.php`.

- Business data that contains a way of joining the sales data to the geographical data. This can be a standardized country name or the two or three letter code that is used in the World Borders shapefile.

Fortunately, your `CarSales` database already contains the three letter ISO country code, so you should have no difficulty joining your geographical data to your sales data. The map that you are trying to create will look something very close to that shown in Figure 5-9.

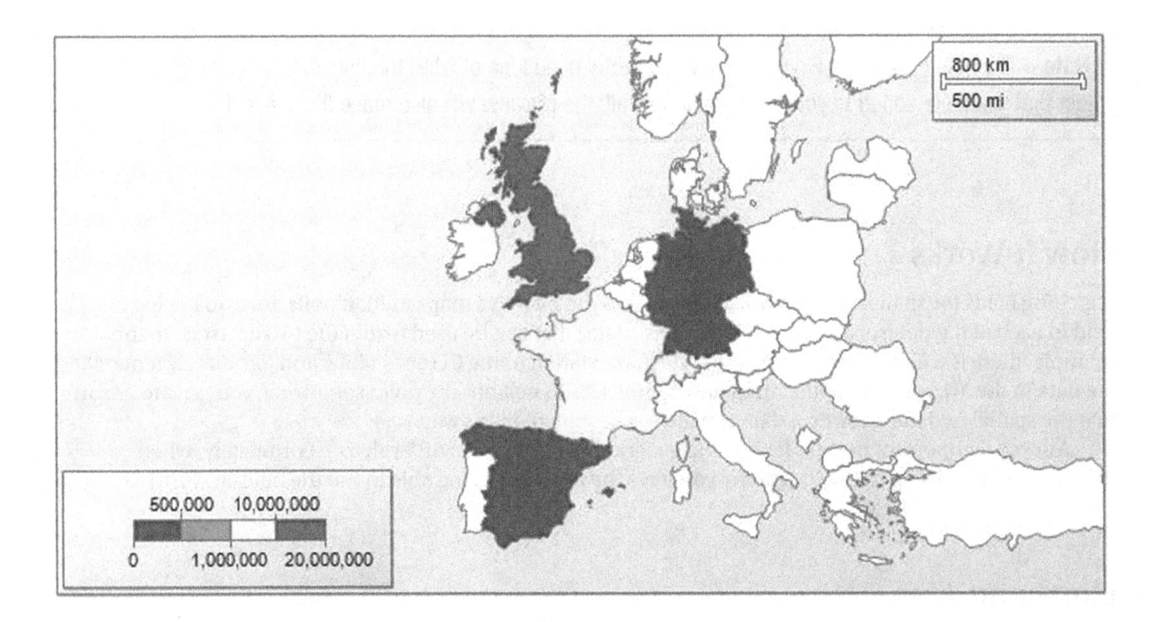

Figure 5-9. *A heat map of sales in European countries*

Of course, in a black and white book you may not get the full effect of the colors, so you may want to load and preview the file from the sample solution to see the map as it appears on a screen.

■ **Note** In the real world, the geographical data that you use may not have a field that allows you to join two fields as easily as you did in this example. So be prepared to either add a reference table of geographical data containing the fields from the two sources that allows you to join the two data sets-and use this in a query when returning the business data-or add an extra column containing the appropriate values to the imported geographical data.

Simplifying the Geographical Data

As I mentioned at the start of this chapter, mapping data can become voluminous, and the more complex a map, the longer it can take to render. Therefore, you will do two things here:

- Load the geographical data from the shapefile into SQL Server.
- Subset the geographical data so that you only display a precise set of countries, Europe in this case.

Loading a Shapefile into SQL Server

Although you could use the source shapefile directly, let's take this opportunity to load the geographical data into SQL Server so that you can output only the data required to draw a map of Europe. This extra effort will speed up the map rendering and thus enhance the user experience.

1. Download (or copy) the shapefile that you will be using for the spatial data into a source folder.
2. Download Shape2SQL and run it using the URL given earlier.
3. Connect to the database into which you want to load the spatial data from the shapefile.
4. Select the source shapefile and unselect any fields that you do not either want or need. The Shape2SQL dialog will look something like Figure 5-10.

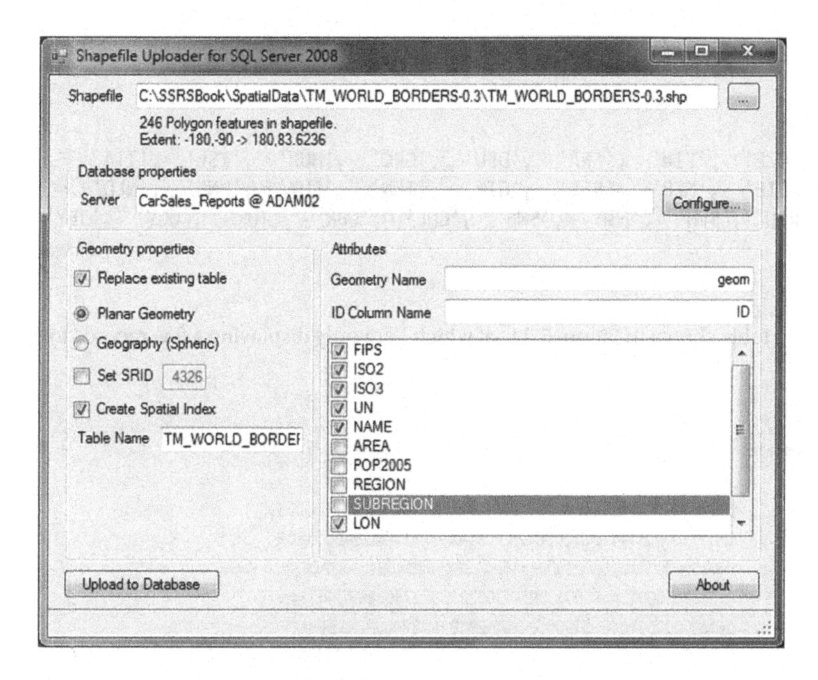

Figure 5-10. *The Shape2SQL dialog*

5. Click Upload to Database.
6. Close the Shape2SQL dialog and rename the table that has been created if necessary. Call it `Reference.WorldMap` in the `CarSales_Reports` database.

That is all that you have to do. You now have spatial (geographical) data in your database rather than in a shapefile.

The Source Data

To create this map, you will need two datasets: the geographical data and the business data. Let's look at each in turn.

The Geographical Data

Now that you have loaded the geographical data into SQL Server, you can define a query that only returns the rows and columns that you need. This query follows (`Code.pr_MAPEurope`):

```
SELECT
 ID
,ISO2
,ISO3
,NAME
,geom
  FROM Reference.WorldMap
  WHERE ISO3 IN
  (
   'DNK'  ,'IRL'  ,'AUT'  ,'CZE'  ,'FIN'  ,'FRA'  ,'DEU'  ,'GRC'  ,'HUN'  ,'ISL'  ,'ITA'
  ,'LVA'  ,'LTU'  ,'SVK'  ,'LIE'  ,'MKD'  ,'BEL'  ,'GIB'  ,'IMN'  ,'LUX'  ,'MNE'  ,'NLD'
  ,'NOR'  ,'POL'  ,'PRT'  ,'ROU' ,'SVN'  ,'ESP'  ,'SWE'  ,'CHE'  ,'GBR'  ,'TUR'  ,'GGY'  ,'JEY'
  )
  ORDER BY NAME
```

Running this code returns the table shown in Figure 5-11, of which I am only displaying a few records to save space.

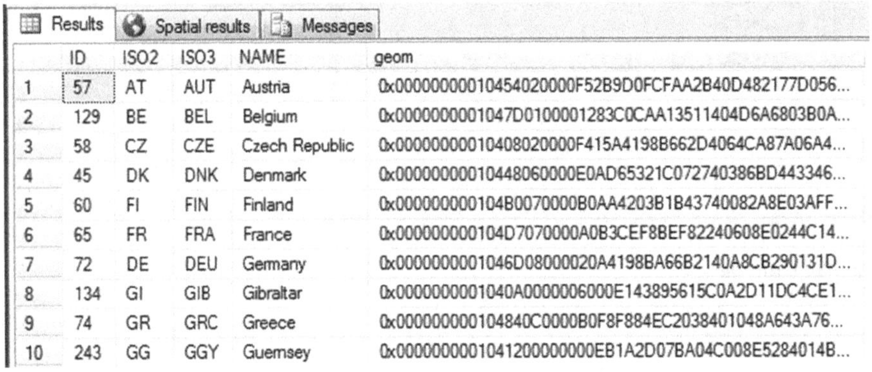

Results | Spatial results | Messages

	ID	ISO2	ISO3	NAME	geom
1	57	AT	AUT	Austria	0x00000000010454020000F52B9D0FCFAA2B40D482177D056...
2	129	BE	BEL	Belgium	0x0000000001047D0100001283C0CAA13511404D6A6803B0A...
3	58	CZ	CZE	Czech Republic	0x00000000010408020000F415A4198B662D4064CA87A06A4...
4	45	DK	DNK	Denmark	0x00000000010448060000E0AD65321C072740386BD443346...
5	60	FI	FIN	Finland	0x0000000001048007000080AA4203B1B43740082A8E03AFF...
6	65	FR	FRA	France	0x000000000104D7070000A0B3CEF8BEF82240608E0244C14...
7	72	DE	DEU	Germany	0x0000000001046D08000020A4198BA66B2140A8CB290131D...
8	134	GI	GIB	Gibraltar	0x0000000001040A0000006000E143895615C0A2D11DC4CE1...
9	74	GR	GRC	Greece	0x000000000104840C0000B0F8F884EC2038401048A643A76...
10	243	GG	GGY	Guernsey	0x00000000010412000000000EB1A2D07BA04C008E5284014B...

Figure 5-11. *Some of the data used to generate the map of European countries*

Generating the Business Data

Now all you need is the metrics for sales by country to date. You can obtain this by running the following code (`Code.pr_MAPEuropeSalesToDate`):

```
SELECT
 CountryISOCode
,SUM(SalePrice) AS SalesToDate
FROM            Reports.CarSalesData
WHERE           CountryISOCode NOT IN ('USA')
GROUP BY        CountryISOCode
ORDER BY        CountryISOCode
```

Running this snippet gives the output in Figure 5-12.

	CountryISOCode	SalesToDate
1	CHE	1440970.00
2	DEU	145750.00
3	ESP	207750.00
4	FRA	2524510.00
5	GBR	15725000.00

Figure 5-12. *The data for European sales over all years*

How the Code Works

The first piece of code selects the geographical data for the countries of Europe from the available world data in the table that you previously loaded into the database. As you can see, the actual geographical data is in a binary format. The second code snippet simply returns the sales figures for all sales to date. It is worth noting that the data is for all sales to date, and so you are not using parameters for the year or month.

Creating the Map

With all the building blocks in place, you can create the map. As you already saw many of the dialogs in the previous map, I will not show them again here.

1. Create a new SSRS report named `_MapOfEuropeWithSales.rdl`. Resize the report so that it is sufficiently large enough to hold a map.
2. Add the shared data source `CarSales_Reports`. Name it `CarSales_Reports`.
3. Create a dataset named `MapOfEuropeWithSales`. Have it use the `CarSales_Reports` data source and the query `Code.pr_MAPEuropeSalesToDate` that you created earlier.
4. Create a dataset named `MAPEurope`. Set it to use the `CarSales_Reports` data source and the query `Code.pr_MAPEurope` that you created earlier.
5. Click the Map item in the SSRS toolbox and drag the mouse in the report area to define the map. Ensure that it covers most of the available report. The New Map Layer dialog will appear showing the Choose a source of spatial data pane.
6. Select SQL Server spatial query and click Next. The Choose dataset with SQL Server spatial data pane will appear.
7. Click the dataset that you created earlier named `MapEurope`. This dialog will look like Figure 5-13.

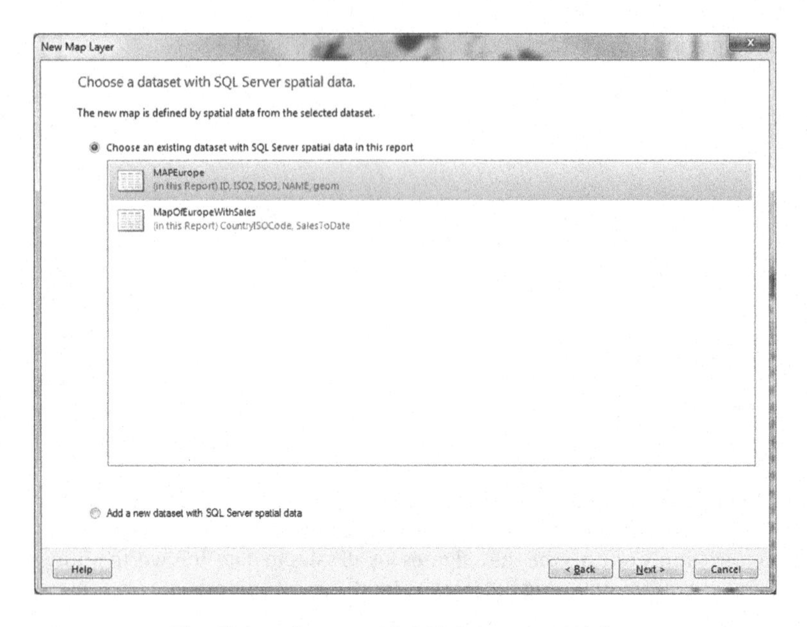

Figure 5-13. *The Choose dataset with SQL Server spatial data pane*

8. Click Next. The Map dialog will display the Choose spatial data and map view options pane.
9. Adjust the size and placement of the map using the buttons on the left of the dialog if you wish. The dialog will look like Figure 5-14. Note that it displays only the countries from the query that you defined previously.

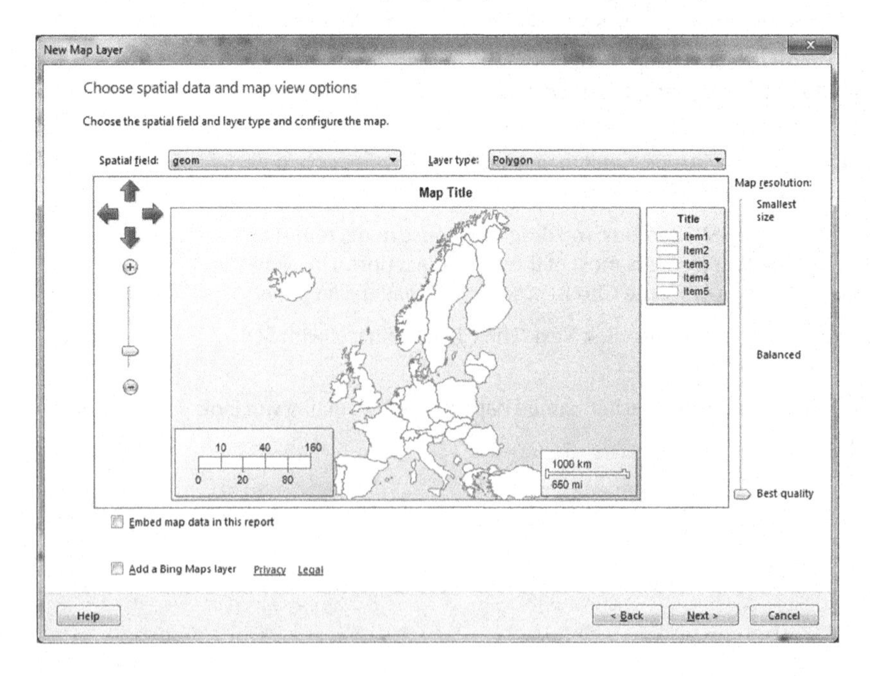

Figure 5-14. *The Choose spatial data and map view options pane*

10. Click Next. The Choose map visualization pane will be displayed.
11. Select Color Analytical Map and click Next.
12. In the Choose the analytical dataset dialog, select the dataset `MapOfEuropeWithSales`. Click Next.
13. Check the box next to the ISO3 field in the upper section of the dialog, and then select the CountryISOCode field from the pop-up menu. Click Next.
14. Select `[SUM(SalesToDate)]` as the field to visualize, and try out any of the themes and color rules you like! Click Finish.
15. Right-click the legend (at the top right) and click Delete Legend.
16. Drag the Distance scale from the bottom right to the top right.
17. Click twice on the map. The MapLayer pane will appear to the right of the map.
18. Click the context menu triangle (at the top right beside the eye icon). Select Polygon color rule. The Map Color Rule Properties dialog will appear.
19. Select Distribution on the left.
20. Choose Custom and add the following data ranges:
    a. 0 to 500,000
    b. 500,000 to 1,000,000
    c. 1,000,000 to 10,000,000
    d. 10,000,000 to 20,000,000
21. The dialog will look like Figure 5-15.

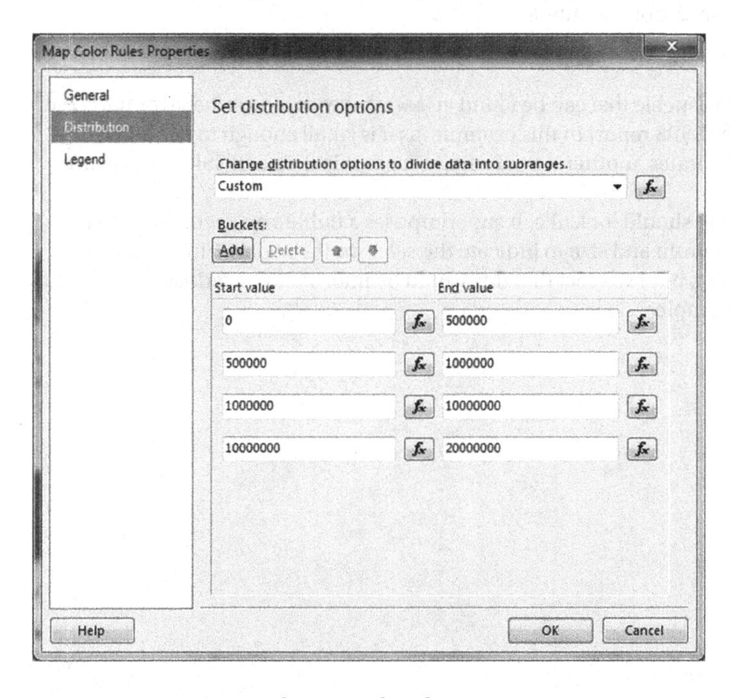

Figure 5-15. *Setting a color range distribution*

22. Click OK. You may want to resize and recenter the map for the best effect.

That's it. You should now have a map that looks like Figure 5-9.

How It Works

The key to the success of this map (as is the case for virtually any map) is to have the correct geographical data that can join accurately to the business data. In this example, you have an ISO country name, which is an ideal field when dealing with data at the country level.

Once the tables have been joined (in step 13), the rest is essentially a question of aesthetics. This covers sizing and placing the map inside the map viewport and configuring the color rule, which is really a legend by another name.

■ **Note** You can resize and reposition maps in two main ways. You can use the direction arrow and slider that are in the Map Layers pane, which appear when you select the map. Alternatively, you can select the map layer and use the mouse to move the map to the position that you want or the scroll wheel to adjust the size of the map.

A Bubble Chart Map of English Regions

You just saw the basics of map creation in SSRS. Now let's push the envelope a little. Let's create a map of the English regions that shows both sales and profits. As the objective is to show two data elements, you will not create a heat map this time, but a bubble map. Using points (or bubbles) to represent the data will allow you to represent two metrics:

- Sales, which will generate the *size* of the bubble
- Profits, which will give the bubble its *color*

This map will use the Regions (UK) shapefile that can be found at `www.sharegeo.ac.uk/handle/10672/50`. You will read the shapefile directly in the SSRS report in this example, as it is small enough to use efficiently and does not need to be subset. This illustrates another way of using geographical data in SSRS for business intelligence.

Figure 5-16 shows what the final map should look like. It superimposes a bubble on each region that changes color to show the percentage of profit and size to indicate the sales for the region. This map could also add a third metric for the region color, but I suspect that this would be information overload, so let's will stick to using only the bubbles in this example.

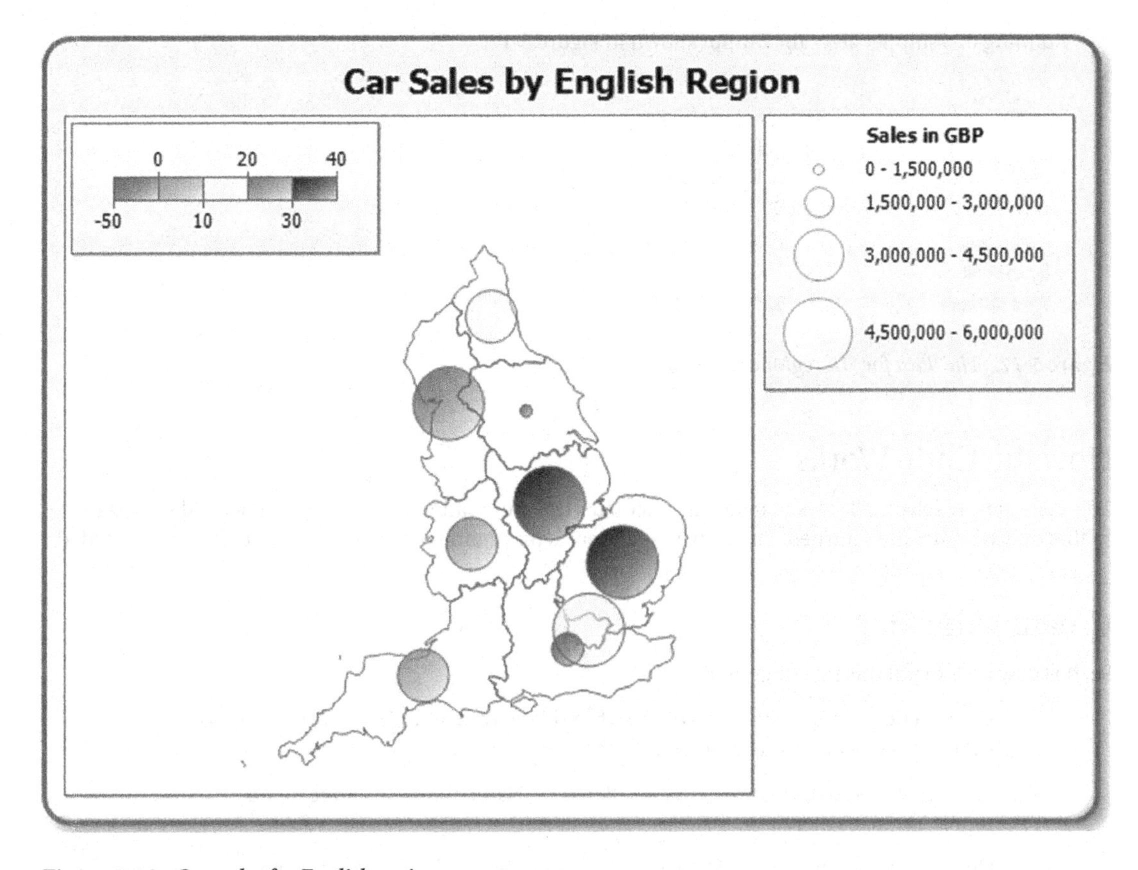

Figure 5-16. *Car sales for English regions*

The Source Data

The business data will give you the area, the sales for each area, and the percentage of profit for sales in each area. You can return this data by running the following code (Code.pr_CarSalesByRegion):

```
SELECT
Region
,SUM(SalePrice) AS TotalSales
,((SUM(SalePrice) - (SUM(CostPrice) + SUM(TotalDiscount) +SUM(SpareParts) +
SUM(DeliveryCharge)
 +SUM(LaborCost))) / SUM(SalePrice)) * 100 AS ProfitRatio
FROM        Reports.CarSalesData
WHERE       CountryISOCode = 'GBR'
            AND Region IS NOT NULL
GROUP BY    Region
```

Running this snippet gives the output shown in Figure 5-17.

	Region	TotalSales	ProfitRatio
1	East Midlands	2876250.00	0.388448
2	Greater London Authority	5839500.00	0.327602
3	North East	85250.00	-0.105865
4	North West	3008250.00	0.262666
5	South West	214750.00	0.030849
6	West Midlands	3310250.00	0.300856

Figure 5-17. *The data for UK regional sales over all years*

How the Code Works

This code simply selects all sales to data and calculates the profit ratio for each region. It also filters the data so that only UK sales are returned, ensuring that any sales that cannot be attributed to a region are excluded.

Creating the Map

So, it is now time to put this map together.

1. Create a new SSRS report named `_MapOfEnglishCarSales.rdl`. Resize the report so that it is sufficiently large to hold a map.
2. Add the shared data source `CarSales_Reports`. Name the data source in the report `CarSales_Reports`.
3. Create a dataset named `CarSalesByRegion`. Have it use the `CarSales_Reports` data source and the query `Code.pr_CarSalesByRegion` that you created earlier.
4. Click the Map item in the SSRS toolbox and drag the mouse in the report area to define the map. Ensure that it covers most of the available report. The New Map Layer dialog will appear showing the Choose a source of spatial data pane.
5. Select ESRI Shapefile as the spatial data source.
6. Browse for the source file (`C:\BIWithSSRS\SpatialData\English Regions\Regions.shp` in this example). The dialog will look like Figure 5-18.

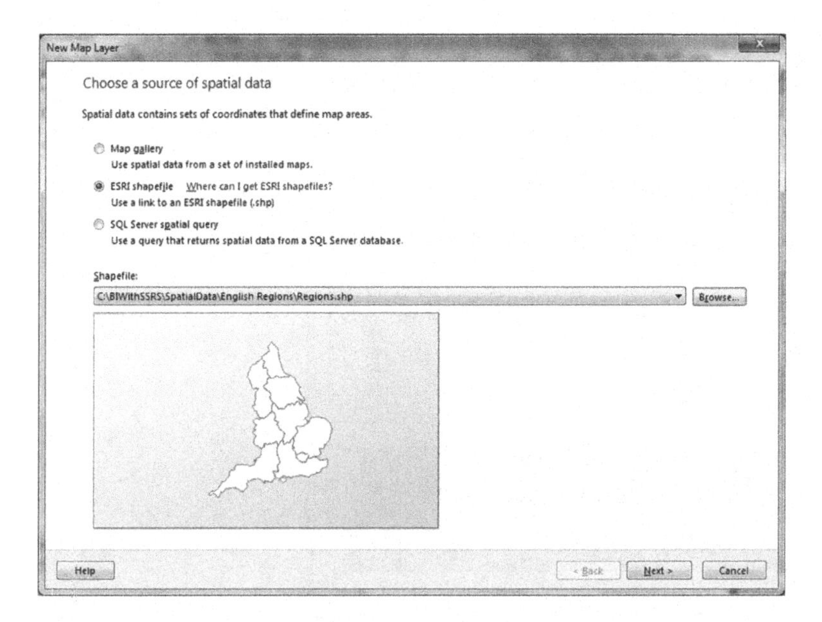

Figure 5-18. *Selecting a shapefile as the source for spatial data*

7. Click Next. The Choose spatial data and map view options pane will be displayed; in it you can resize and reposition the map if you wish.
8. Click Next. The Choose map visualization pane will be displayed. Select Bubble Map as the Visualization type.
9. Click Next. The Choose the analytical dataset pane will appear.
10. Select CarSalesByRegion as the dataset containing the metrics that you want to display on the map.
11. Click Next. The Specify the match fields for spatial and analytical data will be displayed.
12. Check the box next to the Name field in the upper section of the dialog, and then select Region from the pop-up menu. The dialog will look like Figure 5-19.

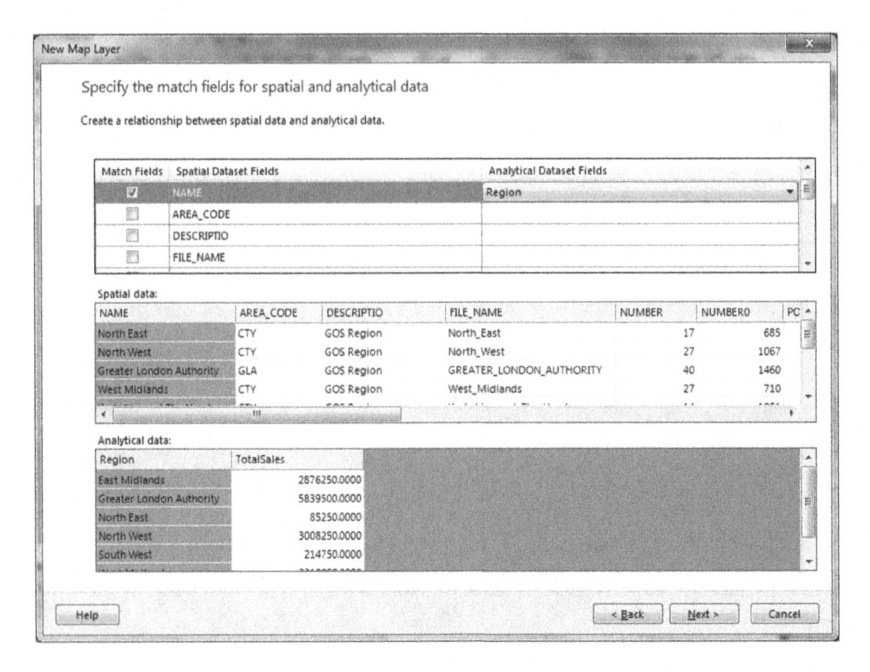

__Figure 5-19.__ The Specify the match fields for spatial and analytical data pane

13. Click Next. The Choose color theme and data visualization pane will appear.
14. Select a theme (I am using Corporate in this example).
15. Ensure that Use bubble size to visualize data is checked, and select `[Sum(TotalSales)]` as the data field.
16. Click Finish. The map will be displayed.

You still have some work to do to make this map into an effective visualization. You need to remove the region colors and apply colors to the bubbles. You also need to set the ranges for both the size and colors of the bubbles and tweak the legends. Let's move on, then, to the second part of this example.

17. Right-click anywhere in the blank map canvas and uncheck both Show distance scale and Show color scale.
18. Click twice on the map. The MapLayer pane will appear to the right of the map.
19. Select Polygon color rule. The Map color rules properties dialog will appear.
20. Select Visualize data by using custom colors.
21. Click Add.
22. Select White as the custom color.
23. Click OK. This will set all the background color for all the regions to white.
24. In the Map layer pane, click the context menu triangle (at the top right beside the eye icon). Select Center point color rule. The Center point Color Rule Properties dialog will appear.

25. Select Visualize data by using color ranges.
26. Select `[Sum(ProfitRatio)]` as the data field to use.
27. Set the Start color to red, the Middle color to yellow, and the End color to dark blue.
28. Click Distribution on the left of this dialog. The Set distribution options pane will be displayed.
29. Select Custom as the distribution option and add four ranges. Set their values as follows:
    a. –50 to 0
    b. 0 to 10
    c. 10 to 20
    d. 20 to 30
    e. 30 to 40.
30. The dialog will look like Figure 5-20.

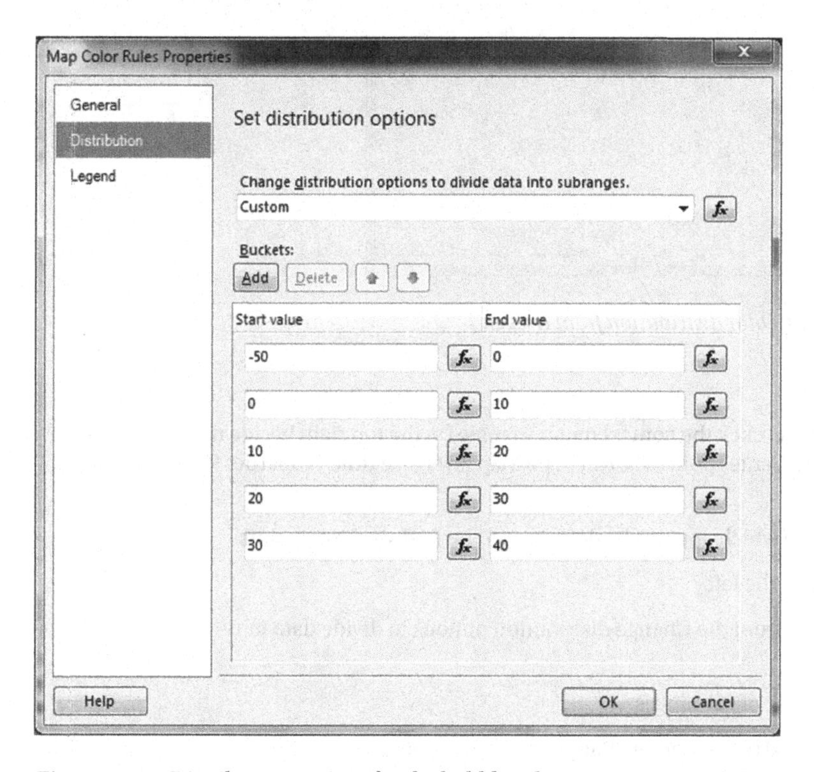

Figure 5-20. *Distribution options for the bubble colors*

31. Select Legend on the left. The Change legend options pane will appear.
32. Leave Show in color scale selected, but select the blank legend from the Show in this legend pop-up. The dialog will look like Figure 5-21.

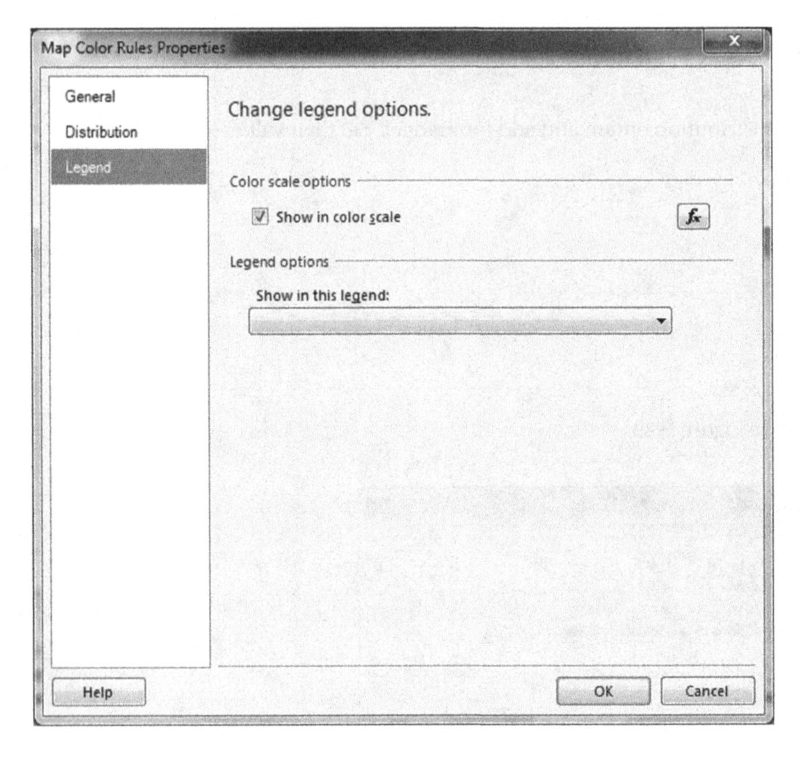

Figure 5-21. *Adding or removing color distribution from a legend*

33. Click OK.
34. In the Map layer pane, click the context menu triangle (at the top right beside the eye icon). Select the Center point size rule. The Map layer size Rule Properties dialog will appear.
35. In the General pane (which appears by default), set the End size to 40 points.
36. Click Distribution on the left.
37. Select Equal interval from the Change distribution options to divide data into subranges pop-up.
38. Select 4 as the number of subranges.
39. Set the Range start at 0 (zero) and the Range end at 6,000,000.

40. Click OK.

41. Add Sales in GBP as the title of the legend, and Car Sales by English Region as the title for the map.

42. Tweak the placement of the color scale and the map itself (as described previously) until you are happy with the final output. It should look more or less like Figure 5-16.

How It Works

This was the most complicated map you have created-or will create-in this chapter, and it required many aspects of the map to be adjusted. In essence, however, you saw how to tweak the following:

- *The map polygon color rule*: This means the regions in this map. These are the properties that you will want to look at if you want to set the colors and range settings for the spatial data representation.
- *The center point color rule*: This is where you can define the colors and the way in which the source data is distributed on the color scale.
- *The center point size rule*: This is where you set the limits and properties of the sizes of the bubbles.

While I cannot explain all the possible properties that you can use when creating maps in SSRS, there are a few other aspects that you might find useful. These include the following:

- *The center point marker type rule*: This is also available from the Map Layers pane. It lets you select the bubble shape from a range of options.
- *Applying data to legends*: You can add multiple legends to a map, and specify which field is displayed in which legend.
- *Tweaking the data sources once the map has been created*: Should you need to make adjustments to the way in which the business and spatial data are used (without rebuilding the map), selecting Layer data from the Map Layers pane will allow you to make changes.

■ **Note** The Specify the match fields for spatial and analytical data dialog in the map wizard gives you a fairly good idea of what data is contained in the spatial and business metric datasets. However, in practice, it's best to prepare the business data carefully, after a trial run to look at the spatial data, so that the two datasets can be joined correctly.

Conclusion

This chapter provided a short introduction on the use of maps and spatial data for business intelligence. The subject is so vast, and the SSRS mapping options so extensive, that you were only able to see a small part of all the available options. I hope that this has whetted your appetite for using maps in your own business intelligence output with SSRS.

CHAPTER 6

Images in Business Intelligence

Business intelligence is not just about metrics. It is also about accentuating and prioritizing information visually and intuitively. While figures and visual elements such as charts and gauges can do much of the work, nothing quite adds pizazz like the judicious use of images in business intelligence reports.

Using images does not necessarily mean just adding logos or decoration. Indeed, many people including me believe that images in BI should not *distract* from the information but *enhance* it. This means appreciating how images can complement the delivery of your data and guide the user as they analyze the information, while making the output both easy and pleasant to read.

In this chapter, you will learn how to use images to attract and retain your users' attention. To achieve this, you will see how to use images via

- Rectangles
- Text boxes
- Chart backgrounds
- Gauge backgrounds
- Table backgrounds
- Chart columns and bars
- Dynamic borders for tables
- Titles

This list is not exhaustive by any means. You will doubtless find many other ways of using images to enhance the power and impact of your BI delivery with SSRS. Before beginning, however, I want to re-emphasize that images have the potential to encumber reports needlessly if they are not used with discretion and taste. Of course, in saying this I am straying into the minefield of design choices, but it has to be said nonetheless. So I will show a variety of stylistic application of images, from the subtle to the almost brutal, and let you decide where the limits of effectiveness lie.

This chapter inevitably relies on many images to create the example reports that are used to explain some of the ways to use images in BI. In order to follow along, you should download the sample data (as described in Appendix A) and place the image files in the directory `C:\BIWithSSRS\Images`.

Image Types

SSRS can accept images in a handful of the most common bitmap file formats. These are the following:

- JPG (pronounced "JayPeg"): A standard (lossy) compressed file format.
- PNG (pronounced "ping"): A modern lossless file format that supports high resolution color.
- BMP: A classic file format frequently associated with Windows Paint.
- GIF: A venerable file format that is perceived as more limited that JPG or PNG. It can, however, display animated images.

You may choose to convert images to a standardized format, or use images in a mixture of these formats in your reports. It is entirely up to you. If you receive or inherit images in other formats, you will have to convert them to one of the four formats listed above. This may require using a package that is able to convert image formats or exporting the image in a different format using the application that created it. As this is a vast and specialized subject, I will let you dig for resources concerning image conversion on your own.

Image Editing Software

If you are lucky enough to have a design department that produces images for your dashboards to order, then you will never need to use image editing software. Should this not be the case, you could well have to tweak or even create your own images.

Few software developers are graphic artists. So I will presume that most readers will not be at home with professional-grade image editors. Obviously, if you are (or are prepared to invest the time to become) a Photoshop® guru, you will not need to use "lesser" products. If, like me, you only need to create images occasionally, or adjust or resize images from time to time, then you could consider trying some of the following applications:

- *Paint*: Bundled with all versions of Windows, it can perform many core bitmap image modification tasks quickly and well.
- *Visio*: As a key Microsoft product, you could well find that your company uses Visio. The big advantages of Visio are the following:
    - It comes with templates containing thousands of images that you can use.
    - It is a vector program, which means that you can group and resize objects before exporting them in a bitmap file format.
- *Pixlr*: A free web-based image editor with lots of available image effects.

This list is not exhaustive by any means and is only intended to give you some pointers. The only thing that counts is that you get the result you want.

Image Size and Resolution

There is one inescapable rule when using images in reports. The larger the image, the slower the report. The report will be slower to load and render, as well as being larger, and consequently will take up more network bandwidth. This will also mean that it will occupy more space in a report server as a snapshot or cached report.

This is where matters can get complicated. Some report users want reports looking like the front cover of a fashion magazine. Others are more reasonable, and accept that there are trade-offs. As the report developer, it is your (thankless) task to convince your users that SSRS is a corporate report creation and delivery tool, and consequently is not really designed to deliver high-resolution images more suited to glossy publications.

So you will need to be prepared to adjust the resolution of any images that you are using. Exactly what is an optimum resolution is open to debate, but in my experience 300 dpi is sufficient for most reports, even those that will be exported as PDF files or printed on 600 dpi office printers.

In practice, setting the resolution of an image nearly always means either

- Starting with a higher resolution image and saving it in a lower resolution, possibly using an image conversion application or even MS Paint.
- Exporting the image in the required resolution from the application used to create it.

This topic can rapidly become a subject of intense and passionate discussion. I have no intention in participating in this debate and will consequently leave you to find a suitable solution among those available.

Image Location

If you are using images in SSRS you will have to decide where your images are stored. Essentially you have three choices: they can be

- Embedded in the report.
- Stored as files on the report server.
- Located in a database table or tables.

The relative merits of each approach are outlined in Table 6-1.

Table 6-1. *Image Location Advantages and Disadvantages*

	Advantages	Disadvantages
Embedded	Instantly and easily accessible in the report.	Much manual effort required to alter (need to delete and reimport). Tricky to export.
Report Server	Centralized storage. Easy to organize.	Potential security questions.
Database	Centralized storage and backup. Easy to categorize.	Slightly trickier to apply. Need to update using code.

I cannot disguise that I am a great believer in storing images in a database. Better still, if you have a reporting database (such as a database where you place the data used in the reports) you potentially have a single place where all BI reporting objects except the reports themselves can be stored and reused. Having said this, I will be using embedded images in nearly all the examples in this book as they are simpler to use and thus will make the learning curve shorter.

Exporting Embedded Images from Reports

To be practical, let's consider the case of inherited reports that have embedded images that you need to tweak or reuse in other reports. In cases like this, it is inevitable that no one will know where the original images are stored; consequently, you will need to retrieve the images from inside the report. There are several ways to do this, but the one that I find the easiest is the following:

1. Preview the report in SSDT, or display the report in a browser.
2. Click the Export menu pop-up in the toolbar and select Word, as shown in Figure 6-1.

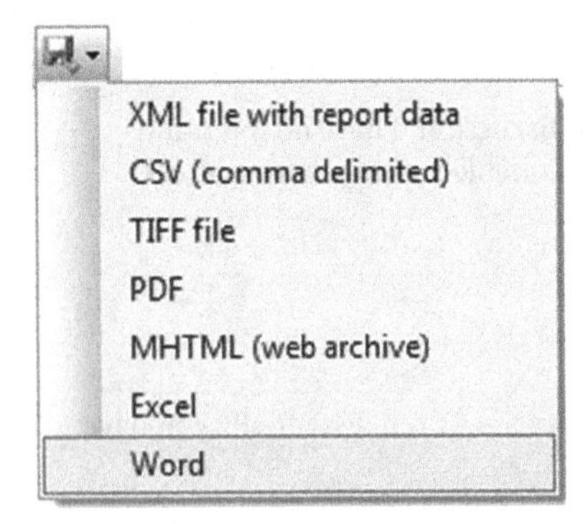

Figure 6-1. *Exporting a report to MS Word*

3. Browse to the destination directory and enter a file name, then confirm by clicking Save.
4. Open the Word document that you just saved and right-click the image. Select Save as Picture from the context menu and then choose the format that you want to use from those on offer (which by a fortunate coincidence include all four recognized by SSRS). Enter a name and browse for a directory for your image file and click Save.

You now have a useable image file that is an exact copy of the image in the report. You can now adjust and resize it, store it in an SQL Server database, and reuse it in other reports. The downside to this approach is that you have no choice but to carry out this process with every unique image in every report.

■ **Note** If you are faced with dozens of reports and hundreds of images, you may prefer to export the reports as PDF files. There are packages (starting with Adobe Acrobat) that can export all the images from a PDF file simultaneously.

Managing images coherently and efficiently for a suite of BI reports can require a little effort. Chapter 11 contains some methods for managing multiple images, using database tables to store images, and copying sets of images between reports.

Image Backgrounds for Text Boxes

Some metrics are more important than others. This is why you will often need to highlight certain sets of figures. Here again, the inherent border and background limitations of plain-vanilla text boxes in SSRS can prove to be something that you want to overcome.

Making a row or column of figures more presentable and readable as well as aesthetically more pleasing is easily achieved with a suitable background image. Indeed, you could well use more than one image to obtain a more striking effect, as Figure 6-2 shows. This visualization uses one image for the darker top part of the box and another for the lighter, lower part of the box. The images also have rounded borders to differentiate them visually from the standard rectangular text box borders in SSRS. This data is for September 2013 in the sample database.

Figure 6-2. *Using images to provide the background for a text box*

The Source Data

Here is the code that is in the stored procedures that you use to deliver this KPI (the data is for September 2013), `Code.pr_WD_MonthlySalesAndProfit`:

```
DECLARE @ReportingYear INT = 2013
DECLARE @ReportingMonth TINYINT = 9

SELECT
(
SELECT
SUM(CostPrice)
FROM    Reports.CarSalesData
WHERE   ReportingYear = @ReportingYear
        AND ReportingMonth = @ReportingMonth
) AS CostPrice
,(
SELECT
SUM(DeliveryCharge)
FROM    Reports.CarSalesData
WHERE   ReportingYear = @ReportingYear
        AND ReportingMonth = @ReportingMonth
) AS DeliveryCharge
,(
SELECT
SUM(CostPrice) - SUM(TotalDiscount)
FROM    Reports.CarSalesData
WHERE   ReportingYear = @ReportingYear
        AND ReportingMonth = @ReportingMonth
) AS GrossMargin
,(
SELECT
SUM(CostPrice) - SUM(TotalDiscount) - SUM(DeliveryCharge) - SUM(SpareParts) - SUM(LaborCost)
```

```
FROM     Reports.CarSalesData
WHERE    ReportingYear = @ReportingYear
         AND ReportingMonth = @ReportingMonth
) AS NetMargin
,(
SELECT
SUM(SalePrice)
FROM     Reports.CarSalesData
WHERE    ReportingYear = @ReportingYear
         AND ReportingMonth = @ReportingMonth
) AS Sales
```

This code returns the data shown in Figure 6-3.

	CostPrice	DeliveryCharge	GrossMargin	NetMargin	Sales
1	363390.00	1875.00	359790.00	345044.00	534690.00

Figure 6-3. *The data output for a text-based metric*

How the Code Works

This stored procedure uses a series of subqueries to return a single row of data for five key metrics.

Building the Visualization

This technique is not difficult in itself, and it will serve to introduce some of the core ideas that you will have to apply in SSRS when using images. So let's see how to add these image backgrounds.

1. Create a new SSRS report named `_MonthlySalesAndProfit.rdl`, and add the shared data source `CarSales_Reports`. Name it `CarSales_Reports`.
2. Add the following four shared datasets (ensuring that you also use the same name for the dataset in the report): `CurrentYear`, `ReportingYear`, `CurrentMonth`, and `ReportingMonth`. Set the parameter properties for `ReportingYear` and `ReportingMonth` as defined at the start of Chapter 1.
3. Add a new dataset named `MonthlySalesAndProfit` using the stored procedure `Code.pr_WD_MonthlySalesAndProfit` based on the T-SQL shown above.
4. Add a new table from the SSRS toolbox and add six new columns (or as many as are required to give you a table containing nine columns).
5. Set the borders for all the text boxes and the table itself to none.
6. Delete the detail row (you will have to confirm this), and add a new row. Your table should end up with two rows only.
7. Set the width of columns 1,3,5,7, and 9 to 3.13 inches. Use the `Size` property in the Properties window to do this.
8. Set the width of columns 2, 4, 6, and 8 (these are the columns used as separators between the metrics) to 0.15 inches.

9. Set the height of the first row to 0.25 inches, and the height of the second row to 0.73 inches.
10. In the Report Data window, right window, right-click Images, then Add Image. Navigate to the directory `C:\BIWithSSRS\Images`, select the file `BoxBottom.png`, and click Open. Do the same to embed the image `BoxTop.png`. You will see that these images are now available in the `Images` folder of the SSRS toolbox.
11. In the first row, expand BackgroundImage in the Properties window. Set the following properties for columns 1,3,5,7, and 9. You can do this cell by cell, or you can preselect the relevant cells by Ctrl-clicking on them and then setting the properties.
    a. Source: Embedded
    b. Value: BoxTop (you can select this using the pop-up menu for this property once the source is set)
    c. MIMEType: image/png
    d. BackgroundRepeat: Clip
12. In the second row, expand BackgroundImage in the Properties window. Set the following properties for columns 1,3,5,7, and 9:
    a. Source: Embedded
    b. Value: BoxBottom
    c. MIMEType: image/png
    d. BackgroundRepeat: Clip
13. Still in the second row, add the following data fields:
    a. Column 1: Sales
    b. Column 3: CostPrice
    c. Column 5: GrossMargin
    d. Column 7: DeliveryCharge
    e. Column 9: NetMargin
14. Format the font of the text boxes in which you added the data fields to Arial 24 point black. Center them vertically and horizontally.
15. Add the titles as seen in Figure 6-2. These are the same text as the data fields, but with spaces between the words. Set the text to be centered vertically and horizontally. Set the font to Arial 11 point italic. Set the Color to white.

This is all that you have to do. Rather than a classic (not to say banal) table of sales metrics, you now have a much more attractive and original presentation to bring your data to the user's attention.

How It Works

This visualization extends the use of the table to shape the presentation by adding separator columns: columns 2, 4, 6, and 8. It then uses tailored images and font formatting to customize the presentation of the metrics.

The tricky part here is, inevitably perhaps, the creation of the images that will be used as backgrounds. If you are not fortunate enough to have ready-made backgrounds, you will have to be prepared to spend a few minutes learning how to cut and recolor images that you find either in the enterprise or from other sources.

In this example, I indicate the exact size for row height and width, so that the images fitted perfectly. In practice, you can always start with an approximate size, add the image, and then manually resize the cell. Once you have the precise dimensions, you then copy and paste them into the size properties for all the other cells of the same type.

■ **Note** It is far too easy to "overcook" the end result when applying images to text boxes. Remember that you want to enhance the readability and give an overall impression of clarity and limpidity. So my advice is to keep the images literally and figuratively in the background.

Report and Table Backgrounds

SSRS does a great job of presenting tabular data. Unfortunately, too many tables in SSRS are simple to the point of being boring. So, to redress the balance and add some life to otherwise mundane tables of figures, you can add an image to an entire table to add a little visual differentiation to your dashboards.

Figure 6-4 shows the scorecard of quarterly sales by make for 2013. If the black and white display medium of a printed book does not make it clear (and you are not reading a color e-book), you can always open the file `_ScoreCardCountryIndicators.rdl` in the sample SSDT application to appreciate the full effect in color.

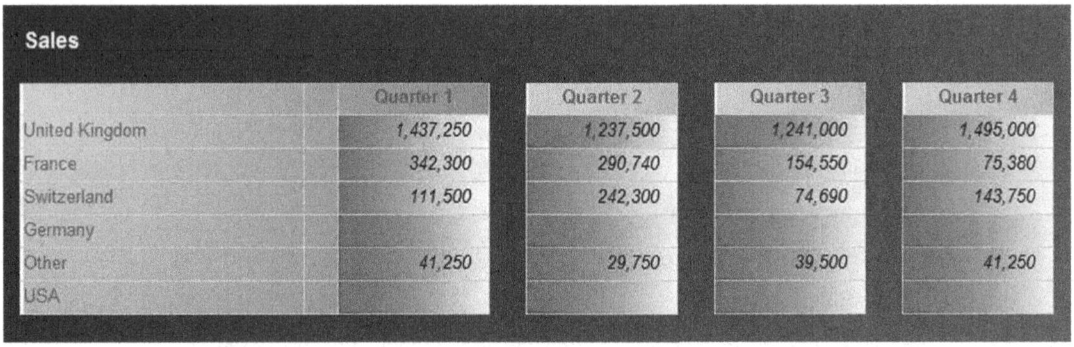

Sales

	Quarter 1	Quarter 2	Quarter 3	Quarter 4
United Kingdom	1,437,250	1,237,500	1,241,000	1,495,000
France	342,300	290,740	154,550	75,380
Switzerland	111,500	242,300	74,690	143,750
Germany				
Other	41,250	29,750	39,500	41,250
USA				

Figure 6-4. *Adding a background to a report and a table*

To make the scorecard more interesting, I added an image background to the report itself. Then there is a gradient image, which has been added to the table. As you can also see, "cells" (or text boxes, to give them their real name) that have been attributed a specific color will obscure the table background.

This way, you can see how the images layer in the order Report ➤ Table ➤ Text Box. You may feel (and I certainly do) that this may be a certain level of visual overload. So let's not discuss the aesthetics but concentrate on mastering the techniques. Hopefully you will find a real-world requirement where they will be useful.

The Source Data

The data for this scorecard is produced by the following code snippet, which is in the stored procedure Code. pr_ScorecardCountryIndicatorsSalesAndBudget:

```
DECLARE @ReportingYear INT = 2013

IF OBJECT_ID('Tempdb..#Tmp_KPIOutput') IS NOT NULL DROP TABLE Tempdb..#Tmp_KPIOutput

CREATE TABLE #Tmp_KPIOutput
(
Qtr TINYINT
,Country NVARCHAR(100) COLLATE DATABASE_DEFAULT
,Sales NUMERIC(18,6)
,SalesBudget NUMERIC(18,6)
,Delta NUMERIC(18,6)
,DeltaPercent NUMERIC(18,6)
,SortOrder TINYINT
)

-- Budget elements

INSERT INTO #Tmp_KPIOutput
(
Country
,Qtr
,SalesBudget
)

SELECT
BudgetDetail AS Country
,CASE
WHEN Month IN (1,2,3) THEN 1
WHEN Month IN (4,5,6) THEN 2
WHEN Month IN (7,8,9) THEN 3
WHEN Month IN (10,11,12) THEN 4
END AS Qtr
,SUM(BudgetValue) AS BudgetValue

FROM      Reference.Budget

WHERE      Year = @ReportingYear
        AND BudgetElement = 'Countries'

GROUP BY      BudgetDetail
                ,CASE
                    WHEN Month IN (1,2,3) THEN 1
                    WHEN Month IN (4,5,6) THEN 2
                    WHEN Month IN (7,8,9) THEN 3
                    WHEN Month IN (10,11,12) THEN 4
                END
```

```
-- Sales

;
WITH Sales_CTE
AS
(
SELECT
CountryName
,CASE
      WHEN MONTH(InvoiceDate) IN (1,2,3) THEN 1
      WHEN MONTH(InvoiceDate) IN (4,5,6) THEN 2
      WHEN MONTH(InvoiceDate) IN (7,8,9) THEN 3
      WHEN MONTH(InvoiceDate) IN (10,11,12) THEN 4
 END AS Qtr
,SUM(SalePrice) AS SalesValue

FROM          Reports.CarSalesData

WHERE         YEAR(InvoiceDate) = @ReportingYear

GROUP BY     CountryName
               ,CASE
                    WHEN MONTH(InvoiceDate) IN (1,2,3) THEN 1
                    WHEN MONTH(InvoiceDate) IN (4,5,6) THEN 2
                    WHEN MONTH(InvoiceDate) IN (7,8,9) THEN 3
                    WHEN MONTH(InvoiceDate) IN (10,11,12) THEN 4
               END
)

UPDATE       Tmp
SET          Tmp.Sales = CTE.SalesValue
FROM         #Tmp_KPIOutput Tmp
             INNER JOIN Sales_CTE CTE
             ON CTE.CountryName = Tmp.Country
             AND CTE.Qtr = Tmp.Qtr

-- Calculations

UPDATE    #Tmp_KPIOutput
SET       Delta = Sales - SalesBudget
          ,DeltaPercent =
                         CASE
                           WHEN Sales IS NOT NULL THEN ((Sales - SalesBudget) / SalesBudget) + 1
                           ELSE NULL
                         END
```

```
UPDATE      #Tmp_KPIOutput
SET         SortOrder =
                         CASE
                              WHEN Country = 'United Kingdom' THEN 1
                              WHEN Country = 'France' THEN 2
                              WHEN Country = 'Switzerland' THEN 3
                              WHEN Country = 'Germany' THEN 4
                              WHEN Country = 'Other' THEN 5
                              ELSE 4
                         END
-- Output
 SELECT Qtr, Country, Sales, DeltaPercent, SortOrder from #Tmp_KPIOutput
```

Running this code for 2013 gives the output shown in Figure 6-5.

	Qtr	Country	Sales	DeltaPercent	SortOrder
1	1	France	342300.000000	4.564000	2
2	1	Germany	41250.000000	0.275000	4
3	1	Other	NULL	NULL	4
4	1	Switzerland	111500.000000	1.486667	3
5	1	United Kingdom	1437250.000000	0.958167	1
6	1	USA	NULL	NULL	5
7	2	France	290740.000000	3.876533	2
8	2	Germany	29750.000000	0.198333	4
9	2	Other	NULL	NULL	4
10	2	Switzerland	242300.000000	3.230667	3
11	2	United Kingdom	1237500.000000	0.825000	1
12	2	USA	69250.000000	0.461667	5
13	3	France	154550.000000	2.060667	2
14	3	Germany	NULL	NULL	4
15	3	Other	NULL	NULL	4
16	3	Switzerland	74690.000000	0.995867	3
17	3	United Kingdom	1241000.000000	0.827333	1
18	3	USA	39500.000000	0.263333	5
19	4	France	75380.000000	1.005067	2
20	4	Germany	NULL	NULL	4
21	4	Other	NULL	NULL	4
22	4	Switzerland	143750.000000	1.916667	3
23	4	United Kingdom	1495000.000000	0.996667	1
24	4	USA	41250.000000	0.275000	5

Figure 6-5. *The output for the shaded sales scorecard*

How the Code Works

This code snippet works much like those that you saw in Chapter 2. A temporary table collects sales data as well as the budgetary information (where available). Then the process calculates the percentage difference between the two, and sets a custom sort order. As you can see from the sample output, you do not need all the columns to create the scorecard.

Building the Scorecard

So, armed with your source data, you can create the scorecard.

1. Create a new SSRS report named `_ScoreCardCountryIndicators.rdl`. Resize the report so that it is sufficiently large enough to hold a large trellis chart. Add the shared data source `CarSales_Reports`. Name it `CarSales_Reports`.
2. Create a dataset named `SalesTrend`. Have it use the `CarSales_Reports` data source and the stored procedure code `pr_ScorecardCountryIndicatorsSalesAndBudget` (which you saw above).
3. Add the shared datasets `CurrentYear` and `ReportingYear` to the report, ensuring that you also use the same name for the dataset in the report. Set the parameter properties for `ReportingYear` as defined at the start of Chapter 1.
4. Add the following images to the report:
    a. `BlueShade.jpg`
    b. `GreenShade.jpg`
    c. `RedShade.jpg`
    d. `GraduatedGrayBackground.png`
    e. `GraduatedGrayBackground_Small.png`
5. Right-click inside the report and select Body Properties. Select Embedded as the image source, and `GraduatedGrayBackground` as the image to use.
6. Leaving the report body selected, display the properties window. Expand BackgroundImage and set the BackgroundRepeat to Clip.
7. Add a table to the report. Ensure that it has a header row and a data row, and three columns. Set `SalesTrend` as the dataset. Set the width of the first column to 2.3 inches, the second to 0.25 inches, and the third to 1.18 inches.
8. Ensure that all the text box borders and the table outer border are set to white solid 1 point.
9. In the properties pane, click the ellipses for the Filters property. Click Add and set the filter to `[Qtr] = 1`.
10. Click Sorting in the Tablix properties (or open it again by clicking on the ellipses for the `SortExpressions` property). Click Add and select `[SortOrder]` as the column to sort by, and the Order as `A to Z`.
11. Add the field `Country` to the left-hand cell of the data row, and the field `Sales` to the right-hand cell of the data row. Add the title Quarter 1 to the right-hand cell of the header row. Format the font to Arial bold 12 point gray.

12. Select the table (it is often easiest to click outside the table and then drag the pointer over part of the table to do this). Display the properties pane. Expand BackgroundImage and set the following properties:

    a. Source: Embedded

    b. Value: GraduatedGrayBackground_Small

    c. BackgroundRepeat: RepeatY

13. Set the custom format for the sales data text box to `#,#` and the font to Italic. Right-align the cell contents.

14. Right-click the sales data text box and select Text Box properties. Set the following properties:

Section	Property	Value
Fill	Fill Color	No Color
	Image Source	Embedded
	Use this image (expression)	`=switch(Fields!DeltaPercent.Value<=0.9, "RedShade",Fields!DeltaPercent.Value>1.1,"BlueShade" ,Fields!DeltaPercent.Value>0.8,"BlueShade")`

15. Copy the table that you just created, then delete the first two columns of the copy. Position the copy to the right of the first table.

16. Set the title of the column in the second table to Quarter 2. Set the filter for this table to `[Qtr] = 2`. This way you will display the data for a different quarter.

17. Duplicate the second column twice more and set the titles and filters to Quarters 3 and 4, respectively. Align the four tables so they look like Figure 6-4.

18. Add a text box to the top left of the report and set "Sales" as the text. Format it to Arial Black white.

■ **Note** In steps 5 and 6, you used both the Body Properties dialog and the Properties window to set the background image properties. You could have used the Properties window only. This is to stress that the dialog does not contain the full range of options for many objects, but it is much easier to use.

How It Works

This report uses images to provide some interesting presentation effects. The report background has a dark-to-light shading from top left to bottom right, while the tables apply a contrary shading from bottom left to top right. The text box background for the figures displays the year-on-year progress as "classic" three color signals. An expression in the report is used to select the image that is used in certain cells in the table rather than using status flags in the source data, as we did in previous chapters.

Using Background Repeat

In steps 6 and 12 you saw the `BackgroundRepeat` property. This defines how an image relates to the size of the object that contains it. This applies specifically to tablixes and text boxes.

- *Default*: This is usually the same as repeat.
- *Repeat*: An image that is smaller than the container object will be repeated (horizontally and vertically) to fill the container.
- *RepeatX*: An image that is narrower than the container object will be duplicated (horizontally) to fill the container.
- *RepeatY*: An image that is shorter than the container object will be duplicated (vertically) to fill the container.
- *Clip*: An image that is bigger than the container will be cropped to fit the container.

■ **Note** When adding images (using the Image item in the SSRS toolbox) you will notice that the image can be resized proportionally with the image object. Unfortunately, this does not happen when you add images as backgrounds to reports, tables, or text boxes.

Rectangle Borders

There may be times when you want to group a set of elements together and to emphasize, visually, the fact that the user is looking at a collection of, say, gauges or charts. This is where using a rectangle as a grouping element can be useful. However, a rectangle only has an extremely limited set of border options (unlike charts) and can only show a rectangular border. None of the more diverse options such as raised, sunken, embossed, or frame (that you can add out of the box to charts, for instance) are available. This can be a slight problem, as it can detract from the visual unity of a dashboard where you want, for example, to present a series of elements such as charts, tables, and gauges inside similar-looking borders.

Figure 6-6 is an example of a series of gauges placed inside a rectangle whose background image provides a more attractive border. Indeed, this border is lifted directly from one of the standard border types available in SSRS charts, thus ensuring visual homogeneity. If you read Chapter 3, you already saw how to create the gauges for this visualization. If you need to refer back to the gauge, take a look at Figure 3-5.

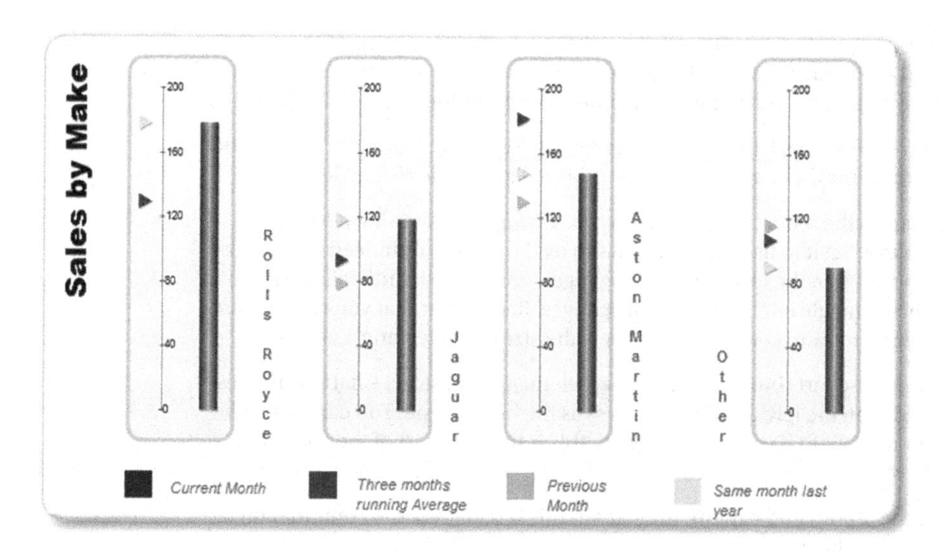

Figure 6-6. *Using an image to provide the border for a rectangle*

Here you will take this core set of gauges as a starting point and encapsulate them, visually anyway, in a custom rectangle with a border. I have to add that I am cheating somewhat by showing you the full effect here, including the title and the legend. You will only see how to add these in Chapter 7.

This technique is nothing other than a trick to use a border design as the background for the collection of elements. Clearly, this depends primarily on having a background image. One fundamental point to note is that this approach presumes that the elements you are placing in a container will *not* grow or shrink when the report is displayed. So it only really suits fairly rigid presentations. However, as this use case covers many types of dashboards, it is in practice really useful to know how to apply varied types of borders to collections of objects in a report.

Using Chart Borders for Gauges

Let's start simply by taking advantage of some of the border styles that SSRS includes that you can add when creating charts. Suppose that your dashboard already has a couple of charts, and that you have used a light gray embossed border for them. Now what you want to do is to use the same border type to enclose the gauges displaying sales by make. You will create a dummy chart that will only serve to provide a border. You will capture the border as an image that you will then use as a background image for a rectangle. This rectangle will group the gauges into a coherent object. Once this is done, you will delete the chart you created.

1. Copy the SSRS report named `_MonthSalesByMakeOverTime.rdl` that you created in Chapter 3 (you can also find this file in the sample data in the `CarSalesReports` project on the Apress web site for this book). Name the copy `_MonthSalesByMakeOverTimeExtended.rdl`.

2. Open the file that you just copied. Increase the height of the report area to approximately twice the height of the gauge visualization.

3. Click the Rectangle item in the SSRS toolbox, then draw a rectangle under the gauges. Ensure that the rectangle is slightly bigger than space taken by the gauges and that it leaves a little whitespace to the top, left, right, and bottom.

4. Select the gauges (and any other items if there are any that complete the collection) and drag them inside the rectangle. This creates a grouped element. Resize the rectangle so that it is only a little taller and wider than the gauges.
5. Drag the rectangle back to the top of the report. Note that it includes all the other items in the report. The gauges are now grouped inside the rectangle.
6. Add a chart to the report. It can be any type, as you are only using it to create a border image. Set it to use the same dataset as the gauges, or indeed any dataset in the report. Place the chart below the gauges, and resize it until it is the same size as the rectangle encompassing the gauges. Remember that you can use the Properties window to get and set precisely the size of the rectangle and chart.
7. Right-click the chart (but not on any chart element) and select Chart Properties. Click Border on the left, and then Emboss as the border type. You can adjust the line color and width if you want. After all, this is the only part of the chart that interests you.
8. Preview the report, and capture the screen (this is usually some combination of keys involving the PrtSc key).
9. Open MS Paint, and paste the screen capture.
10. Using the rectangle selection tool, drag the pointer (or marquee if you prefer) over the interior of the chart, leaving its border untouched, then press Delete.
11. Select the border encompassing the empty space that remains and copy it, leaving the rest of the screen capture behind.
12. Click New to create a new Paint file without saving the current file, and paste the copied border into Paint. Save the file. Call it `C:\BIWithSSRS\Images\EmbossedBorder.png`.
13. Back in SSRS design mode, delete the chart that you made.
14. Decrease the height of the report so that it only contains the rectangle and gauges.
15. In the report data window, right-click Images and then Add Image. Navigate to the image that you just saved (`C:\BIWithSSRS\Images\EmbossedBorder.png`) and click Open.
16. Right-click inside the rectangle, but not on any of the gauges or indeed on any other object, and select Rectangle Properties from the context menu.
17. Click Fill, and select Embedded as the image source. You can then select the image name from the Use this Image pop-up. The dialog will look like Figure 6-7.

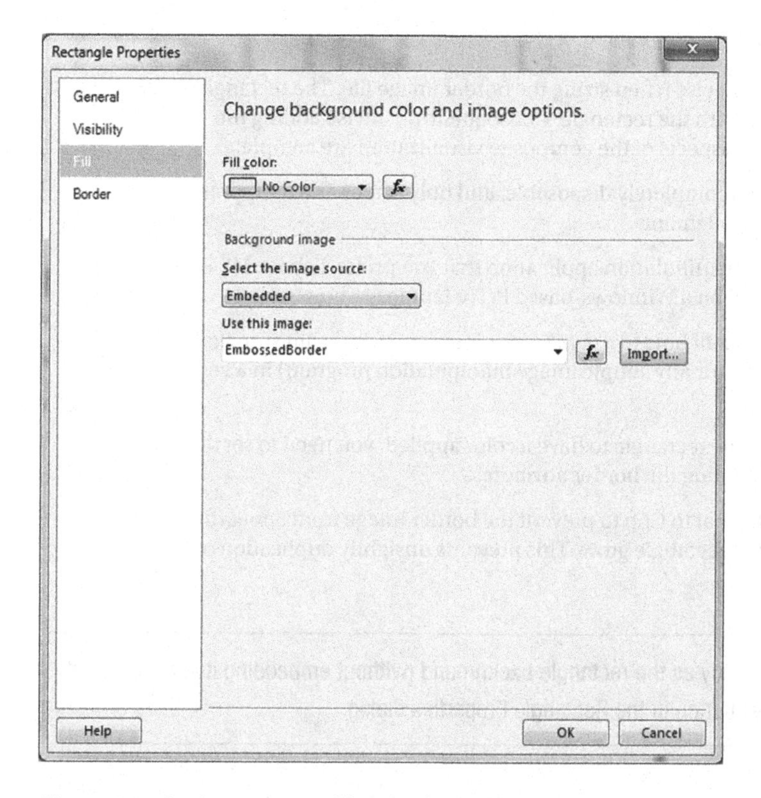

Figure 6-7. *Setting an image file to be the background for a rectangle*

18. Click OK.
19. Adjust the height and width of the rectangle if required to show all the borders-and nothing but the borders. You may need to tweak the placement of the other objects relative to the rectangle to obtain the best look for your gauges.
20. With the rectangle selected, display the properties window and expand BackgroundImage. Set the BackgroundRepeat to Clip.

The gauges (with titles that are explained at the end of this chapter, and a legend that is described in Chapter 7) now have an embossed border, identical to the border that you can apply to SSRS charts, and should look like Figure 6-5.

How It Works

This example boiled down to creating an object with a border that was the same size as the group of objects that you wanted to enclose, and then capturing the border only as an image file. Then all you had to do was to use the image file as the background for the rectangle.

There are a few things to note:

- You need to be extremely precise when sizing the border image file. The rectangle background cannot resize with the rectangle. Consequently, I advise adding the border only when all other aspects of the composite visualization are complete.
- The chart that you create is completely disposable, and only serves as a temporary source of a border for other elements.
- Feel free to use any image manipulation application that you prefer. I chose MS Paint because it is always at hand on a Windows-based PC or laptop.
- I realize that the explanation of how to use MS Paint is succinct, but I am sure that you will get the hang of this (or any simple image manipulation program) in a very short time.
- If you want the interior of the rectangle to have a color applied, you need to set this fill color in step 8 when defining the border attributes.
- You set the BackgroundRepeat to Clip to prevent the border image from appearing more than once should the rectangle grow. This prevents unsightly duplication of the border.

■ **Note** You can import the image directly as the rectangle background (without embedding it into the report) if you prefer. To do this, click the Import button in the Rectangle Properties dialog.

Using a Custom Image

This example showed how to use SSRS to generate the image that becomes the border for a group of objects. If you are proficient with other ways of generating image files, you can use any way that you choose to create the background image that adds a border to a rectangle. Just remember to make the size of the image as accurate as you can when creating it. This way you will avoid much to-ing and fro-ing between your image manipulation application and SSRS as you continually tweak the size of the image in the source application.

Borders for Tables

While SSRS does produce efficient tables to deliver tabular output, the available border options for tables are extremely limited. As was the case with rectangles, you cannot display embossed, raised, or sunken borders in just a couple of clicks as you can with charts.

However, all is not lost, because there is a way to add enhanced borders to tables using images. If your tables are likely always to be the same height, then there is a method to add a background image pretty much as you did previously with a rectangle. If your table is going to vary in height (because the number of detail rows varies), there is another technique that you can use. Let's look at each of these in turn.

Fixed-Size Table Borders

If your table will always contain the same number of rows and columns, it could look something like Figure 6-8.

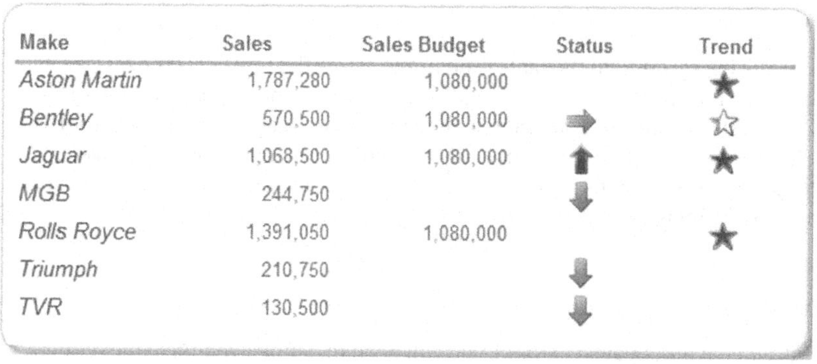

Make	Sales	Sales Budget	Status	Trend
Aston Martin	1,787,280	1,080,000		
Bentley	570,500	1,080,000		
Jaguar	1,068,500	1,080,000		
MGB	244,750			
Rolls Royce	1,391,050	1,080,000		
Triumph	210,750			
TVR	130,500			

Figure 6-8. *Adding an image border to a table*

Here is how to add the border. This technique is very like the approach taken in the previous example, only applied to a table instead of to a rectangle.

1. Open the file containing the table to which you want to add a fixed-size background. This example will take a copy of the file `KPI_Basic` that you created in Chapter 2. I have named the copy `KPI_Basic_Border.rdl`.
2. Add a new row at the top of the table. Add another below the bottom rows. Both of these new rows must be outside the existing group (if there is one) for the table. Set the font size for all the cells in the row to 9 points.
3. Add a new column to the left and a new column to the right of the table. Set them to 0.125 inches wide (this is roughly 9 points).
4. Remove all borders from the newly inserted columns and rows. Ensure that none of the cells have a background color or any content.
5. Add the image `C:\BIWithSSRS\Images\EmbossedBorderKPI.png` to the report. The technique for this is explained in the previous example.
6. Select the table and display the Properties window. Set the BackgroundImage properties to the following:
   a. Source: Embedded
   b. Value: EmbossedBorderKPI
   c. MIMEType: image/png
   d. BackgroundRepeat: Clip
7. Preview the report. As you will not see the full image until you do this, you can then adjust the image size (and reimport it) and/or tweak the detail row height until the table has an apparently flawless border.

How It Works

Apart from adding the image as the background to the table (and not to any of the table's text boxes), all you have to do is add extra rows and columns to the outside of the table that provide the requisite empty space needed by the table border. Adding these "spacer" rows merely means ensuring that they are not part of any row group, and are the same height (and width, in the case of the columns) as the border in the image. You may want to add a little white space as well. In most cases, there may be a certain amount of trial and error until you have balanced the size of the image and the size of the row.

Dynamic Table Borders

If your table will grow or shrink when it is displayed in SSRS, you will need to define dynamic table borders. An example of a table with a dynamic border is shown in Figure 6-9. It builds on the _ColorSales report that you saw in Figure 2-5 in Chapter 2. This report has a number of detail rows that varies depending on the selected month and year, as the number of different colors of vehicle sold will vary. Consequently, in this example, the vertical borders will extend proportionately as the table changes in height. This would definitely *not* be the case for a table with an image background, as you saw in the previous example.

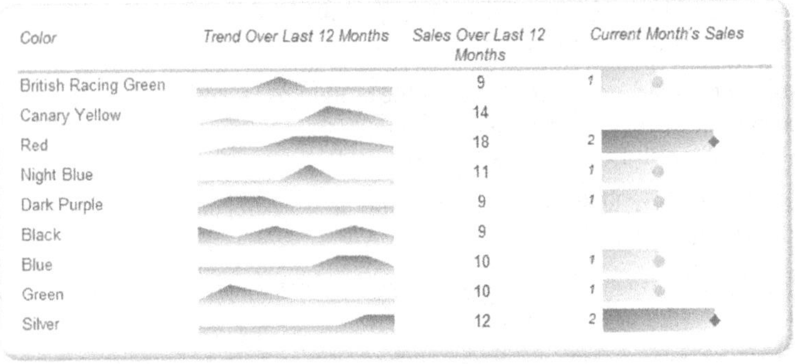

Color	Trend Over Last 12 Months	Sales Over Last 12 Months	Current Month's Sales
British Racing Green		9	1
Canary Yellow		14	
Red		18	2
Night Blue		11	1
Dark Purple		9	1
Black		9	
Blue		10	1
Green		10	1
Silver		12	2

***Figure 6-9.** Using images to add dynamic borders to a table*

Here is how to add a dynamic border.

1. Open the file containing the table to which you want to add a fixed-size background. This example will take a copy of the file `KPI_Basic` that you created in Chapter 2. I have named the copy `_ColorSales.rdl`.

2. Add new blank rows and columns outside the existing table, as described for the previous example. Set the row heights and the column widths to 0.2 inches. Make sure that the newly created cells have nothing in them and that no background color is applied.

3. In your image editing software (such as MS Paint), open an existing background image file containing a border, such as `C:\BIWithSSRS\Images\EmbossedBorderKPI.png`.

4. Copy and save independently the following eight image files:

    a. The top left-hand corner angle of the border, 20 pixels square. Name it `EmbossedBorder1_TopLeft.png`.

    b. The top right-hand corner angle of the border, 20 pixels square. Name it `EmbossedBorder1_TopRight.png`.

    c. The bottom left-hand corner angle of the border, 20 pixels square. Name it `EmbossedBorder1_BottomLeft.png`.

    d. The bottom right-hand corner angle of the border, 20 pixels square. Name it `EmbossedBorder1_BottomRight.png`.

    e. A 20-pixel square from the top border. Name it `EmbossedBorder1_Top.png`.

    f. A 20-pixel square from the bottom border. Name it `EmbossedBorder1_Bottom.png`.

    g. A 20-pixel square from the left border. Name it `EmbossedBorder1_Left.png`.

    h. A 20-pixel square from the right border. Name it `EmbossedBorder1_Right.png`.

5. Embed these eight images in the report, using the method described in previous examples.

6. Display the Properties window, then click on the top left cell of the table. Set the BackgroundImage properties to the following:

    a. Source: Embedded

    b. Value: EmbossedBorderl_TopLeft

    c. MIMEType: image/png

    d. BackgroundRepeat: Clip

7. Do the same for the other three corners of the table as far as the Source, MIMEType and BackgroundRepeat properties are soncerned. Ensure that you use the following files for the Value property:

    a. Top right corner: `TopRight.png`

    b. Bottom left corner: `BottomLeft.png`

    c. Bottom right corner: `BottomRight.png`

8. Add the image `EmbossedBorder1_Top` to all the cells for the top row except the first and last. Set the `BackgroundRepeat` property to RepeatX. Do the same for the bottom row using the image `EmbossedBorder1_Bottom`.

9. Add the image `EmbossedBorder1_Left` to all the cells for the leftmost column except the top and bottom cells. Set the `BackgroundRepeat` property to RepeatY. Do the same for the rightmost column using the image `EmbossedBorder1_Right`.

You're done! This approach may be a little laborious, especially when it comes to copying and pasting the image files that compose the borders, but it works well in practice. If you are lucky, you can define a standard border style across your BI reports, and reuse these files many, many times, thus maximizing your initial investment in time.

How It Works

The interesting thing about this approach is that you are not using one image to provide the border, but eight. This is because you need the following elements:

- Images for each of the *four corners*.
- Images for a section of the *left* and *right* vertical borders.
- Images for a section of the *top* and *bottom* horizontal borders.

If you take a look at Figure 6-10, I hope that this idea will become clearer.

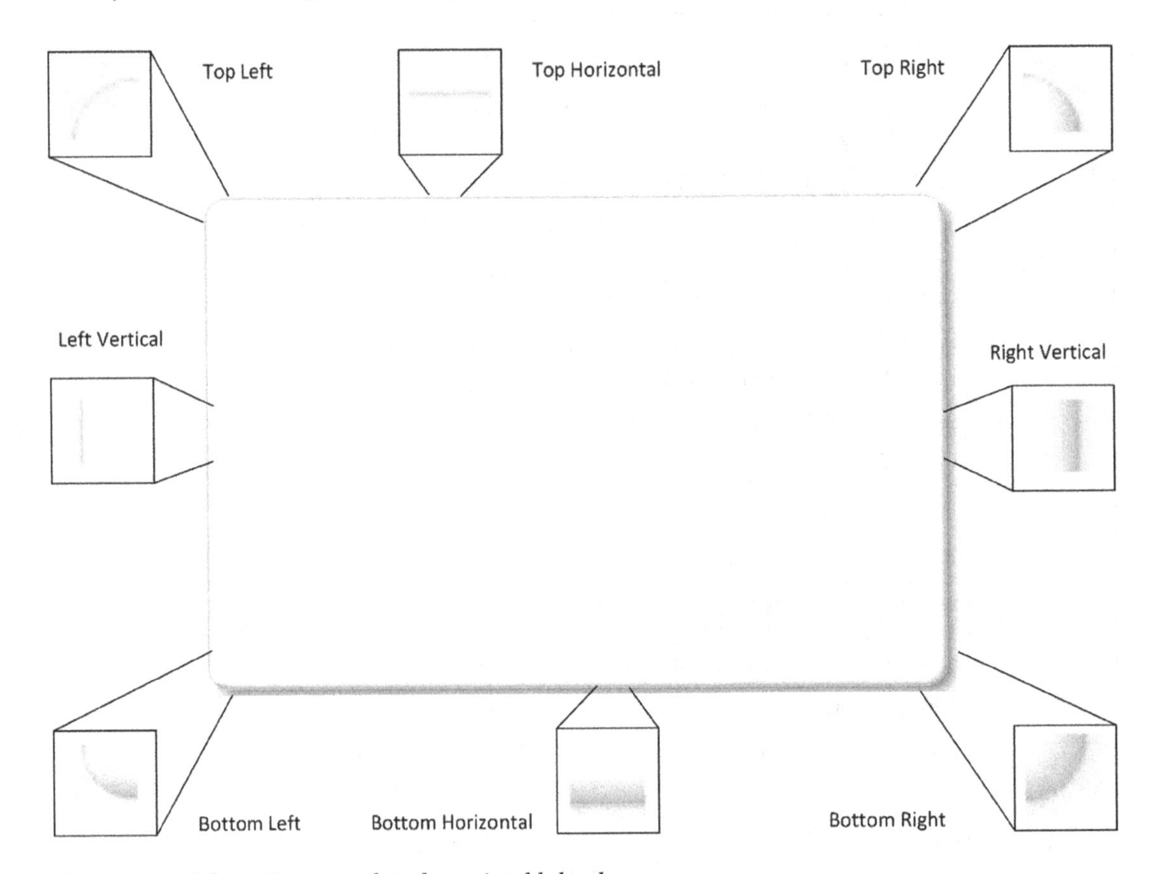

Figure 6-10. *Schematic approach to dynamic table borders*

Once you have the image files, you then need placeholder rows and columns around the table. When creating a border like this the arduous part can be setting the size of the rows and columns to match exactly the size of the image files. Fortunately, SSRS will convert pixels to inches (or centimeters) if you enter nnPx (where nn is the size of the image height or width in pixels) as the height or width in the Properties window. So I advise you to concentrate on getting the images right first, and then set the SSRS properties as a function of the image files.

To get your dynamic borders to work, you must set the `BackgroundRepeat` property to RepeatX (for all the border images that make up the main horizontal borders in the top and bottom rows) and RepeatY (for all the border images that make up the main vertical borders in the left and right columns). This way SSRS will extend the image correctly to fill up the cells for the rows and columns. Conversely, the corner image must be set to Clip, as you do not want them repeating under any circumstances.

■ **Note** If your table will always (and I mean always) be the same size, you can simply use a background image for the border. You may nonetheless need to add an empty top row and empty left column to "center" the table in the image.

Gauge Backgrounds

SSRS provides a complete and varied toolkit for creating gauges, including a swathe of different gauge backgrounds. On certain rare occasions, you could nonetheless find yourself needing to override the standard gauge backgrounds with a custom image. Figure 6-11 demonstrates a set of gauges showing sales by country designed for a smartphone. In this case, screen space is so limited that even removing the country name helps. So, to indicate which dial relates to which country's sales figures, I am using the country flag as the background for the gauge.

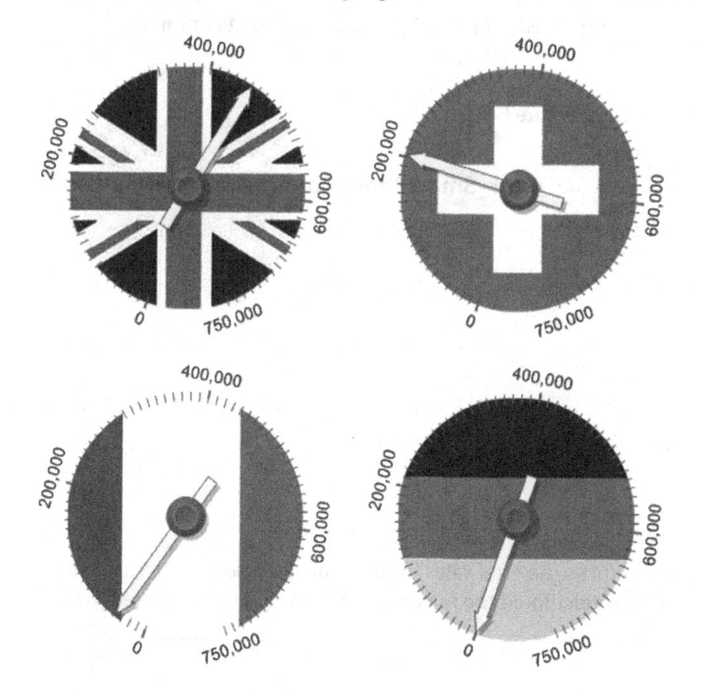

Figure 6-11. *Adding a background to a gauge*

The Source Data

First, you need some data to work with. The following code snippets (available as `Code.pr_SmartPhoneCountryGauges` and `Code.pr_SmartPhoneCountryGaugesMaxValue` in the `CarSales_Report` database) will give you the data you need for May 2013:

```
DECLARE @ReportingYear INT = 2013
DECLARE @ReportingMonth TINYINT = 5

-- Code.pr_SmartPhoneCountryGaugesMaxValue
```

```
SELECT
Code.fn_ScaleQuartile(SUM(SalePrice)) AS CurrentMonthSalesMax
FROM            Reports.CarSalesData
WHERE           ReportingYear = @ReportingYear
                AND ReportingMonth = @ReportingMonth
                AND CountryName IN ('United Kingdom','France','Germany','Switzerland')

-- Code.pr_SmartPhoneCountryGauges

SELECT
SUM(SalePrice) AS CurrentMonthSales
,CountryName AS Country

FROM            Reports.CarSalesData
WHERE           ReportingYear = @ReportingYear
                AND ReportingMonth = @ReportingMonth
                AND CountryName IN ('United Kingdom','France','Germany','Switzerland')
GROUP BY        CountryName
```

Running these snippets gives the output shown in Figure 6-12.

pr_ScorecardCostsGauges

	CurrentMonthSales	Country
1	37690.00	France
2	202800.00	Switzerland
3	450750.00	United Kingdom

pr_SmartPhoneCountryGaugesMaxValue

	CurrentMonthSalesMax
1	750000

Figure 6-12. *The data for a simple gauges with an image as the background*

How the Code Works

This T-SQL snippet is incredibly simple, really: it returns the total sales for the four selected countries for the chosen year and month. A second stored procedure calculated the maximum sales value to be applied to all the gauges so that the scales are identical.

Building the Visualization

Let's see how this visualization is built. It is, to a certain extent, a revision of some of the gauge techniques you saw in Chapter 3, so I will not explain everything in detail.

1. Copy the SSRS report named `__BaseReport.rdl` and rename the copy `SmartPhone_CarSalesByCountryWithFlagDials.rdl`. This will set the parameters and core datasets.
2. Add two datasets: `SmartPhoneCountryGauges` and `SmartPhoneCountryGaugesMaxValue` using the appropriate stored procedures shown above. Delete the range on the gauge.

3. Embed the following four images into the report (all from the directory `C:\BIWithSSRS\Images`): `FrenchFlag_Round`, `UKFlag_Round`, `GermanFlag_Round`, and `SwissFlag_Round`.
4. Place a radial gauge on the report area and set `SmartPhoneCountryGauges` as the dataset. Filter this first gauge to use the United Kingdom as the Country to display.
5. Set the pointer to use `CurrentMonthSales` as its source data field. Set the following pointer properties:

Section	Property	Value
Pointer Options	Pointer type	Needle
	Needle style	Stealth arrow with tail
	Placement (relative to scale)	Inside
	Distance from scale (percent)	0
	Width (percent)	15
Pointer Fill	Fill style	Solid
	Color	Yellow
Cap Options	Cap style	Rounded with wide indentation
	Cap width (percent)	33

6. Right-click the gauge itself and display the Properties window. Expand BackFrame, then FrameImage. Set the following properties:

Property	Value
ClipImage	True
MIMEType	Image/jpeg
Source	Embedded
Value	UKFlag_Round

7. Right-click the scale and select Scale Properties. Set the following properties:

Section	Property	Value
General	Minimum	0
	Maximum (Expression)	`=Sum(Fields!CurrentMonthSalesMax.Value, "SmartPhoneCountryGaugesMaxValue")`
Layout	Start angle (degrees)	50
	Sweep angle (degrees)	320
	Scale bar width (percent)	0
Labels	Placement (relative to scale)	Outside
	Distance from scale (percent)	3
Label Font	Font	Arial
	Size	14 point
	Color	Dark Blue
Major Tick Marks	Hide major tick marks	Unchecked
	Major tick mark shape	Rectangle
	Major tick mark placement	Outside
	Width (percent)	2
	Length (percent)	7
Minor Tick Marks	Hide minor tick marks	Checked

8. Make three copies of the gauge and filter them to use the three other countries (France, Switzerland, and Germany). Set the BackFrame FrameImage to the appropriate country flag for each gauge as described in step 6.

How It Works

The effect produced by these gauges depends purely and simply on the use of a Frame Image for the gauge Back Frame. The images also had to be prepared and had to be exactly the same shape as the gauge, round in this case. Once again, you may need a little practice with some image editing software (and a mastery of your favorite search engine) to learn how to cut circles out of images.

Chart Columns and Bars

I am the first to admit that using images in chart bars and columns is perhaps overkill. After all, it was "new" in Excel nearly 20 years ago. Any presentation purist will immediately sniff at such a tacky gimmick. Yet there may be one or two cases (such as in a chart on a smartphone, for instance) where using an image can make the chart more readable, and possibly more arresting. Indeed, on a tablet or smartphone this technique can do the following:

- Save space, since you no longer need a legend
- Grab the user's attention, and stop them flicking to the latest game app that they recently downloaded

An example of using images in chart series is given in Figure 6-13.

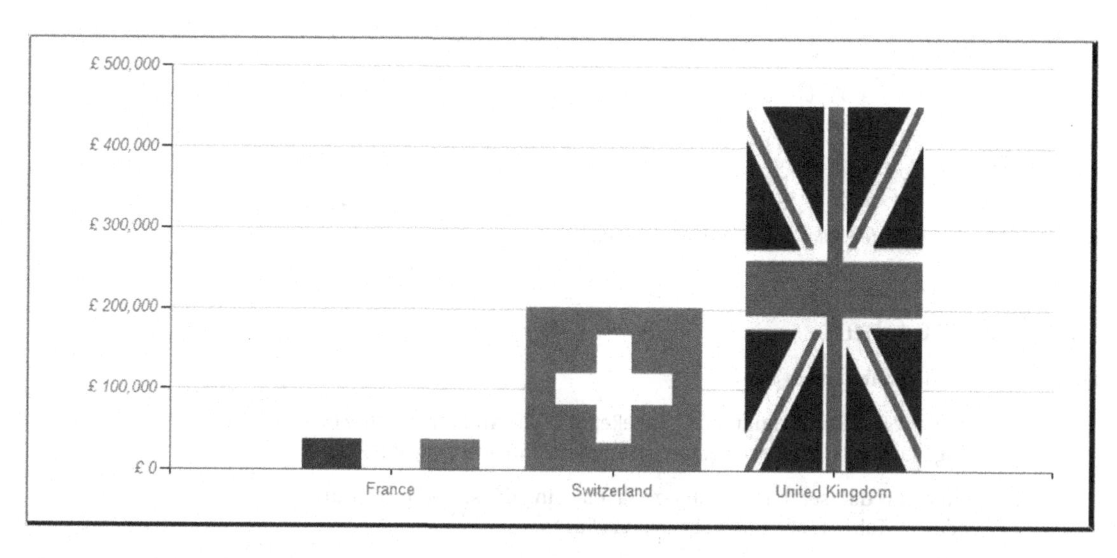

Figure 6-13. *Using images for the columns in a chart*

The Source Data

The source data for this chart, for May 2013, can be returned using the following code that can be found in the sample data as `Code.pr_TabletTabbedReportCountrySales`:

```
DECLARE @ReportingYear INT = 2013
DECLARE @ReportingMonth TINYINT = 5

SELECT
CASE
WHEN CountryName IN ('France','United Kingdom','Switzerland', 'United States', 'Spain')
THEN CountryName
ELSE 'OTHER'
END AS CountryName
,SUM(SalePrice) AS Sales
FROM            Reports.CarSalesData
WHERE           ReportingYear = @ReportingYear
                AND ReportingMonth = @ReportingMonth
GROUP BY        CountryName
```

The data that is output looks like Figure 6-14.

	CountryName	Sales
1	France	79000.00
2	Switzerland	37690.00
3	United Kingdom	418000.00

Figure 6-14. *The data for sales per country per month*

Building the Chart

Let's get straight down to business.

1. Copy the SSRS report named __BaseReport.rdl and rename the copy FlagChart.rdl. This will set the parameters and core datasets.

2. Add a new dataset named CountrySales using the stored procedure Code.pr_TabletTabbedReportCountrySales.

3. Embed the following six images into the report (all .png files from the directory C:\BIWithSSRS\Images): EuropeFlag, GermanFlag, USAFlag, GBFlag, SpainFlag, FranceFlag, and SwissFlag.

4. Place a column chart on the report area and set CountrySales as the dataset. Set Sales as the ∑ Values and CountryName as the Category Groups.

5. Click any of the columns and display the Properties window. Expand BackgroundImage and set the following properties:

   a. Source: Embedded

   b. Value (expression):

```
=Switch (
        Fields!CountryName.Value= "United Kingdom", "GBFlag"
        ,Fields!CountryName.Value= "France", "FranceFlag"
        ,Fields!CountryName.Value= "UnitedStates", "USAFlag"
        ,Fields!CountryName.Value= "Germany", "GermanFlag"
        ,Fields!CountryName.Value= "Spain", "SpainFlag"
        ,Fields!CountryName.Value= "Switzerland", "SwissFlag"
        ,Fields!CountryName.Value= "OTHER", "EuropeFlag"
        )
```

   c. MIMEType: Image/png

   d. BackgroundRepeat: Fit

6. Set the font color and line color for both axes to appealing shades of gray, and format the numbers on the vertical axis as you wish. Refer back to Chapter 4 if you need any help with how to do this.

How It Works

As you can see in step 5, it is the expression that is used for the background image of the chart column that does the real work here. Specifically, it tests the field that sets the column data and chooses the appropriate image using a switch statement. This way the flag for the relevant country is displayed as the column.

■ **Note** You will not see the flags in the report design. You will have to preview (or deploy) the report to get the full effect.

Images as Titles

If you have spent any time in SSRS creating charts, you will doubtless have come to appreciate the flexibility with which you can adjust chart titles. Most notably, you can pivot the text used as a title to produce vertical titles.

Unfortunately, this is not possible in other circumstances, such as a title to accompany a series of gauges, which you saw in Figure 6-5. This is where creating a title as an image and then adding the image as the background to a rectangle can work wonders. As an example, look at Figure 6-12. This takes the set of gauges used previously and adds a rectangle containing an image file that contains the title. This time the rectangle is inside the outer "container," which has the border image.

1. Re-open the file `_MonthSalesByMakeOverTimeExtended` that you used earlier in this chapter.
2. Create an image file containing the text "Sales by Make," pivoted vertically. You can do this in certain image editing packages, in MS Visio (when you then select the pivoted text and save it as a `.png` file), or by creating a dummy chart containing the pivoted text and then carrying out a screen capture to grab the image of the text, as you saw earlier in this chapter.
3. Embed the image file (here it is `C:\BIWithSSRS\Images\SalesByMakeTitle.png`) in the report.
4. Add a rectangle to the report where you want the title to appear.
5. Set the following properties for the BackgroundImage of the rectangle:
    a. Source: Embedded
    b. Value: SalesByMakeTitle
    c. MIMEType: image/png
    d. BackgroundRepeat: Clip
6. Adjust the size and position of the rectangle so that the image contents are visible and the title appears where you want it.

Once again, as is the case with any image file, the hard part is getting the image to be the right size and proportion. So be prepared for a little trial and error, but this is the price of the creative freedom that you now have in your SSRS visualizations.

Conclusion

This chapter took you on a rapid tour of some of the ways that you can use images in SSRS for your dashboards, scorecards, and business intelligence delivery. You saw how images can be used as the background for tables and gauges as well as ways in which images can remove the need for legends and titles in certain cases.

Used with discretion, images can enhance the way that you present metrics in your BI reports. They can add variety as well as ensure a coherent and standardized look and feel for your corporate business intelligence.

CHAPTER 7

■ ■ ■

Assembling Dashboards and Presentations

Over the course of the last few chapters you have seen how to create many different types of visualizations. As business intelligence is all about delivering multiple views of the organization's data, you now need to see how some of these individual elements can be combined to deliver insights that are even more powerful for being coalesced into a single coherent view of specific information.

This chapter will give you a series of hints and tips on ways to assemble dashboards and other BI presentations in a structured and controlled way. First, you will see how to use existing widgets (which I also tend to call *components* in this context) to assemble dashboard-style reports. Then you will see how to align objects, group and contain objects, and control the way that certain elements grow. Nearly all that I will cover concerns the presentation aspect of business intelligence with SSRS. There is a simple reason for this: your audience will not only be distracted by a lopsided final output, they will also lose belief in the validity of your information. So the final presentation tweaks can be nearly as important as the accuracy of your data.

This chapter presumes a certain way of working with BI reports. As you have probably noticed, I tend to build reports up starting with the smallest object, which is then assembled into bigger visualizations and then-and this is one of the aims of this chapter-a series of visualizations finally becomes a complete report. Of course, there are other ways of working, and I do not for one second want to imply that this is the only method. It just happens to be a technique that suits the learning process that a structure like a book imposes, and it can reap rewards in practice. Once again, you are free to adopt any style of dashboard creation that gives you the results that you want.

This chapter builds on the work done in the preceding chapters. This does not mean that you have to have built all the components that you will be using. You can download the sample SSDT application and take certain "widgets" that you find there as the basis for building the examples in this chapter, if you prefer.

Remember above all that dashboards, scorecards, and most other business intelligence reports are designed to be relatively immobile. They are not normally designed to grow or shrink. Consequently, you will probably spend a large amount of time setting the size and alignment of objects in a report. This is perfectly normal when developing SSRS BI visualizations. This chapter aims to give you some ideas as to how to do this efficiently and productively.

A First Dashboard

To begin with, I will show you how to use the series of visualizations that you created in some of the previous chapters to assemble a first dashboard. This dashboard will illustrate the principles of dashboard assembly in SSRS.

Figure 7-1 shows what your first dashboard will look like. As you can see, it is composed of five widgets:

- `_SalesPercentageByMake.rdl`: This is the top left-hand set of four gauges that shows how a sales map targets for four main makes of cars. This component is described in Chapter 3.
- `_MapOfEnglishCarSales.rdl`: This visualization shows the sales to date in England by region. This component is described in Chapter 5.
- `_ThermometerYTDSales.rdl`: This gauge displays the sales up to and including the selected month along with the target for the year. This component is explained in Chapter 3.
- `_BubbleChartMarginByAgeAndMake.rdl`: This chart shows the percentage profit by make of car for vehicle age ranges. This component is described in Chapter 4.
- `KeyFiguresForMonth.rdl`: This text-based KPI gives key information for five essential metrics. This component is described in Chapter 2.

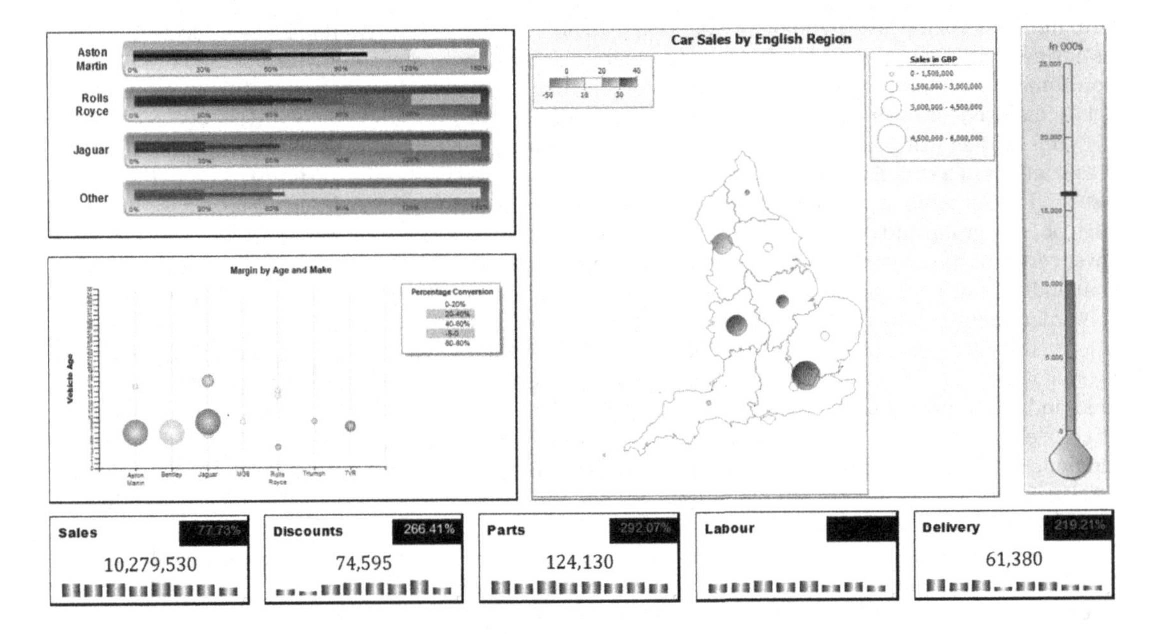

Figure 7-1. A simple dashboard

That, then, is the theory. Now let's move on to the practice.

1. Make a copy of the report `__BaseReport.rdl` and rename the copy `Dashboard_Yearly`. If you are starting from scratch, follow steps 1-5 from the first example in Chapter 3, or indeed the initial steps of many of the examples that you have seen so far in this book.
2. Resize the report area so that it is approximately large enough to hold all of the objects. I suggest about 16 inches by 10 inches to start.
3. Open the report `_SalesPercentageByMake.rdl` and take a look at the datasets that it uses. In this case, there is only one, `SalesPercentageByMake`. Recreate this dataset (paying attention to capitalization) in the new dashboard file that you created in step 1.

4. Select everything in the report `_SalesPercentageByMake.rdl` (Ctrl-A can be useful). Copy and paste the widget elements into the dashboard. Place it on the top left of the report area, possibly leaving a little space above and to the left.

5. Open the four remaining reports listed above and recreate their datasets, then copy and paste the widgets into the dashboard. Place the different elements so that the dashboard looks reasonably similar to the one in Figure 7-1.

6. Tweak the sizes of the five component elements until the proportions suit your aesthetic requirements.

7. Adjust or add any borders that are needed to ensure that the component parts share a common look and feel. In this example, for instance, this means setting the map border to Line rather than Embossed (as it was in the source file). You may in practice have to set all the borders to be the same color and weight to ensure a standardized presentation across all the elements of the dashboard.

8. Select the three components at the top of the dashboard that share a common top edge. Then select Format ➤ Align ➤ Tops.

9. Ensure that any other elements that need to be aligned vertically are aligned using the same approach. This will specifically apply the KPIs copied from the `KeyFiguresForMonth` file.

10. Adjust any fonts and colors to ensure a harmonious aesthetic for the dashboard. You saw several examples of how this is done in Chapters 2 to 5, so I will not repeat the techniques here; instead I refer you back to those chapters.

11. Select all the dashboard elements and position them towards the top and left of the report area, leaving a small amount of white space above and to the left of the components that make up the dashboard.

12. Reduce the height and width of the report area so that there is no empty space at the edges of the dashboard.

■ **Note** I always advise making the initial report as wide and as high as you can before assembling a dashboard. This gives you plenty of room to experiment with different presentations. After all, reducing the height and width of the report area takes only a couple of seconds.

How It Works

This dashboard presents no arcane difficulties, so is a good example to learn how to assemble multiple components into a single report. First, you need a "core" dashboard to contain the datasources and datasets that are common to all of the widgets. Then you add the datasets and widgets from the sub-elements that will be used to assemble the dashboard. You can add the datasets all together, or add them one at a time as you copy and paste the individual widgets into the dashboard; it is entirely up to you.

You may find that a major part of dashboard assembly is the time spent tweaking the fonts and colors of your components. This will be inevitable, but it does emphasize the point that applying corporate standards to the creation of individual reporting elements that are assembled into more complex presentations can save you time in the long run.

Hints and Tips

These tips and tricks will help you with this example.

- Remember that simple components like charts and maps are easy to resize. More complex composite elements like multiple gauge widgets can take longer, as you may have to resize individual elements inside the component. Anything based on a table can take a while to resize, as you will have to adjust the width and height of columns and rows, possibly across several elements.
- Aligning the top and left edge of certain objects is not just for aesthetics. This will ensure that a slight (and possibly invisible) horizontal discrepancy does not cause one element to appear on the final output *below* another, where it has been "pushed" by the rendering engine.
- When fine-tuning the relative sizes of component parts in a dashboard, remember that you can use the size and location properties to configure the position on the report to a fraction of an inch.
- Although I did not mention it during the process above, remember that SSRS also gives you the possibility to set the horizontal and vertical spacing automatically. These options are in the Format menu, and can be extremely useful when assembling complex visualizations such as dashboards.

Grouping Objects

As you probably know by now, SSRS does not have a function that enables you to group objects so that multiple items can be manipulated as a single object. There are, however, a few workarounds to overcome this limitation.

At the start of Chapter 3, you saw some slightly "edgy" gauges of sales by make of car. However, you did not group these gauges into a coherent visualization, the sort of output that is shown in Figure 7-2 (where, incidentally, the output is for January 2013). I want to take this as an initial example of using tables to group objects.

Figure 7-2. *Assembling a set of gauges inside a table*

You have already created initial gauge for this tiny dashboard (which is destined to be displayed on a smartphone) in Chapter 3. Even if you did not, it exists in the sample application, so here is how you can make a single gauge into a coherent presentation.

1. Make a copy of the report `CarSalesGauges.rdl`. Name it `SmartPhone_CarSalesGauges.rdl`.
2. Extend the size of the report area to give yourself plenty of room to work.
3. Add a table to the report. Set the dataset to be the same dataset as the one that is used for the charts (or gauges) that it will contain. In this example, it is `SmartPhoneCarSalesGauges` using the stored procedure `Code.pr_SmartPhoneCarSalesGauges`.
4. Delete the detail row (and confirm the deletion).
5. Set the text box borders to None.
6. Add five new rows, or as many as are required to end up with a six-row table. The table should contain three columns. If not, add or delete columns until the table is five rows by three columns.
7. Select the table. Display the properties window and set the BackgroundColor property to Black.
8. Set the borders for all of the text boxes and the table itself to none.

9. Select the left-hand text box in the first row and set its size to 4 centimeters wide by 0.6 centimeters high. Do the same for the right-hand text box on the first row. Then do the same for the text boxes on the third and fifth rows.

10. Select the left-hand text box in the second row and set its size to 4 centimeters by 4 centimeters. Do the same for the right-hand text box on the second row. Then do the same for the text boxes on the fourth and sixth rows.

11. Select any text box in the second (middle) column and set the width to 0.6 centimeters.

12. Resize the existing gauge to 4 centimeters by 4 centimeters. Drag the gauge inside the left-hand cell on the second row.

13. Copy this gauge and paste five copies as follows: two in the first column in: the fourth and sixth rows and the other three in the second, fourth, and sixth rows of the third column.

14. Set the filter for each of the gauges to the make of car that is displayed above each gauge in Figure 7-1. You do this by right-clicking each gauge and selecting Gauge Panel Properties followed by Filters. The existing gauge is filtered to display only Aston Martin, so all you have to do is change this to Rolls Royce, Jaguar, etc.

15. Add the title for each gauge in the cell above the gauge. Center the title and set the font to Arial 10-point white.

You now have a set of gauges that is contained and grouped in a table.

How It Works

As you can see, you are concentrating your efforts principally on creating a table that will contain the required number of gauges. Nearly all of your time is spent on defining column widths and row heights. Then you ensure that any table borders and backgrounds (or their absence) map to your presentation requirements. Finally, you place copies of the widget that you want to display inside the table and tweak the filters to ensure that you place the precise gauge that you want in the right part of the table.

■ **Note** If you prefer, you can add the dataset to the table and then place the first gauge in a textbox in the table. This way the gauge will "inherit" the dataset from the table.

Hints and Tips

Although this technique is not difficult, there are a few things that you need to know when using tables to control object alignment.

- Objects cannot be added to row groups. So you need to remove row groups first.
- Use added rows for spacing/blank white space. This is a frequently-used technique, and it can seem a little laborious at first. However, it is well worth the effort as it makes it easy to define the spacing accurately by setting the row height for the "spacer" rows.
- The table needs a dataset, so it seems easier to use the same dataset as the objects in it. However, you can use an empty (or "dummy") dataset if you prefer, as the charts or gauges can use their own datasets independently of the table dataset. Using a dummy dataset is described a little further on in this chapter.

A More Complex Dashboard

Now that you have seen some of the basic techniques needed to control object placement, I suggest that you make another dashboard to learn how to present more complex business intelligence reports using SSRS. This dashboard is `DashboardMonthly` in the sample application, and it looks like Figure 7-3.

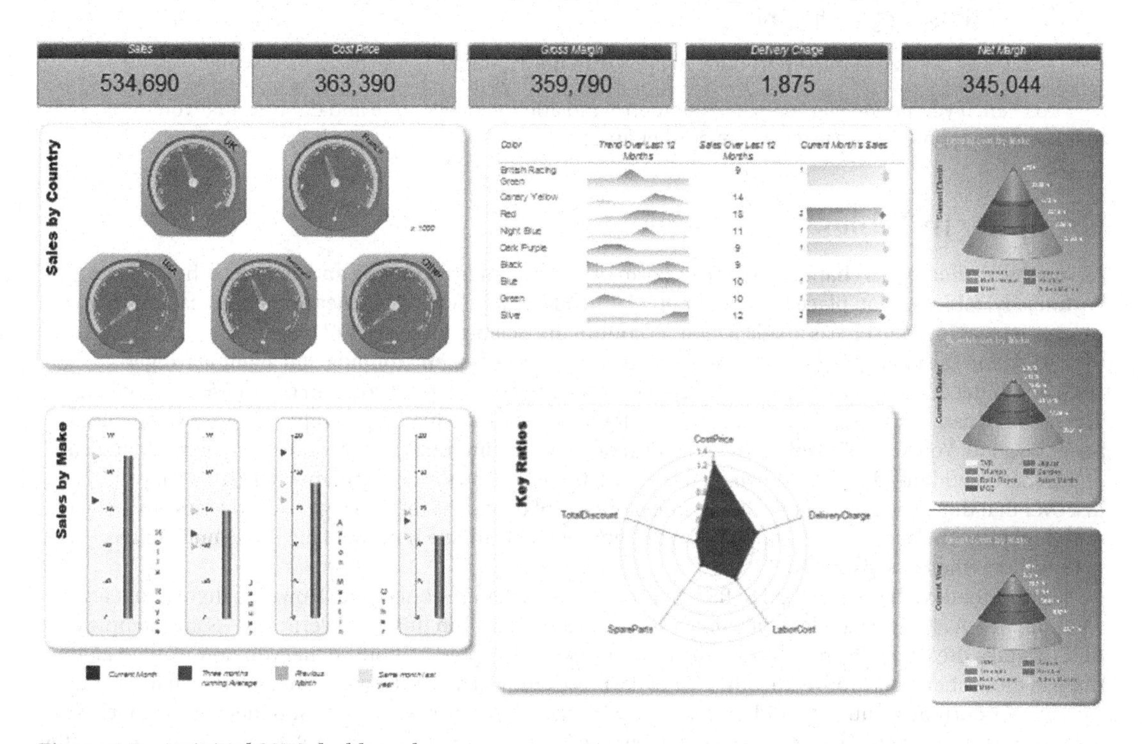

***Figure 7-3.** An initial SSRS dashboard*

This dashboard uses the six source components that you saw in previous chapters. They are

- `_MonthlySalesAndProfit.rdl`
- `_GaugeSalesByCountry.rdl`
- `_ColorSales.rdl`
- `_PyramidSales.rdl`
- `_MonthSalesByMakeOverTime.rdl`
- `_RatioAnalysis.rdl`

Creating the Basic Dashboard

Before you can add the components, you need a report to place them in. You can create this by following these steps.

1. Make a copy of the report `__BaseReport.rdl`. Name the copy `Dashboard_Monthly.rdl`.
2. Set the size of the report body to approximately 17 inches wide by 12 inches high.

You can, if you prefer, start with a blank report and add the datasource and datasets that you have used in most of the reports you created in previous chapters.

Grouping in Tables

As you saw earlier in this chapter, tables can be an excellent tool for grouping and aligning other objects. Fortunately, tables let you place most other objects inside a table "cell." This means that you can place gauges, images, or even charts inside a table to control the position of a series of objects.

To extend the concept a little, tables can also be very useful when grouping and aligning widgets *vertically*. Indeed, this not only groups a set of charts or gauges, but it can solve an otherwise tricky issue. The problem can be that you have a set of, say, charts that you have placed in a report one above the other. So far, so good. You have aligned them perfectly, and all looks fine-until you preview the report. Then some of the charts suddenly drop way down on the report, far below the others. After much frustration, you discover that the problem is a table at one side of the set of charts. As this table grows, it pushes any object that was below the bottom of the table in the design view beneath the bottom of the (now much longer) table in the report when it is displayed.

Unfortunately, this is standard SSRS behavior where tables are concerned. However, there is an elegant solution, which is to use a table to position a set of objects (charts, in this case) vertically. As an example of this, let's take the chart `_PyramidSales.rdl` that you saw in Chapter 4. Suppose that you now want to align three of these charts one above the other. To make this example clearer, I suggest that the top chart displays sales for the current month, the middle chart shows sales for the current quarter, and the bottom chart sales for the current year. I do not want to reiterate how to create charts here, so I will refer you to Chapter 4 for the principles, and the stored procedure `Code.pr_WD_PyramidSales` in the `CarSales_Reports` database for the code that returns the data for the three pyramid charts. To create this type of presentation, follow these steps.

1. Drag a table on to the report. Set the dataset to be the same dataset as the one that is used for the charts (or gauges) that it will contain. In this example, it is `PyramidSales` using the stored procedure `Code.pr_WD_PyramidSales`.
2. Delete all of the columns but one. Then delete the data row.
3. Set all the text box borders to None. Ensure that the table border is also set to none.
4. Add four more rows. Set the width of the column to 2.7 inches.
5. For the second and fourth cells in the column, set the height to 0.25 inches. Set the font size to 3 points; otherwise the height of the row will stretch to the (larger) point size.
6. Copy and paste the chart `_MonthPyramid` into the top cell. Set the dataset to `PyramidSales` and the chart filter in the chart properties to `ChartType = MonthlySales`.

7. Copy this chart into the third cell and set the filter to `ChartType = QuarterlySales`.
8. Copy this chart into the fifth cell and set the filter to `ChartType = YearlySales`.
9. Tweak the vertical titles for the charts to read Current Month, Current Quarter, and Current Year (from top to bottom).

This set of charts will now look like the rightmost visualization (the three pyramid charts one above the other) in Figure 7-3.

How It Works

As you can see, this dashboard has a table to the left of the pyramid charts. If the pyramid charts were not contained-and constrained-by a table, then their vertical alignment would be skewed when the table of color sales grew, and the second pyramid chart (of sales for the quarter) would have been placed under the bottom row of the table to its left. As it is, the fact that the charts are enclosed in a table controls the way that the SSRS rendering engine displays the final output.

Grouping Inside a Rectangle

I will now use this dashboard to show some further techniques for aligning visualizations-and elements in visualizations. A first trick is to use a rectangle as a container for other elements.

This dashboard contains two examples of using rectangles as containers to group inner objects. The first one is probably self-evident. On the left is a set of gauges inside a rectangle, and then the rectangle provides the border-only it is supplied by a background image, as described in Chapter 6. The only real tricks to know when placing objects inside a rectangle are the following:

- Always draw the rectangle outside the objects that you want to enclose. If necessary, extend the size of the report provisionally to create the space needed for this.
- Ensure that the rectangle is at the back (or bottom, if you prefer) of the object "layer." To define this property, select the rectangle and choose Format ➤ Order ➤ Send to Back.
- Drag the objects inside the rectangle once the rectangle attributes (and background image) are set.

Now that you understand this, let's see the steps required to assemble the Sales by Country set of gauges in this dashboard. To make things clearer, Figure 7-4 shows you a close-up of the widget.

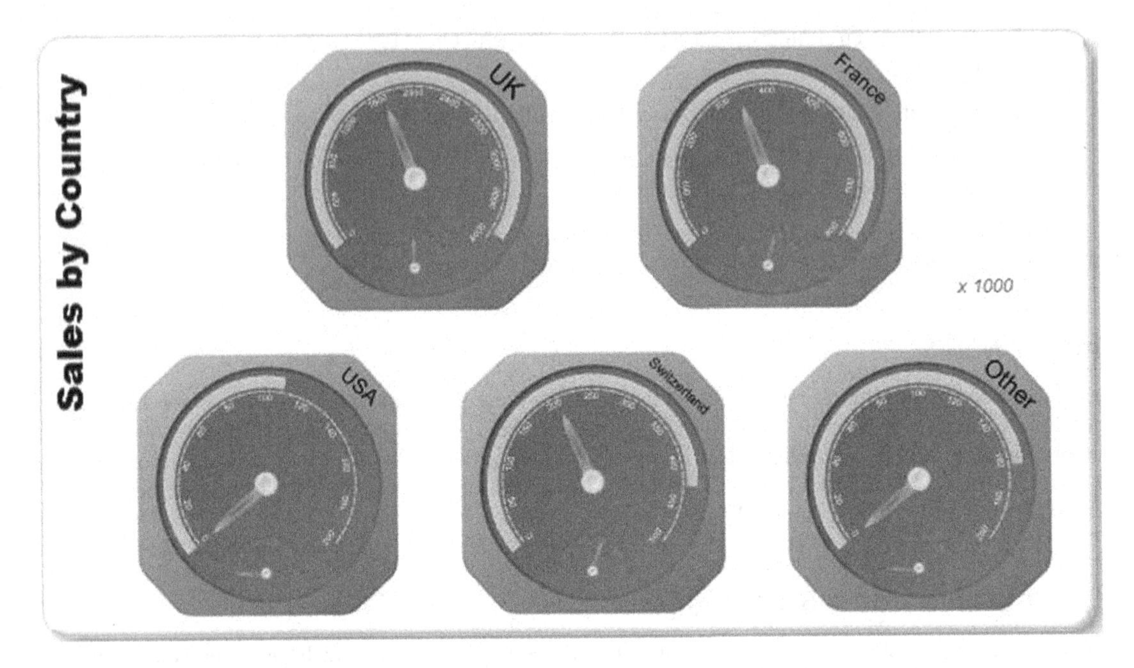

Figure 7-4. *A rectangle container for gauges and texts*

1. Drag a rectangle onto an empty area on the report. Resize the rectangle so that it will be approximately big enough to hold all the elements that you want to place inside the gauge.
2. Add the image file `EmbossedBordeFull66By37.png` to the report.
3. Set this image as the background for the rectangle, and set the BackGroundRepeat as Clip. If you need to revise this, it is described in Chapter 6.
4. Open the report(s) containing the object(s) that you will be adding to the dashboard. In this case, it is `_GaugeSalesUK.rdl`.
5. Add any datasets used by the objects that you will be placing inside the rectangle to the report. You can see these in the file containing the "source" object. Be very careful to use exactly the same names for the datasets, and remember that SSRS is case-sensitive for these names. If you are building the objects directly inside the final dashboard, then this will not apply.
6. Resize the base gauge to suit your requirements, and place it inside the rectangle.
7. Make the number of copies that you need of the source gauge (four in this example) and place them inside the rectangle.
8. Set the filter for each gauge to the relevant element. In this example, it means selecting a country for the filter. The country names are indicated in Figure 7-3 on the top right of each gauge.
9. Alter the country name in the copies of the gauges to display the country used for the filter.
10. If necessary, adjust the font size of the country name at the top right of each gauge.

11. Prepare an image containing the rotated text "Sales by country" as described in Chapter 6 and add the image to the report.
12. Drag the image onto the report inside the rectangle containing the gauges. Resize the image and place it to the left of the gauges.
13. Add a text box to the right of the first row of gauges. Set its text to read "x 1000". Set the font to Arial 10 gray.

How It Works

Assembling this particular visualization is not desperately difficult. The key is to remember that once the elements are placed inside the rectangle they can be moved, copied, or deleted as a single unit. Adding the rectangle border was to remind you of one of the techniques you saw in Chapter 6. This has the added advantage of ensuring a certain visual conformity as far as the borders of certain components of the dashboard are concerned.

Adding Custom Elements

Sometimes a component widget needs a little enhancement before it can be used effectively in a dashboard. The fourth of your component parts in this dashboard (`_MonthSalesByMake`) is a case in point. It needs both a border and a legend. To minimize risk and to ensure a fallback position, I suggest first duplicating the source file. Then you add any new elements, and finally copy the new, improved source file into the dashboard. To give you a more readable view of what you are doing, Figure 7-5 shows the progression from the initial set of four gauges to the complete component that you will copy into the dashboard.

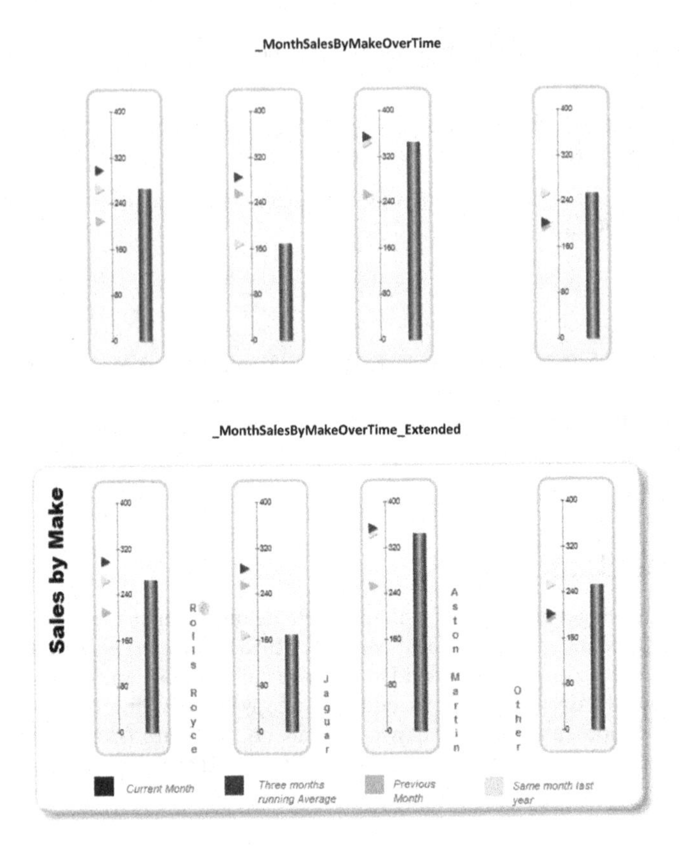

Figure 7-5. *Enhancing multiple gauges with common elements*

1. Make a copy of the file `_MonthSalesByMakeOverTime.rdl`. Name the copy `_MonthSalesByMakeOverTime_Extended`.
2. Open the copy.
3. Create an image file for the title Sales by Make. Import this image into the report and then drag the image from the toolbox into the report. Resize and place the image, and set its BackgroundRepeat property to Clip.
4. Add a text box to the right of the first gauge. Set it to be tall and thin (more precisely, 0.19 inches wide by 1.65 inches high). Enter the text "Rolls Royce" but add a return between each letter or space.
5. Copy this text box to the right of the two central gauges, and to the left of the right-hand gauge. Set the texts to Jaguar, Aston Martin, and Other as you see them in Figure 7-5.
6. Add four rectangles under the gauges. Make them into squares and space them out as in Figure 7-5.
7. Set the colors of the four text boxes to correspond to the colors of the pointers in the gauges. In this example, they are Dark Blue, Blue, Light Steel Blue, and Aqua.

8. Add a text box to the right of the first colored rectangle. Ensure that it is large enough to hold the longest text that will be used, once you have set the font (Arial, 9 point) and font color (grey). Set the text alignment in the Properties window to Middle (for VerticalAlign) and Left for TextAlign.

9. Copy the text box three times and place the copies to the right of the colored rectangles. Change the texts to those shown in Figure 7-5.

10. Select the four colored rectangles and their text boxes, and align all the objects vertically to Middles. Leaving the eight items selected, click Format ➤ Horizontal Spacing ➤ Make Equal.

11. Add the same image file to the report (`EmbossedBordeFull66By37.png`) that you used for the previous component.

12. Add a rectangle to the report, away from the existing objects and in a large clear area of the report. Add the background image that you specified in the previous step and adjust the rectangle properties as you did for the previous component.

13. After any final tweak to perfect the component, save it and copy it in its entirety to the dashboard.

14. Add the `MakeSalesComparedOverTime` dataset from the component report to the dashboard.

15. Place the component in the dashboard and finalize it by tweaking the size and proportions as required for the overall aesthetic.

■ **Note** I realize that when I say that some tweaking is required I am guilty of serious understatement. You may have to spend some time resizing and aligning objects. When developing dashboards and more advanced visualizations, this is part of the challenge–and the fun.

How It Works

In this case, you add the decorative elements first, and then set the frame afterwards. This is a conscious choice because it can be easier to manipulate objects without the irritation of selecting the background rectangle by mistake. Equally, selecting multiple objects-the better to align them, for instance-is much easier without a background rectangle getting in the way.

Adding the Radar Chart

Finally you come to a component that can probably be used as-is. This is the radar chart that you saw in Chapter 4. To add it, follow these steps.

1. Open the file `_RatioAnalysis.rdl`.

2. Find the dataset that it uses (`RatioAnalysis`) and add this to the dashboard.

3. Copy the chart into the dashboard.

4. Position and resize the chart.

Grouping Two Components

A rectangle that you are using to group elements in this dashboard is probably less obvious. In fact, there is an invisible rectangle around the Sales by Make and Key Ratios visualizations. You might be wondering why this is necessary, as it is blissfully easy to align the two widgets and assume that the work is done. Here, as well, the table is the villain of the piece. Because the Key Ratios radar chart is under the table, it will be pushed downwards as the table grows. If, however, you place the two visualizations (Sales by Make and Key Ratios) inside a rectangle, and then set the rectangle's border to None, you have created a container that will force the two visualizations to move together. This is altogether more pleasing visually.

This is what you need to do.

1. Enlarge the report so that there is some empty space at the bottom.
2. Drag a rectangle from the toolbox into the space that you just created. Make the rectangle slightly larger than the two components that you want to group.
3. Place the two components inside the rectangle.
4. Select the two components and set them to align vertically at the top (Format ➤ Align ➤ Tops).

■ **Note** As an alternative to using a rectangle, you can use a horizontal line to force all the objects below the line to move down in a report together. In this case, you will need to make sure that the line's color is the same as the background.

Finishing the Dashboard

A final element that needs to be added is the dynamic table of color sales. Fortunately, adding this table is easy.

1. Open the file `_ColorSalesWithBorder.rdl`.
2. Find the datasets that it uses (there are several of them) and add them to the dashboard.
3. Add all the images that the component file contains (one by one!) to the dashboard.
4. Copy the table into the dashboard.
5. Position and resize the table. This can mean adjusting the width of multiple columns.

Now you can align any elements-such as aligning the top edges of the Sales by Country rectangle, the color sales table, and the pyramid charts table, for instance-that need it. You have completed your dashboard and can now test the result.

Hints and Tips

Make use of these tips and hints for this example.

- I always advise previewing a complex dashboard every time you add a new component. In case of error it can help to have some context as to what the (often cryptic) SSRS error messages mean, or at least an idea which objects it refers to.
- You must add any image file(s) to the destination report (the dashboard) if the source file (the component) contains images. Otherwise, the images will simply not appear in the dashboard. Of course, if you have used an SQL Server source for images, then providing that you have set up the appropriate dataset in the dashboard, your images will appear perfectly.

Creating a Legend

Sometimes you need to create your own legend for a set of charts because the legend has to be common across several charts. An example is the trellis chart that you saw at the end of Chapter 4. To provide a solution to this issue, let's create the legend that you see in Figure 7-6 (using the file `_TrellisChartCostsPerMonth.rdl` from Chapter 4).

Figure 7-6. *An independent legend*

1. Insert a table below the trellis chart table. Add the datasource named `Dummy` (if this does not exist, then it is simply a datasource whose SQL reads `SELECT ''`). Add three rows outside the existing row group and then delete the existing row group and header row.
2. Set all the BorderStyle to None for all the borders in the table.
3. Add columns until the table has ten columns.
4. Set the width of the first and last columns to 0.125 inches.
5. Set the width of the second, fourth, sixth, and eighth columns to 0.25 inches.
6. On the second (middle) row set the BackgroundColor of the textbox to the following:
    a. Second: Aqua
    b. Fourth: Gold
    c. Sixth: Lime
    d. Eighth: Tan
7. Set the font size for all textboxes in the top and bottom rows to 2 points. Set the row height of these rows to 0.125 inches.
8. Select the entire table and add an outside border.

Once this is done, your trellis chart should look like Figure 4-14 in Chapter 4.

Hints and Tips

These hints and tips will be helpful for this example.

- The `Dummy` dataset only exists to prevent SSRS complaining that you have a table without valid data attached.
- Adding empty rows and columns to create spacing from the outside border in the legend is a little long-winded, but it allows you to create a much more aesthetically pleasing custom legend.
- You may want to add a few spaces before the text in the legend to produce a pleasing balanced effect and distance the text from the colored textboxes.

A Few Random Tips

To finish this chapter, I have added a few tips that you may find useful. As they do not really fit into any of the previous examples, I will give them without any attempt at categorization.

- You can use an outer table as a placeholder for other tables. This means first adding the "placeholder" table and then adding tables inside the cells of the outer table. However, if you do this, you are limited to using a single dataset-the one attached to the outer table.
- You can place multiple textboxes into a single cell of a table. Just remember to add a rectangle to the textbox first, and then add any nested textboxes inside the rectangle.
- You can use the cursor keys to move objects around the report area rather than the mouse. This has the advantage of keeping objects on the vertical or horizontal plane where you have already positioned them. If you find that the cursor keys move objects in increments that are too large, you can use Ctrl-Cursor to move in tiny increments.

Conclusion

There are many, many ways of assembling complex reports. Users will have many varied requirements and also, each of us has our own preferred techniques and favorite ways of getting things done. It follows that it is not possible to anticipate all the requirements that you will have to address. I hope however that you will find these methods and techniques useful when you assemble your own dashboards.

CHAPTER 8

Interface Enhancements for Business Intelligence Delivery

The kindest thing that anyone can say about the SSRS interface is that it is showing its age. Parameter selection can be more than a little laborious. Having to confirm your choices to refresh the report just seems downright old-fashioned. The parameter section of a report is intrusive and ungainly. Indeed, if you are using SSRS reports on a mobile device, the built-in interface elements can be annoyingly clunky to use.

Fortunately, help is at hand. With a little applied ingenuity you can replace access to most (and hopefully all) of the parameters in a report with more intuitive interface elements. Indeed, parameter selection can be integrated into the report design as tabs or slicers-or even pop-up choices.

I am not suggesting that the SSRS interface can be replaced entirely. What I hope to show is that it can be "revamped" to freshen up the way in which user interaction is enabled. This is nothing less than a shameless attempt to imitate, in a limited way, the more intuitive interactivity of self-service BI products such as Power View. It is also an attempt to declutter the user experience and allow report users to select, slice, and dice their data more easily.

The interface tweaks that you will look at in this chapter include

- Interactive parameter selection
- Imitating pop-up menus to select filter elements
- Using slicers to filter data

You will also be adding a couple of charts to the visualizations in this chapter. These charts are largely there to show you the new and improved parameter interface works. However, they will also extend your charting knowledge when using SSRS. Given that I have already covered many aspects of chart creation in Chapter 4, I will assume that you are already familiar with charting essentials in SSRS, and so I will not be explaining all the steps in laborious detail. Nonetheless, I hope that the chart ideas will still be useful for your projects.

Report Refresh and Parameter Postback

All the interface tweaks that you will look at in this chapter are based on one simple technique. I can only describe it as *parameter postback* for want of a better term. In its purest form, it consists of

- Hiding the report parameter
- Using the `Action` property of a report element (be it a textbox, chart series, or even map viewport) to refresh the current report and "post back" the parameter with a new value-and all other parameters with their current value. I used the term *refresh* here, where in fact what SSRS is doing is to jump to the report as if it was being opened anew.

This simplified overview requires a number of comments to make you aware that this approach does have its downsides.

- *All* hidden parameters-and not just the one whose value you are changing-have to be defined as part of the Action property that you are setting. In other words, there is no "viewstate" in SSRS, and unless you want a parameter to revert to its default value, you have to pass its current value to the refreshed report in every action property.
- As all the hidden parameters have to be posted back to the report when it is refreshed as a part of *every* action property, it follows that the more the interface is revamped to allow for multiple interactive selections, the more you will be setting multiple parameters to return their existing value. Remember that if you do not pass back the current value of a parameter, it will be reset to its default value. This can rapidly lead to a lot of time spent tweaking action properties when you add new parameters to a report.
- You no longer have the option to refresh a report once all parameters are set, as the report will refresh every time a single parameter is changed.
- Refreshing a report is not instantaneous. Consequently, you may have to pay attention to the size of the dataset(s) used and the best way to optimize and/or cache your reports. Some techniques for doing this are described in Chapter 12.

However, I do not want for a moment to suggest that this approach is difficult or laborious. I only want you to understand that it is not a magic bullet, and that a little work can be required. Fortunately, most users (and most report developers) seem to find that the result is well worth the effort.

■ **Note** Should you really need to customize the SSRS interface to a high level, you can always develop a .NET interface around the `ReportViewer` control. Alternatively, you can build SharePoint parts that allow for an extended interface. However, building BI reports with either of these techniques can prove to be an immense amount of work, and both are outside the scope of this book.

So let's move on to seeing what can be done to enhance and rejuvenate the SSRS interface to deliver a more intuitive and friendly user experience. Once again, you will be reusing some of the widgets that you created in previous chapters. All these widgets are available in the sample SSDT application that is available for download on the Apress web site should you wish to use them.

Interactive Parameter Selection

Let's start with a simple example. Suppose that you have a report that displays a chart of sales to date. Users have expressed a clear opinion that the SSRS user interface is too boring and clunky, and they are tired of clicking on the "View Report" button-or they forget to click it. They complain that changing the choice of month in Excel or even (horror of horrors) a rival reporting tool updates a chart or table quickly and easily. Consequently, they want to have this level of interactivity in SSRS reports.

No problem! In this example, you will replace the year and month parameters with an interactive and visual selection of the parameters. A single click (or tap if you are using a touch-sensitive screen) will update the report and refresh all the data.

To see where you are going, an example of a year and month selector is given in Figure 8-1. In the case of this widget, the contents are deliberately simple so that you can concentrate on the interactive date selector; all that is displayed is the sales by make for 2013 up to and including the month of October. However, as you can see, the choice of year and month is visible in the report. Moreover, if you look at the sample file (`Tablet_YearAndMonthSelectord.rdl`), the parameters are not visible at the top of the report as they would be normally.

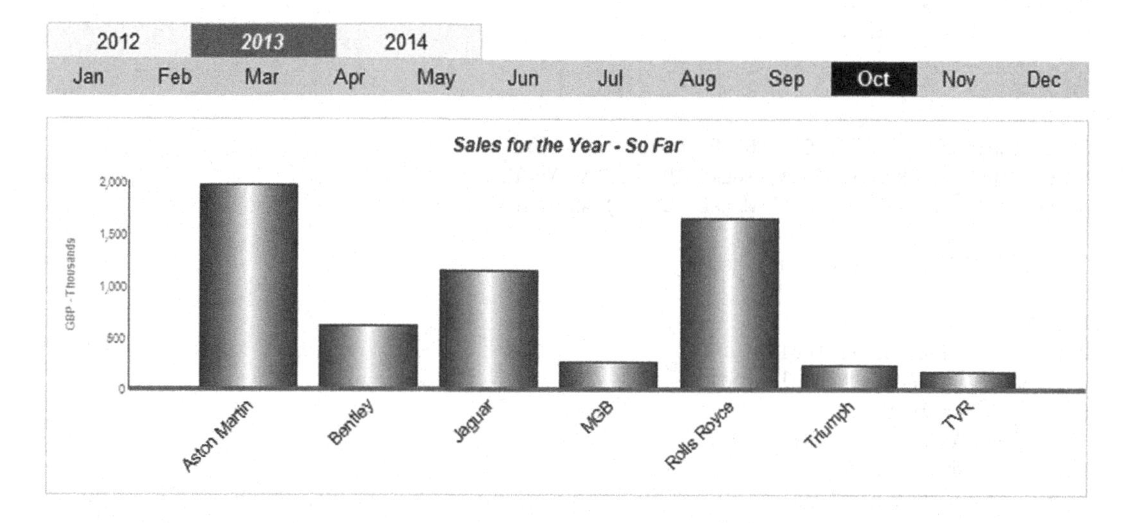

Figure 8-1. *A year and month selector in a report*

The Source Data

This report actually requires four datasets over and above the four used to return the year and month that you have used in nearly every report that you have built in this book so far. The datasets are

- `DateWidgetYear`: This is a shared dataset that returns the last three years from the source sales data. This is based on the stored procedure `Code.pr_WD_DateWidget`.
- `DateWidgetColourScheme`: This is also a shared dataset. Using the stored procedure `Code.pr_WD_DateWidgetColourScheme`, it sets the colors used in the custom date selector.
- A dummy dataset: This is needed for the tables that contain the year and month elements in the date selector to prevent SSRS returning an error.
- `ChartHighlight_SalesYTD`: Finally, a normal, run-of-the-mill dataset that returns data that is displayed in the report. This data is used in the chart.

The reason that two of these datasets are shared is simple. You will probably want to reuse your year and month selector in several reports. So it is more logical to design this widget as a reusable element-and that also means ensuring that the data it is based on can be reapplied easily. This means using a shared dataset that you can develop once, then reapply to multiple reports.

Here, then, is the code itself for these elements (the dummy dataset was explained in the previous chapter):

```
DECLARE @ReportingYear INT = 2013
DECLARE @ReportingMonth TINYINT = 10

-- Code.pr_WD_DateWidget

WITH YEAR_CTE
AS
(
SELECT TOP (3)
YEAR(InvoiceDate) AS theYear, ROW_NUMBER() OVER (ORDER BY YEAR(InvoiceDate)) AS ID
FROM Reports.CarSalesData
GROUP BY YEAR(InvoiceDate)
ORDER BY YEAR(InvoiceDate)
)

SELECT
(SELECT theYear FROM YEAR_CTE WHERE ID = 1) AS Year1
,(SELECT theYear FROM YEAR_CTE WHERE ID = 2) AS Year2
,(SELECT theYear FROM YEAR_CTE WHERE ID = 3) AS Year3

-- Code.pr_WD_DateWidgetColourScheme

SELECT
 'DimGray' AS BckGrdSelected1
,'WhiteSmoke' AS BckGrdNonSelected1
,'Black' AS BckGrdSelected2
,'LightGrey' AS BckGrdNonSelected2
,'White' AS TextSelected1
,'Black' AS TextNonSelected1

--Code.pr_Tablet_ChartHighlight_SalesYTD

SELECT      Make
           ,SUM(SalePrice) / 1000 AS SalesYTD
FROM       Reports.CarSalesData
WHERE      ReportingYear = @ReportingYear
           AND  ReportingMonth <= @ReportingMonth
GROUP BY   Make
```

The results returned by these three code snippets are shown in Figure 8-2.

Code.pr_Tablet_ChartHighlight_SalesYTD

	Make	SalesYTD
1	Aston Martin	1509.59
2	Bentley	489.75
3	Jaguar	871.00
4	MGB	197.00
5	Rolls Royce	1082.55
6	Triumph	163.00
7	TVR	100.75

Code.pr_WD_DateWidget

	Year1	Year2	Year3
1	2012	2013	2014

Code.pr_WD_DateWidgetColourScheme

	BckGrdSelected1	BckGrdNonSelected1	BckGrdSelected2	BckGrdNonSelected2	TextSelected1	TextNonSelected1
1	DimGray	WhiteSmoke	Black	LightGrey	White	Black

Figure 8-2. *The output for the stored procedures behind the year and month selector*

How the Code Works

This code works like this:

- DateWidgetYear: This stored procedure finds the last three years for which there are sales in the CarSalesData view. It then pivots the results (here using simple sub-queries) so that the output is ready to display horizontally in a table.
- DateWidgetColourScheme: This is nothing more than a hard-coded color selection for the backgrounds and texts of the year and month tables that are used for selecting these elements.
- ChartHighlight_SalesYTD: This procedure returns the data for the chart. The result is predivided by 1,000, which is one way of reducing the width of the figures on the vertical axis.

Building the Report

Now that you have prepared your data, it is time to create the gauge itself. Here is how.

1. Create a new SSRS report named Tablet_YearAndMonthSelector.rdl. Add the shared data source CarSales_Reports. Ensure that its name in the report is CarSales_Reports.
2. Add the following four shared datasets (ensuring that you also use the same name for the dataset in the report): CurrentYear, CurrentMonth, ReportingYear, and ReportingMonth. Set the parameter properties for ReportingYear and ReportingMonth as defined at the start of Chapter 1.
3. Create a dataset named ChartHighlight_SalesYTD. Have it use the CarSales_ Reports data source and the query Code.pr_Tablet_ChartHighlight_SalesYTD that you created earlier.

4. Add the following two shared datasets:

   a. `DateWidgetYear` (using the shared dataset `DateWidgetYear.rsd`)

   b. `DateWidgetColorScheme` (using the shared dataset `DateWidgetColorScheme.rsd`)

5. Add a table to the report and delete its header row. Ensure that it has three columns, and configure it to use the dataset `DateWidgetYear`. This will be the year selector table.

6. Set all the text box borders to Light Gray, 1 point.

7. Set each of the text boxes to 1 inch wide, and apply the following fields:

   a. Leftmost column: Year1

   b. Center column: Year2

   c. Rightmost column: Year3

8. Click the leftmost text box and set the `BackgroundColor` property to the following expression:

```
=IIF(Parameters!ReportingYear.Value = Fields!Year1.Value,First(Fields!BckGrdSelected1.Value,
"ColorScheme"),First(Fields!BckGrdNonSelected1.Value, "ColorScheme"))
```

9. Set the `Color` property for this text box to the following expression:

```
=IIF(Parameters!ReportingYear.Value = Fields!Year1.Value,First(Fields!TextSelected1.Value,
"ColorScheme"),First(Fields!TextNonSelected1.Value, "ColorScheme"))
```

10. Carry out steps 8 and 9 for the other two text boxes in this table, only replace Year1 in the expression with Year2 for the center text box and with Year3 for the rightmost text box.

11. Add a table to the report and delete its header row. Ensure that it has 12 columns, and configure it to use the dummy dataset. This will be the month selector table.

12. Set all the text box borders to Light Gray, 1 point.

13. Enter the texts Jan through Dec in the 12 text boxes to display the 12 months of the year.

14. Click the leftmost text box and set the `BackgroundColor` property to the following expression:

```
=IIF(Parameters!ReportingMonth.Value = 1,First(Fields!BckGrdSelected2.Value, "ColorScheme"),
First(Fields!BckGrdNonSelected2.Value, "ColorScheme"))
```

15. Set the `Color` property for this text box to the following expression:

```
=IIF(Parameters!ReportingMonth.Value = 1,First(Fields!TextSelected1.Value, "ColorScheme"),Fi
rst(Fields!TextNonSelected1.Value, "ColorScheme"))
```

16. Do the same for the other eleven text boxes, only replace `Value = 1` in the expression with `Value = 2` (for February), `Value = 3` (for March), etc.

17. Right-click the Year1 text box in the upper table, and select Text Box Properties. Click Action on the left.

18. Select Go to report as the enabled action.

19. In the Specify a report pop-up, do not select the name of the report, but enter `=Globals!ReportName` as the report to jump to. The dialog will display `[&ReportName]`.

20. Click twice on the Add button and define the two following parameters as expressions (so click the Fx button to display the Expression dialog):

    a. Name: ReportingYear - Value: `=Fields!Year1.Value`. The dialog will show `[Year1]` after you click OK in the Expression dialog.

    b. Name: ReportingMonth - Value: `=Parameters!ReportingMonth.Value`. The dialog will display `[@ReportingMonth]` after you click OK in the Expression dialog.

21. Check that the dialog looks like Figure 8-3, then click OK.

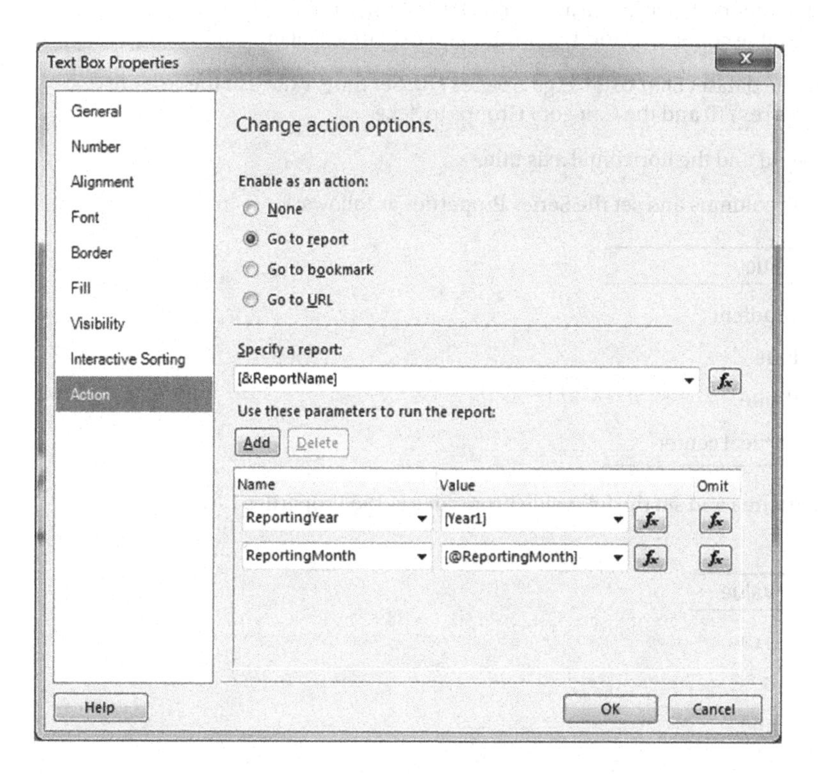

Figure 8-3. *The Action dialog*

22. Set actions for the two other text boxes in the table containing the years. They must be identical to those that you set for Year 1, with the only differences being that you set the `ReportingYear` value to be Year2 for the center text box and Year3 for the rightmost text box.

23. Set actions for all 12 text boxes in the month selector table. This is done as for the years, except that the ReportingMonth values are 1 (for January), 2 (for February), etc. These are *not* expressions, and are hard-coded.

24. The ReportingYear parameter is set for the textboxes for all 12 months to the expression =Parameters!ReportingYear.Value, which will show as [@ReportingYear] in the Text Box Properties dialog. The widget should look like Figure 8-4.

[Year1]		[Year2]		[Year3]							
Jan	Feb	Mar	Apr	May	Jun	Jul	Aug	Sep	Oct	Nov	Dec

Figure 8-4. *A year and month selector widget*

25. Drag a chart onto the report area. Choose a column chart. Resize it so that is about as wide as the table containing the 12 months, and 8-10 times taller.

26. Set the chart to use the dataset ChartHighlight_SalesYTD. Set the ∑ Values in the Chart Data pane to SalesYTD and the Category Groups to Make.

27. Delete the chart legend and the horizontal axis title.

28. Right-click any of the columns and set the Series Properties as follows:

Section	Property	Value
Fill	Fill style	Gradient
	Color	Blue
	Secondary color	White
	Gradient style	Vertical center

29. Click inside the chart area and set the following properties in the Properties window:

Property	Element	Value
CustomPosition	Enabled:	True
	Left	0
	Right	100

30. Right-click the Chart Title, select Chart Title Properties, and set the following properties:

Section	Property	Value
General	Title text	Sales for the Year - So Far
	Title position	Top center
Font	Font	Arial
	Size	9 point
	Color	Dim Gray
	Bold	Checked
	Italic	Checked
Border	Line style	None

31. Right-click the Vertical Axis Title, select Vertical Axis Properties, and set the following properties:

Section	Property	Value
General	Title text	GBP - Thousands
	Title alignment	Center
Font	Font	Arial
	Size	7 point
	Color	Dim Gray

32. Right-click the Vertical Axis, select Axis Title Properties, and set the following properties:

Section	Property	Value
Axis Options	Always include zero	Checked
Labels	Disable auto-fit	Selected
	Label rotation angle (degrees)	0
Label Font	Font	Arial
	Size	7 point
Number	Category	Number
	Decimal places	0
	Use 1000 separator	Checked
Major Tick Marks	Hide major tick marks	Checked
Minor Tick Marks	Hide minor tick marks	Checked
Line	Line color	Dim Gray

33. Right-click the Horizontal Axis and set the following properties:

Section	Property	Value
Labels	Disable auto-fit	Selected
	Label rotation angle (degrees)	-45
Label Font	Font	Arial
	Size	8 point
	Color	Gray
Major Tick Marks	Hide major tick marks	Checked
Minor Tick Marks	Hide minor tick marks	Checked
Line	Line width	2 point
	Line color	Dim Gray

You can now preview the chart. All you have to do is to click a year or month to update the chart with data for the selected parameters.

How It Works

In this report, the focal point-the chart-is the easy part. The chart merely reflects the sales per make for the year to date, as set by the parameters `ReportingYear` and `ReportingMonth`. It is *how* these two parameters are set that is interesting. What happens is that you have one table that displays the last three years, and another table that displays a hard-coded list of months. If you click or tap on a year or a month in either of the tables, the text box's action property is triggered. This property sets the year parameter (if you click a text box in the year table) or the month parameter (if you click a text box in the month table) and then refreshes the report by jumping to itself as the destination report.

This is, in effect, a kind of "postback" for the report, which ends up refreshing itself, only with a changed parameter.

Hints and Tips

These tips and tricks will help with this example.

- In step 19 you can, alternatively, click the Fx (function) button and then expand Built-in Fields and select `ReportName`, instead of entering `[&ReportName]` or `=Globals!ReportName`.
- When it comes to selecting the parameter values (as in step 20) you have three options:
    - Type in the value as the dialog will display it. So if you know the parameter name, you can type in `[@ReportingYear]`, for example.

- Type in the expression as it would appear in the Expressions dialog, such as =Parameters!ReportingYear.Value.
- Click the function button (Fx) and select a parameter where one exists. So, for instance, instead of typing [@ReportingYear], you can display the Expression dialog, expand Parameters as the Category, and then select a parameter, such as ReportingYear (in this example). The dialog will display =Parameters!ReportingYear.Value, but once you click OK only [@ReportingYear] will be displayed in the Text Box Properties dialog.

- In an attempt to make the report a little more dynamic I set the years to display only the last three years for which there are data. You can simplify the process and hard-code the years if you prefer. However, the extra effort to have greater automation seemed trivial, so I wanted to show you how to do it.

■ **Note** Some interface elements are bound to be reused over and over again. Date parameter selection is such a case. If you find that you are reusing an interface element such as this, then it is probably worth spending some time to perfect it, and then save it as a reusable "widget" in a separate report. You can then copy and paste it into any report that requires it. You should also probably define any datasets a widget used as shared datasets, which you can then add to new reports in a couple of clicks.

Creating a Year and Month Selector Widget

As you have gone to a certain amount of trouble to create the interface element that allows users to select the year and month for the data that they wish to display, it is a good idea to "capture" these elements as a coherent set of items that can be reused in other reports. Here is how you can do this.

1. Make a copy of the report Tablet_YearAndMonthSelector.rdl. Name the copy __DateSelector.rdl.
2. Delete the dataset ChartHighlight_SalesYTD.
3. Delete the chart from the report.
4. Resize the report so that only the year and month selector are visible.

The widget should look like Figure 8-4.

This report can become the basis for other reports as it contains not only the widget itself, but all the required datasets for displaying and selecting the year and month for your reports using the CarSales_Reports database.

Creating Interactive Menus to Select Filter Elements

Sometimes when creating dashboards or mobile output in SSRS, space on the screen is at a premium. This can mean that ranges of interactive parameters are not an option. Well, if this is the case, there is an alternative. The solution consists of imitating a set of menu choices in a table inside the report. It is not, perhaps, the most visually arresting technique, but it works!

An interactive menu in SSRS is nothing more than a sly trick. It uses two elements:

- A text box to display the selected element (months, in this example).
- A table to display the list of "pop-up" elements (the 12 months of the year, in this example). When a user chooses one of these (in essence he or she clicks on a text box), the action value for the text box displays the same report and passes back any selected values. This table will be invisible until toggled by the text box described above.

To give you an idea of where you are going, Figure 8-5 shows an example of the text box and table (both visible and hidden).

Menu Hidden

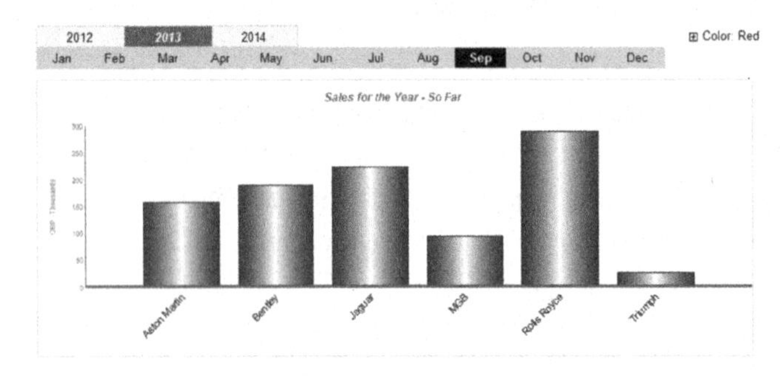

Menu Visible

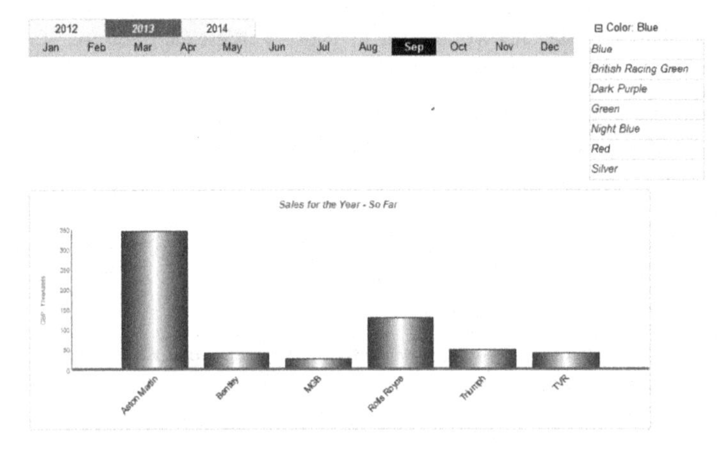

Figure 8-5. *Imitating a pop-up menu*

The Source Data

This report is an extension of the previous one, so you will base the new report on a copy of the previous one. Although the new report uses nearly the same data, there are nonetheless a couple of minor changes that have to be made. The first change is that the stored procedure `pr_Tablet_ChartHighlight_SalesYTD` is now called `pr_Tablet_ChartHighlight_SalesBy Color`, and it accepts another parameter so that the `WHERE` clause can be extended to include `AND Color = @Color` so that the chart will filter not only on year and month, but color as well. Finally, you need a code snippet to provide the list of colors available up to and including the selected month that is displayed in the "menu" that you can see in Figure 8-4. This is provided by the stored procedure `Code.pr_ColorsList`, which looks like this:

```
SELECT    DISTINCT Color
FROM      Reports.CarSalesData
WHERE     ReportingYear = @ReportingYear
          AND ReportingMonth <= @ReportingMonth
```

Building the Report

Here is how you can extend the previous report to include a pseudo pop-up menu.

1. Make a copy of the report `Tablet_YearAndMonthSelector.rdl`. Name the copy `Tablet_YearAndMonthAndColorSelector.rdl`.
2. Add a dataset named `ColorsList` based on the stored procedure `Code.pr_ColorsList`.
3. Delete the dataset named `ChartHighlight_SalesYTD`.
4. Add a dataset named `ChartHighlight_SalesByColor`. Base this on the stored procedure `Code.pr_Tablet_ChartHighlight_SalesByColor` (essentially the same as before, but with a color parameter added).
5. Adjust the `Action` property for all of the year and month text boxes in the year and month selector tables to add the following parameters:
    a. Name: Color
    b. Value: `[@Color]`
6. Add a text box at the top right of the report. Set the text box to contain the following expression:

```
="Color: " & Lookup(Parameters!Color.Value, Fields!Color.Value,
Fields!Color.Value, "ColorsList")
```

7. Right-click the text box and select Text Box Properties from the context menu. In the General tab, set the name to `TxtColorSelector`.
8. Add a table composed of one row (the data row) and one column under the text box that you just added. Set this table to use the dataset `ColorsList`.
9. Add the field Color to the unique text box in the table.

10. Set the table's text box font to the following:
    a. Color: Blue
    b. Font: Arial Narrow - Italic
11. Right-click the text box in the color table and click Action on the left. Specify the report to be `=Globals!ReportName` like the text boxes for the year and month selector tables.
12. Add three parameters:
    a. Name: Color - Value: `=Fields!Color.Value`
    b. Name: ReportingMonth - Value: `=Parameters!ReportingMonth.Value`
    c. Name: ReportingYear - Value: `=Parameters!ReportingYear.Value`
13. Select the text box in the table. Then right-click the table selector (the grey square at the top left of the grey row and column selectors) and select Tablix Properties.
14. Click on Visibility on the left.
15. Select Hide as the option for when the report is initially run.
16. Check the "Display can be toggled by this report item" check box.
17. Select `TxtColorSelector` from the pop-up list.
18. Click OK.

That is all that you have to do.

How It Works

When you display the report, you will see the "classic" SSRS plus symbol next to the text box containing Color: Red (or another color from the list of available colors). Clicking the plus symbol will cause the table of colors to appear because its visibility is controlled by this text box. As it is a table it will push any other objects further down the report. Clicking a color in the table of colors that you are using as a pseudo menu triggers an action to redisplay the report using the selected vehicle color.

Using Slicers

Most report consumers in any business are also Excel users. They probably create their own lists and pivot tables in Excel, nearly always with a considerable degree of skill. This means that they are used to interactive and intuitive ways of interacting with their data. Slicers are one technique that they probably use. Consequently, adding slicers to SSRS is something that can have real appeal to BI report users.

Slicers are nothing more than a visual selection technique. In this section, I will show you how to create

- Dynamic slicers, where a set of slicers is created from the source data. This technique allows the user to select a single slicer element.
- Static slicers, where you, the developer, define a slicer set, but you allow the user to select multiple elements from the set.

Single Slicers

Single slicers are nothing more than a parameter selection presented in a visually appealing way. In this example, you will be using a tablix to become a slicer set. This allows a dynamic set of choices to be returned from the source data. In this specific example, the slicer is the country. An example of the final output is shown in Figure 8-6. You can see here that clicking on a country name has highlighted the sales for that country in the top chart, and displayed the cost metrics for the country in the bottom chart.

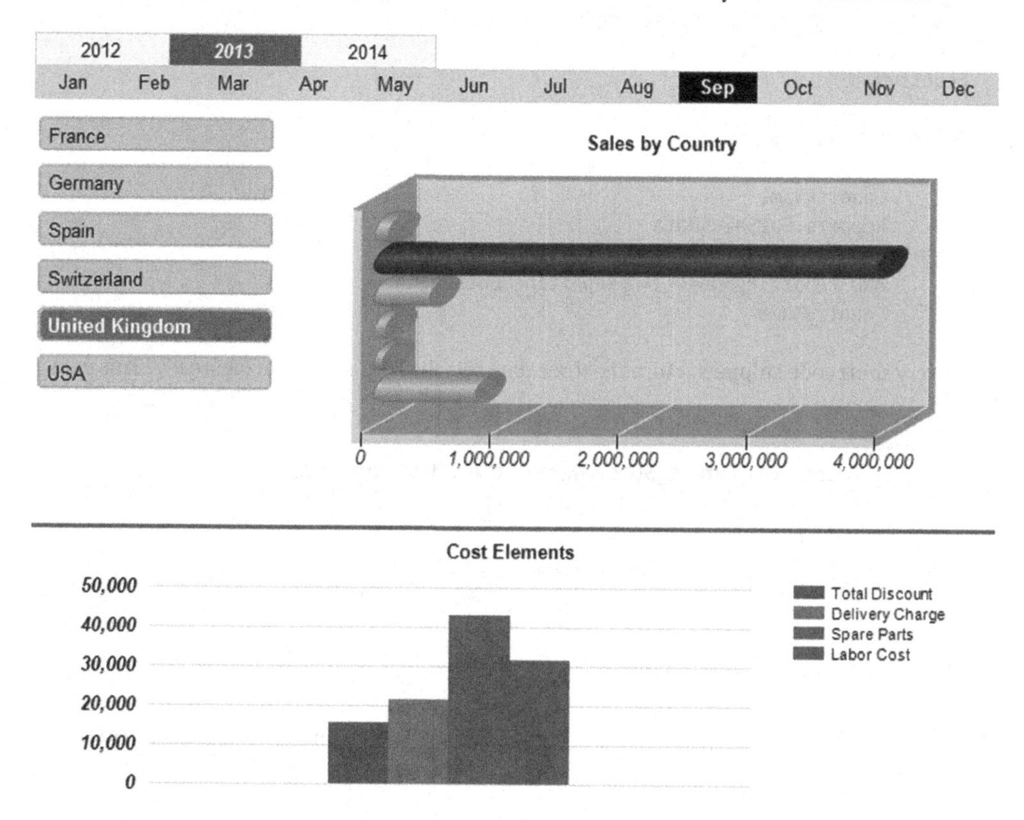

Figure 8-6. *A simple dynamic slicer to select a single element*

The Source Data

This report requires three datasets: one for the list of countries, and two for the two charts. The code for all three follows:

```
-- Code.pr_Tablet_SlicerHighlightCountrySales

SELECT
             CountryName AS CountryName
            ,SUM(SalePrice) AS Sales
FROM        Reports.CarSalesData
WHERE       ReportingYear = @ReportingYear
            AND ReportingMonth <= @ReportingMonth
GROUP BY    CountryName

-- Code.pr_Tablet_SlicerHighlightCountrySalesCosts 2013, 9, 'United Kingdom'
```

```
SELECT
             SUM(TotalDiscount) AS TotalDiscount
            ,SUM(DeliveryCharge) AS DeliveryCharge
            ,SUM(SpareParts) AS SpareParts
            ,SUM(LaborCost) AS LaborCost
FROM        Reports.CarSalesData
WHERE       ReportingYear = @ReportingYear
            AND ReportingMonth <= @ReportingMonth
            AND CountryName = @Country
GROUP BY    CountryName

-- Code.pr_WD_DynamicCountrySlicer

SELECT DISTINCT    CountryName
FROM               Reports.CarSalesData
WHERE              ReportingYear = @ReportingYear
                   AND ReportingMonth <= @ReportingMonth
ORDER BY           CountryName
```

These three very short code snippets return the three data sets that you can see in Figure 8-7. This data is for May 2013.

Code.pr_Tablet_SlicerHighlightCountrySalesCosts

	TotalDiscount	DeliveryCharge	SpareParts	LaborCost
1	20950.01	29365.00	57595.00	38748.00

Code.pr_WD_DynamicCountrySlicer

	CountryName
1	France
2	Germany
3	Spain
4	Switzerland
5	United Kingdom
6	USA

Code.pr_Tablet_SlicerHighlightCountrySales

	CountryName	Sales
1	France	787590.00
2	Germany	71000.00
3	Spain	91750.00
4	Switzerland	428490.00
5	United Kingdom	3915750.00
6	USA	108750.00

Figure 8-7. *The data sets for the tables output using a slicer*

How the Code Works

The list of countries returns the countries where there are sales for the selected month and year. The data for country sales groups sales by country for the selected month and year. Finally, the cost data is returned for the selected country, month, and year.

Building the Report

Now that you have the data in place, you can build the report.

1. Either build another year and month selector, as described in the first example in this chapter, or make a copy of the report `__DateSelector.rdl`. Name the copy `Tablet_SlicerHighlight.rdl` and delete the dataset `ChartHighlight_SalesYTD`. Remember that these reports are available on the Apress web site if you want to use them as starting points.
2. Add three datasets based on the three stored procedures described a little earlier. They should be
   a. `DynamicCountrySlicer` using the stored procedure `Code.pr_WD_DynamicCountrySlicer`.
   b. `SlicerHighlightCountrySales` using the stored procedure `Code.pr_Tablet_SlicerHighlightCountrySales`.
   c. `SlicerHighlightCountrySales` using the stored procedure `Code.pr_Tablet_SlicerHighlightCountrySales`.
3. Verify that you also have the `Country` parameter; if not, add it (it is a string). Right-click the parameter and select Parameter Properties from the context menu.
4. In the General pane, set the parameter visibility to Hidden.
5. In the Default Values pane, select Specify values. Click Add and enter United Kingdom as the default value. Click OK.
6. Add a table to the report area under the year and month selector, on the left. Remove the header row and all columns except one. Set the dataset to `DynamicCountrySlicer`.
7. Add the two images, `SlicerButton1` and `SlicerButtonSelected1`, from the directory `C:\BIWithSSRS\Images` (assuming that you have downloaded the sample data).
8. Right-click inside the text box in the table and select Expression. Enter the following expression: `="    " & Fields!CountryName.Value`.
9. Set the background image for the text box to the following expression: `=IIF(Fields!CountryName.Value = Parameters!Country.Value,"SlicerButtonSelected1","SlicerButton1")`.
10. Add a chart to the right of the table that you just entered. Choose the chart type 3-D clustered horizontal cylinder. Remove both axis titles and the legend. Use `SlicerHighlightCountrySales` as the dataset. Set the chart title to Sales by Country.
11. In the Chart Data pane, set Sales as the ∑ Values and CountryName as the Category Group.
12. Right-click the vertical axis, and in the vertical axis properties, check Hide axis labels and Hide major tick marks.
13. Right-click any of the cylinders and select Series Properties. Click Fill on the left and add the following as the expression for the color: `=IIF(Parameters!Country.Value = Fields!CountryName.Value, "Blue", "Silver")`.

14. Add a second chart to the report under the first chart. Choose the chart type Column. Remove both axis titles. Use `SlicerHighlightCountrySalesCosts` as the dataset. Set the chart title to Cost Elements.

15. In the Chart Data pane, set the Σ Values to

    a. TotalDiscount

    b. DeliveryCharge

    c. SpareParts

    d. LaborCost

16. Right-click the horizontal axis, and in the vertical axis properties, check Hide axis labels and Hide major tick marks.

17. Right-click the vertical axis, and set the following vertical axis properties:

Section	Property	Value
Axis Options	Always include zero	Checked
Labels	Enable auto-fit	Selected
Label Font	Font	Arial
	Size	8 point
	Bold	Checked
	Italic	Checked
Number	Category	Number
	Decimal places	0
	Use 1000 separator	Checked
Major Tick Marks	Hide major tick marks	Checked
Minor Tick Marks	Hide minor tick marks	Checked
Line	Line style	None

18. Select the chart, and in the Chart Properties window, select the Berry Palette.

19. Right-click the chart and select Chart Properties. Click Border on the left. In the chart border options, set the border type to Line. Choose a 2-point Dim Gray border and apply this to the top border only by clicking on the top line in the preview in the dialog. Click OK.

You can now preview the report. Each time you click an element in the slicer, the lower chart will be updated to show only the data for the selected country, and the upper chart will highlight the metric for the chosen country.

How It Works

This report is specifically designed for a tablet. As this is the case, you want to enable users to make selections-and see what they have selected-more easily and intuitively. So you are combining the date selector that you already used with a country selector that resembles an Excel slicer. The "slicer" is dynamic in the sense that it will contain as many elements as there are countries for the selected year and month.

To enhance the visual effect, you will use two image backgrounds for the text box in the tablix that is the slicer. One is applied if the element is selected; the other is applied to all unselected rows. Once again, an expression detects the state and applies the corresponding image.

At a purely stylistic level, I have omitted axis titles in the chart of sales by country because this allows the user to focus on the highlighted bar, which is in the same color as the chosen slicer element. The chart cylinder for the selected country is highlighted using the expression in step 11, which tests the current country against the selected country parameter, and applies a different color if the two are equal.

Multiple Slicers

Users frequently want to combine slicer elements to perform "what-if" analyses. This is perfectly possible with SSRS. Here you will look at how to "hard-code" a predefined set of slicer elements. The user can then select one, several, or indeed all of the slicer elements to change the report output. An example is given in Figure 8-8, where three out of four potential elements are selected. The table and chart will then only display data for the selected elements.

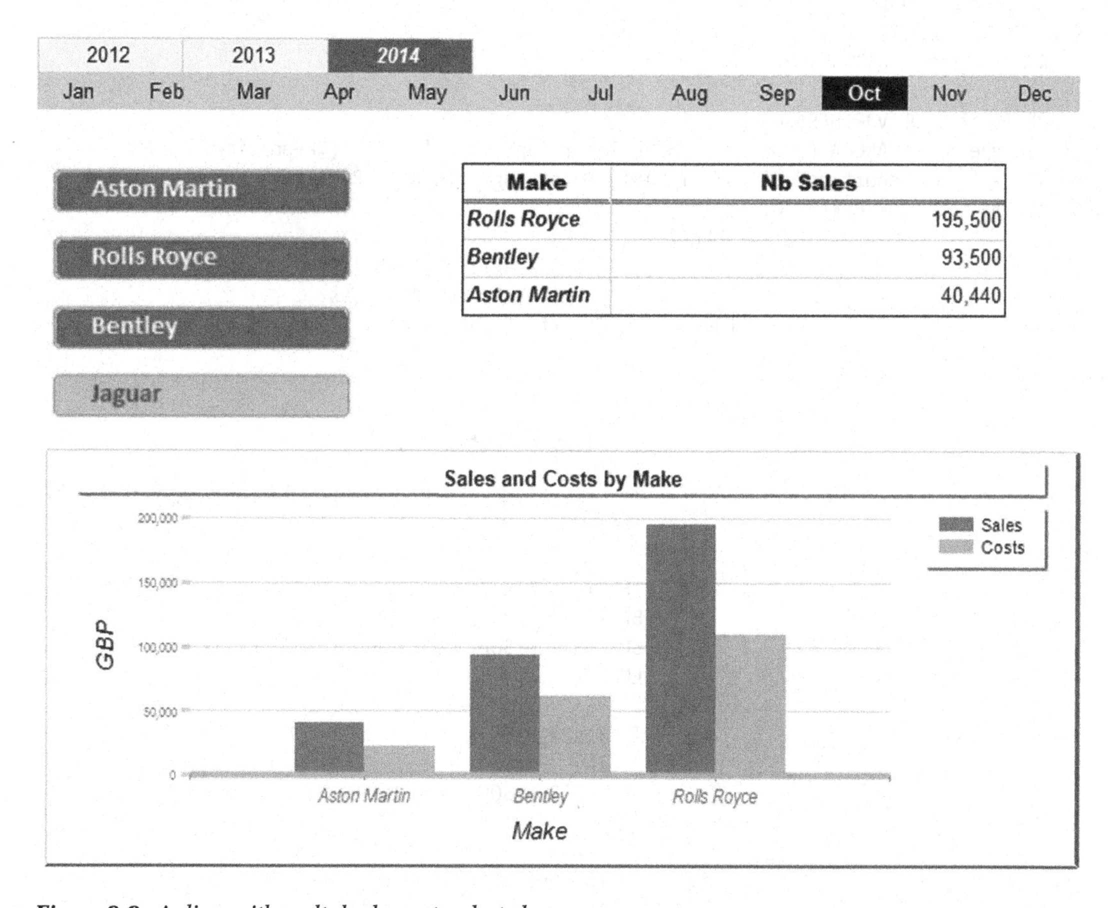

Figure 8-8. *A slicer with multple elements selected*

■ **Note** These examples cover applying a single slicer to a report. You can, of course, add as many slicer sets as you want to a report in order to offer greater analytical flexibility to your users.

The Source Data

In this particular example, you will be using the same data set for the two visualizations, table and chart, in the report. Not only that, but you will be using some dynamic SQL. The slicer sends back multiple parameters, one Boolean flag per make of car, that will be displayed in the report. The code detects the status of the Boolean flag and extends the WHERE clause for the corresponding make if the flag is set to true. This code is available in the stored procedure Code. pr_MonthlyCarSalesSlicer.

```
DECLARE @ReportingYear INT
DECLARE @ReportingMonth TINYINT
DECLARE @IsAstonMartin BIT = True
DECLARE @IsRollsRoyce BIT = True
DECLARE @IsJaguar BIT = True
DECLARE @IsBentley BIT = True

DECLARE @SlicerSQL VARCHAR(8000) = ''
DECLARE @WhereSQL VARCHAR(8000) = ' WHERE ReportingYear = ' + CAST(@ReportingYear AS
CHAR(4)) + ' AND ReportingMonth = ' + CAST(@ReportingMonth AS VARCHAR(2))
DECLARE @MainSQL VARCHAR(8000) =
                              'SELECT      Make
                                           ,SUM(SalePrice) AS Sales
                                           ,SUM(CostPrice) AS Costs
                               FROM        Reports.CarSalesData'

SET @SlicerSQL = ' AND Make IN ('

                                        IF @IsAstonMartin = 1
                                        BEGIN
                                        SET @SlicerSQL = @SlicerSQL + '''Aston Martin'','
                                        END

                                        IF @IsRollsRoyce = 1
                                        BEGIN
                                        SET @SlicerSQL = @SlicerSQL + '''Rolls Royce'','
                                        END

                                        IF @IsJaguar = 1
                                        BEGIN
                                        SET @SlicerSQL = @SlicerSQL + '''Jaguar'','
                                        END

                                        IF @IsBentley = 1
                                        BEGIN
                                        SET @SlicerSQL = @SlicerSQL + '''Bentley'','
                                        END
```

```
SET @SlicerSQL = LEFT(@SlicerSQL, LEN(@SlicerSQL) -1) + ')'

SET @MainSQL = @MainSQL + @WhereSQL + @SlicerSQL + ' GROUP BY Make'

EXEC (@MainSQL)
```

Running this dynamic SQL produces the result shown in Figure 8-9.

	Make	Sales	Costs
1	Aston Martin	40440.00	22500.00
2	Bentley	93500.00	61400.00
3	Rolls Royce	195500.00	110000.00

Figure 8-9. *The output from the dynamic SQL for a muliple selection*

How the Code Works

As the SSRS report can send back a combination of possible Boolean values, dynamic SQL seemed the best way to handle the subtleties of the requisite `WHERE` clause. Otherwise, the code is not difficult; it aggregates the sales and costs for the selected month and year, and groups the result by make. Then the `WHERE` clause is extended to filter the result set by the selected make of vehicle.

■ **Note** In production code you should really use sp_executesql and not exec to run dynamic SQL.

Building the Report

Once the code is in place, you can build the report itself. As this report also uses the interactive year and month selectors, you must add them to the report, or build on a previous report containing these elements.

1. Make a copy of the report `__DateSelector.rdl`. Name the copy `Tablet_SimpleSlicer.rdl`.
2. Add a dataset that you name `MonthlyCarSalesSlicer` using the stored procedure `Code.pr_MonthlyCarSalesSlicer`. This will add four new parameters to the report, one for each of the makes shown in Figure 8-7.
3. Expand the Parameters folder in the Report Data window. Set the four parameters `IsAstonMartin`, `IsRollsRoyce`, `IsJaguar`, and `IsBentley` to be hidden. In the Default Values pane for each of them, select the Specify values button, click Add, and set the value to True.
4. Embed the following images to the report from the folder `C:\BIWithSSRS\Images` (assuming that you have downloaded the sample files from the Apress web site):
    a. `SlicerButton_AstonMartin`
    b. `SlicerButton_Jaguar`

c. `SlicerButton_RollsRoyce`

d. `SlicerButton_Bentley`

e. `SlicerButtonSelected_ AstonMartin`

f. `SlicerButtonSelected_ Jaguar`

g. `SlicerButtonSelected_ RollsRoyce`

h. `SlicerButtonSelected_ Bentley`

5. Add a table to the report area and place it on the left under the interactive year and month selectors. Remove the detail row and delete two of the three columns, leaving a single column. Add six rows (or as many as are necessary to make a total of seven rows).

6. Configure the table to use the `Dummy` dataset (remember that this will stop SSRS complaining that there is no dataset for the table).

7. Set the height of rows 2, 4, and 6 to 0.08 inches, and the font size to 2 points.

8. Set the height of rows 1, 3, 5, and 7 to 0.354 inches.

9. Set the width of any text box, and consequently the entire column, to 2.125 inches.

10. Drag an image item from the toolbox into the top cell of the table. Leave the image source as Embedded, and click the Fx (function) button to the right of the Use this image pop-up, and add the following code:

```
=IIF(Parameters!IsAstonMartin.Value=True,"SlicerButtonSelected_
AstonMartin","SlicerButton_AstonMartin")
```

11. Do the same for the cells in rows 3, 5, and 7-only tweak the code to use the Rolls Royce, Bentley, and Jaguar buttons (in this order).

12. Select the four images that you just placed and display the Properties window. Set the Sizing property to FitProportional.

13. Right-click the image in row 1. Select Image Properties from the context menu. Click Action on the left.

14. Click the Add button six times. Configure the parameters as follows:

Name	Value
IsAstonMartin	`=Iif(Parameters!IsAstonMartin.Value=True,False,True)`
IsBentley	`=Parameters!IsBentley.Value`
IsRollsRoyce	`=Parameters!IsRollsRoyce.Value`
IsJaguar	`=Parameters!IsJaguar.Value`
ReportingYear	`=Parameters!ReportingYear.Value`
ReportingMonth	`=Parameters!ReportingMonth.Value`

The dialog should look like Figure 8-10.

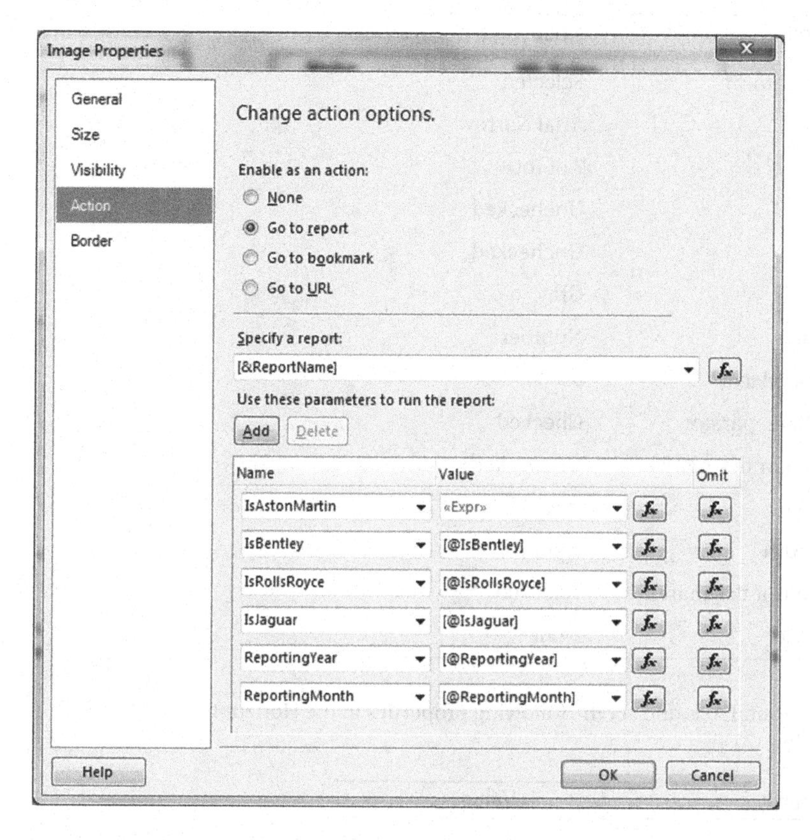

Figure 8-10. *Specifying the actions for each image*

15. Repeat the process for the other three images in the table. Be careful to set the expression that you set for Aston Martin to each of the other makes, adapting the code to the requisite make, of course.
16. Add a table to the report area. Place it to the right of the slicer. Delete one column and set the dataset to be `MonthlyCarSalesSlicer`.
17. Add the `Sales` field to the left column and the `Sales` field to the right column.
18. Right-click the right-hand column and set the Format property in the Properties window to `#,#`. Set the column width to around 2.5 inches.
19. Add a border to the table and below the header row.
20. Drag a chart onto the report area. Place it under the two tables. Set the dataset to be `MonthlyCarSalesSlicer`.
21. Set the following titles:
    a. Chart title: Sales and Costs by Make
    b. Horizontal axis title: Make
    c. Vertical axis title: GBP

22. Right-click the vertical axis, and set the following vertical axis properties:

Section	Property	Value
Labels	Enable auto-fit	Selected
Label Font	Font	Arial Narrow
	Size	7 point
	Bold	Unchecked
	Italic	Unchecked
	Color	Gray
Number	Category	Number
	Decimal places	0
	Use 1000 separator	Checked
Major Tick Marks	Hide major tick marks	Unchecked
	Line color	Gray
	Line width	2 point
Minor Tick Marks	Hide minor tick marks	Checked
Line	Line style	None

23. Right-click the Horizontal Axis and set the following properties in the Horizontal Axis Properties dialog:

Section	Property	Value
Labels	Disable auto-fit	Selected
	Label rotation angle (degrees)	0
Label Font	Font	Arial Narrow
	Size	10 point
	Color	Gray
Major Tick Marks	Hide major tick marks	Unchecked
	Line color	Gray
Minor Tick Marks	Hide minor tick marks	Checked
Line	Line width	3 point
	Line color	Dim Gray

24. Right-click the legend and select Legend Properties from the context menu. Click Shadow on the left and set the Shadow offset to 0.5 point. Click OK.

25. Do the same for the title as you just did for the legend.

26. Right-click the chart and select Chart Properties from the context menu. Click Border on the left and in the preview window add a 1-point light gray border to the top and the left, and a 3-point gray border to the right and bottom. Click OK.

When you have finished building the report, I suggest that you preview it. Then click any of the makes in the slicer, and you will see that the make is unselected and will no longer appear in either the table or the chart.

How It Works

This particular type of report is very strictly controlled, and is indeed quite the opposite of the previous example. This approach is particularly suited to mobile devices where you specifically do *not* want slicers to grow in an uncontrolled fashion. A major upside is that you can apply multiple slicer elements at the same time. A potential downside is that handling multiple Boolean parameters can be laborious in SSRS, and can necessitate dynamic SQL on the server. However, your users will not know this. All they can see is a much more attractive interface and visibly improved interactivity, especially on mobile devices.

It can be slightly disconcerting not to see the background for the slicer buttons appear at all in the design window. This is unfortunately inevitable, as SSRS will only know which text box background to apply when the report is displayed.

In case you were wondering why you used individual images for each make (as opposed to using a simple background, as you did in the previous example), the reason is quite simple. When you use images, you can resize the cells in the table and the image will shrink or grow to fit the cell. This means that you are less constrained by the necessity of creating a perfectly sized image for your report. The downside to this approach is that you cannot superpose a text over an image, so you have to create pressed and unpressed images for each element in the slicer.

It can get a little laborious setting the action for each clickable element in a slicer like this because you have to pass back every parameter. You also have to tweak every one so that it handles the selection that your user will be clicking on. As you can see in step 14, you toggle the state of each slicer element using a simple expression.

Conclusion

In this chapter, you saw how, with a little effort and ingenuity, you can radically transform the SSRS interface. You can replace pop-up selectors with clickable elements in the report itself and even add slicers to give an "Excel-like" feel to a report. If you want, you can also integrate menu-type selections to your BI output. Best of all, these elements can easily become reusable widgets that will not only help you develop reports faster, but will ensure a standardized look and feel across the range of your dashboards and mobile output.

CHAPTER 9

Interface Enhancements

In the previous chapter, you saw some of the basic techniques that you can use to mask parameter selection in SSRS. It is now time to push these techniques a little further and apply them to add some real visual pizazz to your business intelligence delivery with SSRS.

Once again, most of what you will see in this chapter is based on hidden parameters and causing a report to refresh so that it displays a different subset of the data, or indeed a different view of the information. Consequently I recommend that you read the previous chapter before proceeding with this one.

The interface tweaks that you will look at in this chapter include

- Highlighting chart elements for data selection
- Creating tiles to subset data
- Adding a "carousel" effect to filter datasets
- Adding paged recordsets
- Creating "tabbed" reports

This is not all that can be done to add flair and style to your BI delivery with SSRS. Far from it; these ideas are only a starting point. I really hope that they will inspire you to discover further ways to create visually arresting dashboards and scorecards, because with a little time, effort, and ingenuity, SSRS can deliver reports that can rival or surpass most corporate and self-service BI products.

As most of the reports that you will see in this chapter use the interactive year and month selector that you created in the last chapter, I will make life easy for all of us by beginning with a template report that contains the widgets and datasets that are needed for these common elements. This report is called `__DateSelector.rdl` and can be found in the sample SSDT application that is available on the Apress web site.

This chapter suggests several techniques for scrolling through datasets. Inevitably, image buttons are used to allow more intuitive scrolling. There are hundreds of available sample images for this on the Web, but I have created a few extremely simple buttons for use in this chapter; see Table 9-1. You are, of course, free to use them in your own widgets, or to use any buttons that you prefer for this instead.

Table 9-1. Paging Buttons

Button	File Name	Description
◀	`LeftOne.png`	Moves one element to the left
▶	`RightOne.png`	Moves one element to the right
◀◀	`LeftSet.png`	Moves one group of elements to the left
▶▶	`RightSet.png`	Moves one group of elements to the right
|◀	`LeftAll.png`	Jumps to the leftmost element
▶|	`RightAll.png`	Jumps to the rightmost element

Highlighting for Data Selection

Tools like Power View have got users used to interactive highlighting in reports. What this means essentially is that users want to be able to click or tap on a bar in a bar chart (or a column in a column chart or a segment in a pie chart) and have the data in the report filtered by the selected element.

Once again, this is not a problem for SSRS. Indeed, this is nothing more than a form of interactive parameter selection that is easy to implement. The effect is best enhanced by setting a different color for the selected chart element (be it a bar, pie slice, or column) so that the user can see exactly what element is filtering the data. An example of this kind of highlighting is shown in Figure 9-1.

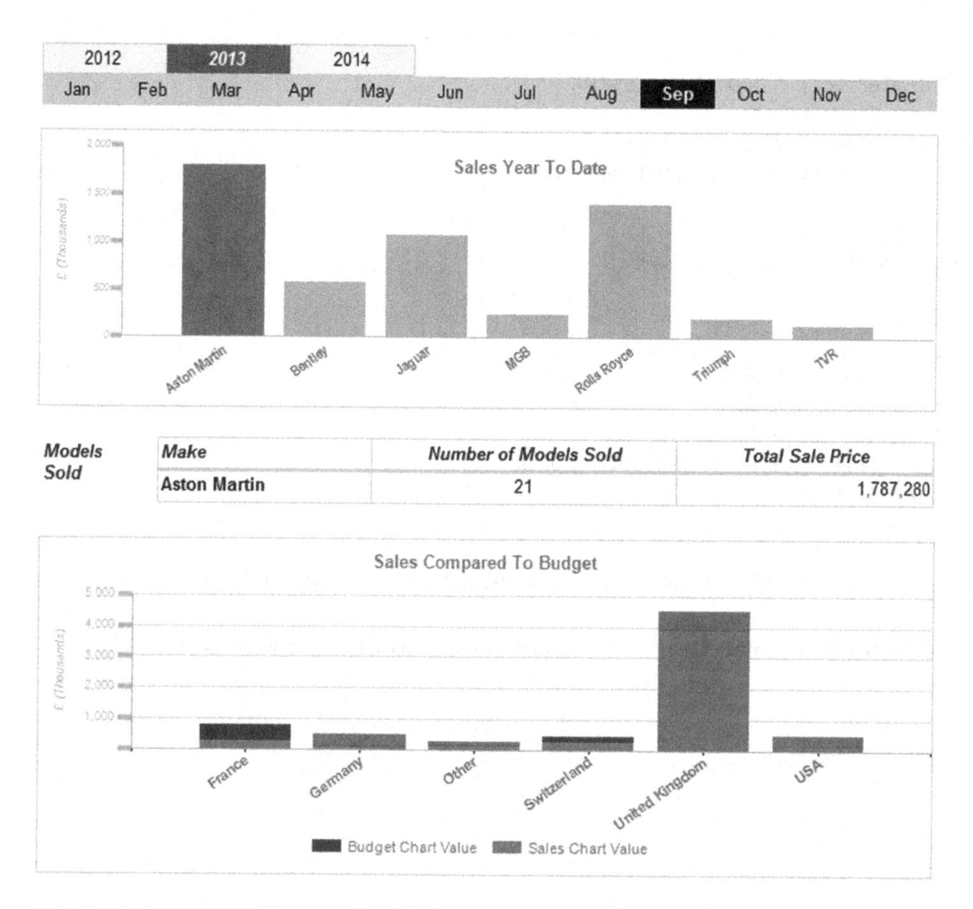

Figure 9-1. *Highlighting report data*

Hopefully the shading in a book is sufficiently clear to allow you to distinguish between the selected and unselected chart columns in the upper chart where the selected make, Aston Martin, is used to filter the data in the table. If the effect is not clear, you can always preview the report `Tablet_ChartHighlight.rdl` in the sample SSDT project to get the full effect.

The Source Data

The source data for the upper chart and the table are shown below. You saw the data for the lower chart in Chapter 4 (`Code.pr_WD_CountrySalesToBudgetRatio`), so I will not repeat it here. What you do need, however, are the code snippets used to display the top chart (`pr_Tablet_ChartHighlight_SalesYTD`) and the table (`pr_Tablet_ChartHighlight_MakeSales`). So here they are:

```
DECLARE @ReportingYear INT = 2013
DECLARE @ReportingMonth TINYINT = 9
DECLARE @Make NVARCHAR(50) = 'Aston Martin'

-- Code.pr_Tablet_ChartHighlight_SalesYTD
```

```
SELECT         Make
               ,SUM(SalePrice) / 1000 AS SalesYTD
FROM           Reports.CarSalesData
WHERE          ReportingYear = @ReportingYear
               AND ReportingMonth <= @ReportingMonth
GROUP BY       Make

-- Code.pr_Tablet_ChartHighlight_MakeSales

SELECT         Make
               ,SUM(SalePrice) AS SalePrice
               ,COUNT(Make) AS NbModels
FROM           Reports.CarSalesData
WHERE          ReportingYear = @ReportingYear
               AND ReportingMonth <= @ReportingMonth
               AND Make = @Make
GROUP BY       Make
```

Running these code snippets gives the results (for September 2013) shown in Figure 9-2.

Code.pr_Tablet_ChartHighlight_SalesYTD

	Make	SalesYTD
1	Aston Martin	1787.28
2	Bentley	570.50
3	Jaguar	1068.50
4	MGB	244.75
5	Rolls Royce	1391.05
6	Triumph	210.75
7	TVR	130.50

Code.pr_Tablet_ChartHighlight_MakeSales

	Make	SalePrice	NbModels
1	Aston Martin	1787280.00	21

Figure 9-2. *Data for the selector chart*

How the Code Works

The first piece of code merely aggregates the total sales for the selected year and month. It divides the result by 1,000 purely for the chart axis. The second piece of code returns the same sales figure (not divided by 1,000) as well as the number of cars sold for the selected (highlighted) color that is passed back using the @Make parameter.

Building the Report

So, with the data in place, here is how to build the report.

1. Make a copy of the .rdl file named __DateSelector. Name the copy Tablet_ChartHighlight.rdl, and open the copy.
2. Add the following datasets:
    a. ChartHighlight_MakeSales, using the stored procedure Code.pr_Tablet_ChartHighlight_MakeSales
    b. CountrySalesToBudgetRatio, using the stored procedure Code.pr_WD_CountrySalesToBudgetRatio
    c. ChartHighlight_SalesYTD, using the stored procedure Code.pr_Tablet_ChartHighlight_SalesYTD
3. Add a column chart to the report. Apply the dataset ChartHighlight_SalesYTD. Widen the chart to match (approximately) the width of the month selector table.
4. Set the following in the Chart Data pane:
    a. ∑ Values: SalesYTD
    b. CategoryGroups: Make
5. Click any of the columns and set the Color property (in the Properties window) to the following expression:

```
=IIF(IsNothing(Parameters!Make.Value),"DarkGray",
IIF(Fields!Make.Value = Parameters!Make.Value,"Red","Silver"))
```

6. Right-click any column and select Series Properties from the context menu. Click Action on the left. Choose Go to report as the action to enable. Set the Specify a report property (after clicking on the Function (Fx) button) to =Globals!ReportName. Remember that this is the ReportName item of the Built-in Fields category.
7. Click the Add button three times to add three parameters. Define them as
    a. Make: =Fields!Make.Value
    b. ReportingYear: =Parameters!ReportingYear.Value
    c. ReportingMonth: =Parameters!ReportingMonth.Value

8. Right-click the vertical axis and select Vertical Axis properties. Set the following properties:

Section	Property	Value
Axis Options	Always include zero	Checked
Labels	Disable auto-fit	Selected
	Label rotation angle (degrees)	0
Label Font	Font	Arial
	Size	6 point
	Color	Silver
	Italic	Checked
Number	Category	Number
	Decimal places	0
	Use 1000 separator	Checked
Major Tick Marks	Hide major tick marks	Unchecked
	Position	Outside
	Length	1.25
	Line color	Silver
	Line width	3 point
Minor Tick Marks	Hide minor tick marks	Checked
Line	Line color	Silver

9. Right-click the horizontal axis and select Horizontal Axis Properties from the context menu. Set the following properties:

Section	Property	Value
Labels	Disable auto-fit	Selected
	Label rotation angle (degrees)	-36
Label Font	Font	Arial
	Size	8 point
	Color	Gray
	Bold	Checked
Major Tick Marks	Hide major tick marks	Checked
Minor Tick Marks	Hide minor tick marks	Checked
Line	Line color	Silver

10. Delete the legend.
11. Right-click the title and select Title Properties from the context menu. Set the following properties:

Section	Property	Value
General	Title text	Sales Year To Date
	Title position	Top center
Font	Font	Arial
	Size	10 point
	Bold	Checked
	Color	Gray
Border	Line style	None

12. Right-click the Vertical Axis Title and select Axis Title Properties from the context menu. Set the following properties:

Section	Property	Value
General	Title text	£ (Thousands)
	Title alignment	Center
Font	Font	Arial
	Size	7 point
	Italic	Checked
	Color	Silver

13. Click the Chart Area, and in the Properties window, expand CustomPosition. Set the following values:
    a. Enabled: True
    b. Height: 95
    c. Width: 100
14. Delete the horizontal axis title.
15. Adjust the chart height and width to suit your aesthetic sensibilities.
16. Add a table to the report. Set it to use the dataset `ChartHighlight_MakeSales`. Leave a header row and three columns. Set all the column widths to around 2.2 inches.
17. Place the field `Make` in the left-hand column, `NbModels` in the center column, and `SalePrice` in the right-hand column.
18. Enter some suitable titles in the table header row. I suggest those that are used in Figure 9-1. Set the titles to be Arial Blue 10 point italic.

19. Add a text box to the left of the table and enter the text "Models Sold" (without the quotes). Make this Arial Blue 10 point italic, too. Adjust its height and width so that the contents flow over two lines.

20. Open the report `_SalesToBudget.rdl` that you created in Chapter 4, or open the example SSDT project where you will find it. Select the chart and copy it onto the report area of the report that you are currently creating. Drag the copied chart below the table. This visualization will use the dataset `CountrySalesToBudgetRatio` that you added in step 2.

21. Apply exactly the same formatting to the chart title, vertical axis title, and horizontal and vertical axes that you applied to the chart you made previously in this report. You can leave the major gridlines, but set their color to silver.

22. Select the chart area for the second chart, and in the Properties window, expand CustomPosition. Set the following values:

    a. Enabled: True

    b. Height: 75

    c. Width: 100

23. Click the vertical axis and (in the Properties window) set the LabelsFormat to `#,#,;(#,#,)`. This will present the output divided by 1,000.

Now that you have your report, you can preview it. The bar colors will change to display those that have been set by an SSRS expression for each chart.

This is a report that you really have to preview to get the full effect. Specifically, the columns will apply the correct colors in the preview, and you will be able to click any column in the top chart to display the sales figures in the table for that specific make.

How It Works

The essential part of this report is the top chart. The data that it displays (sales for the year up to the selected month) is simple. It is the way that clicking a column sets the `@Make` parameter that is interesting. This parameter not only highlights the column (as the column color has been set by the expression that you added in step 5), but also filters the data shown in the table.

Adding the second chart is merely a stylistic flourish. It also serves to make the point that any selection made using a chart as you did here will *not* apply automatically to every element in the report. You have to wire everything up manually. This can be a good thing because it gives you greater control. It can also be a pain when you have to hand craft the code over and over for all the parameters in the various action properties of clickable elements in the report.

The adjustment of the chart area window is to remind you that you are not limited to the default settings for the chart area. The other presentation ideas are only suggestions.

Creating Tiles to Subset Data

Another technique that has become popular in BI reporting is to subset data is using tiles in a report. These are, in essence, a filter that is

- Visual and intuitive
- Well suited to extensive filter sets

It is a little more work to set up a tile set rather than a slicer to allow the reader to select a filter. However, this kind of visualization adds a certain "wow" factor to your reports. So here is how to deliver the report shown in Figure 9-3.

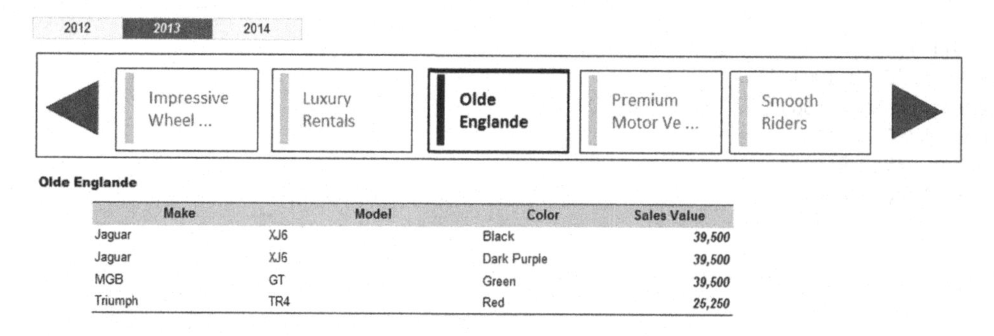

Figure 9-3. *A tiled report*

Clicking any of the vehicle distributor names will display a table of sales for the selected year. The tile containing the client is then highlighted so that you can see which tile is selected (Olde Englande, in this example). Clicking the right-hand button scrolls to the next set of clients, whereas clicking the left-hand button scrolls back to the previous set of clients. Code tweaks are also added to prevent the tile set scrolling past the first or last elements in the set.

The Source Data

This visualization uses three stored procedures, the code for which is given below. The first code block returns the five clients that you can see in the tile set. The second block returns the maximum number of tiled sets that are possible in a given year. This figure is used to prevent empty sets of tiles being displayed through providing a ceiling to the data set. The third block is much simpler, and displays the sales for the selected client.

```
-- Code.pr_TabletTileTiles

DECLARE @ReportingYear INT = 2013
DECLARE @SelectedElement VARCHAR(50) = 'Aston Martin'
DECLARE @BatchRow INT
DECLARE @RowCount INT = 1
DECLARE @BatchCount INT = 1
DECLARE @BatchCountRemainder INT = 1
DECLARE @Counter INT = 1
DECLARE @LowerThreshold INT = 1

SELECT @BatchCount = COUNT(DISTINCT ClientName) FROM Reports.CarSalesData WHERE
CountryName = 'United Kingdom' AND YEAR(InvoiceDate) = @ReportingYear
SELECT @RowCount = @BatchCount / 5
SELECT @BatchCountRemainder = @BatchCount % 5

IF @BatchCountRemainder > 0 SET @RowCount += 1
IF @BatchRow <= 0 SET @BatchRow = 1
```

```
-- Required Iterations
IF OBJECT_ID('tempdb..#Tmp_Output') IS NOT NULL DROP TABLE tempdb..#Tmp_Output

CREATE TABLE #Tmp_Output
(
  ID INT IDENTITY(1,1)
 ,FirstElement VARCHAR(100)
 ,SecondElement VARCHAR(100)
 ,ThirdElement VARCHAR(100)
 ,FourthElement VARCHAR(100)
 ,FifthElement VARCHAR(100)
)

WHILE @Counter <= @RowCount

BEGIN

INSERT INTO #Tmp_Output (FirstElement,SecondElement ,ThirdElement ,FourthElement
,FifthElement)

SELECT [1] AS FirstElement, [2] AS SecondElement ,[3] AS ThirdElement ,[4] AS FourthElement
,[5] AS FifthElement
FROM
     (
      SELECT TOP (5) ROW_NUMBER() OVER (ORDER BY S.ID) AS ID, S.ClientName
      FROM

          (
           SELECT DISTINCT
                    DENSE_RANK() OVER (ORDER BY ClientName) AS ID
                   ,ClientName
           FROM    Reports.CarSalesData
           WHERE   CountryName = 'United Kingdom'
                   AND YEAR(InvoiceDate) = @ReportingYear
          ) S

      WHERE     S.ID BETWEEN ISNULL(@LowerThreshold, 1) AND @LowerThreshold * 5
     ) PT
PIVOT (MAX(ClientName) FOR ID IN ([1],[2],[3],[4],[5])) AS PT

SET @Counter += 1
SET @LowerThreshold = @LowerThreshold + 5
END

IF  @BatchRow > @RowCount
BEGIN
SELECT * FROM #Tmp_Output WHERE ID  = @RowCount
END
ELSE
BEGIN
SELECT * FROM #Tmp_Output WHERE ID  = @BatchRow
END
```

```
-- Code.pr_TabletTileMaxElements

DECLARE @RowCount INT = 1
DECLARE @BatchCount INT = 1
DECLARE @BatchCountRemainder INT = 1

SELECT @BatchCount = COUNT(DISTINCT ClientName)
                     FROM Reports.CarSalesData
                     WHERE CountryName = 'United Kingdom'
                     AND YEAR(InvoiceDate) = @ReportingYear
SELECT @RowCount = @BatchCount / 5
SELECT @BatchCountRemainder = @BatchCount % 5

IF @BatchCountRemainder > 0 SET @RowCount += 1

SELECT @RowCount AS MaxElements

-- Code.pr_TabletTileListOutput

SELECT
Make
,Model
,Color
,SalePrice

FROM       Reports.CarSalesData
WHERE      CountryName = 'United Kingdom'
           AND ReportingYear = @ReportingYear
           AND ClientName = @SelectedElement
ORDER BY   Make, Model, Color
```

The output for the three stored procedures is shown in Figure 9-4.

Code.pr_TabletTileTiles

	ID	FirstElement	SecondElement	ThirdElement	FourthElement	FifthElement
1	2	Impressive Wheels	Luxury Rentals	Olde Englande	Premium Motor Vehicles	Smooth Riders

Code.pr_TabletTileMaxElements

	MaxElements
1	3

Code.pr_TabletTileListOutput

	Make	Model	Color	SalePrice
1	Jaguar	XJ6	Black	39500.00
2	Jaguar	XJ6	Dark Purple	39500.00
3	MGB	GT	Green	39500.00
4	Triumph	TR4	Red	25250.00

Figure 9-4. *The result sets used by a tiled report*

How the Code Works

The procedure `Code.pr_TabletTileTiles` (the core process for this visualization) starts by creating a temporary table where each row returned will be displayed as a set of five tiles. This table is then populated by a PIVOT operation that does the following:

- Selects all the records that correspond to the criteria (a hard-coded country and a parameterized year). These records are sorted in alphabetical order of client name.
- Pivots the rows into sets of five clients (clients) and gives each row a unique ID number.
- Adds these rows, one at a time, to the temporary output table. A WHILE clause adds the sets of five records to the output table.

The code for the stored procedure `Code.pr_TabletTileMaxElements` determines, independently of the previous procedure, the maximum number of records in the output table. Finally, the procedure `Code.pr_TabletTileListOutput` returns all the sales for a client once they have been selected in the tile set.

Building the Report

Once the three stored procedures have been built, you can create the tile visualization. As this report uses the year selector that you have used previously (but not the month selector), I suggest using the `__DateSelector` template, and then removing any references to the reporting month parameter rather than rebuilding the year selector from scratch.

1. Make a copy of the `.rdl` file named `__DateSelector`. Name the copy `Tablet_Tile.rdl`, and open the copy.
2. Delete the table containing the months. Delete the dataset named *Dummy* that is used by this table.
3. For *each* of the three text boxes in the table containing the years, do the following to remove the reference to the `ReportingMonth` parameter:
    a. Right-click the text box.
    b. Click Action on the left.
    c. Click inside the ReportingMonth parameter.
    d. Click Delete.
4. Expand the `Parameters` folder in the Report Data pane, right-click the `ReportingMonth` parameter, and select Delete from the context menu.
5. Add the following datasets:
    a. `Tiles`, using the stored procedure `Code.pr_TabletTileTiles`- this will add the parameter `@BatchRow`
    b. `Maximumelements`, using the stored procedure `Code.pr_TabletTileMaxElements`
    c. `ListOutput`, using the stored procedure `Code.pr_TabletTileListOutput`- this will add the parameter `@SelectedElement`

6. Ensure that importing the datasets has added the following two parameters; otherwise add them, and set their properties as follows:

   a. BatchRow

Section	Property	Value
General	Data type	Integer
	Visible	Hidden
Default Values	Specify values	1

   b. SelectedValue

Section	Property	Value
General	Data type	Text
	Allow null value	Checked
	Visible	Hidden
Default Values	No default value	

7. Add a rectangle to the report. Make it 1.5 inches high by 0.9 inches wide.

8. Place a text box inside the rectangle. Name it TxtFirstElement. Set the following properties in the Properties window:

Property	Value
Color	`=Iif(ReportItems!TxtFirstElement.Value = ReportItems!TxtSelectedElement.Value,"Black","DimGray")`
Font ➤ FontFamily	Calibri
Font ➤ FontSize	14 point
Font ➤ FontWeight	`=Iif(ReportItems!TxtFirstElement.Value = ReportItems!TxtSelectedElement.Value,"Bold","Normal")`
CanGrow	False
CanShrink	False

9. Set the default border width of the rectangle containing the text box to the following expression:

```
=Iif(ReportItems!TxtFirstElement.Value = ReportItems!TxtSelectedElement.
Value,"3pt","1pt")
```

10. Add a rectangle in the shape of a wide vertical line to the left of the text box (as you can see in Figure 9-3). Note that this is purely for visual effect to attract the reader's attention to the selected element. Set its `BackgroundColor` property to the following expression:

```
=Iif(ReportItems!TxtFirstElement.Value = ReportItems!TxtSelectedElement.
Value,"DarkBlue","LightBlue")
```

11. Select the three elements that you just created, which in effect make up a single tile, and make four copies of them. Place the copies on a single row as in Figure 9-3.

12. Change the name of the four copied text boxes inside each set of copied elements to TxtSecondElement, TxtThirdElement, TxtFourthElement, and TxtFifthElement. Make sure that the numbering in the names flows from left to right, so that the last one is TxtFifthElement.

13. Change the code for the *four* expressions that you created in steps 7, 8, and 9 to the name of the corresponding text box for each tile. So, for example, the second text box will contain `=If(ReportItems!Txt`**`Second`**`Element.Value = ReportItems!TxtSelectedElement.Value,"Black","DimGray")` as the expression that sets its color.

14. Embed the following two images from the directory `C:\BIWithSSRS\Images` (assuming that you have downloaded the sample data from the Apress web site):

    a. `LeftOne.png`

    b. `RightOne.png`

15. Add two images to the report: one to the left of the set of elements that make up the tiles, one to the right. Set the left-hand one to use the image LeftOne and the right-hand one to use the image RightOne. Make them both 0.6 inches square.

16. Right-click the leftmost image and select Image Properties from the context menu. Select Action on the left.

17. Select Go to report as the enabled action.

18. In the Specify a report pop-up, do not select the name of the report, but enter `=Globals!ReportName` as the report to jump to. The dialog will display `[&ReportName]`.

19. Click three times on the Add button and define the following expressions as the values for the parameters (so click on the Fx button to display the Expression dialog):

    a. `Name: ReportingYear - Value: =Fields!Year1.Value`

       The dialog will show `[Year1]` after you click OK in the Expression dialog.

    b. `Name: SelectedElement - Value: =Parameters!SelectedElement.Value`

       The dialog will display `[@SelectedElement]` after you click OK in the Expression dialog.

    c. `Name: BatchRow - Value: =IIF(Parameters!BatchRow.Value -1 < 1, 1, Parameters!BatchRow.Value - 1)`

20. Click OK to confirm your modifications to the image properties.

21. Repeat steps 15 through 19 for the right-hand image (RightOne). Set the expression for the BatchRow parameter to

```
=IIF(Sum(Fields!MaxElements.Value, "MaximumElements")
<= Parameters!BatchRow.Value
,Parameters!BatchRow.Value
,Parameters!BatchRow.Value + 1
)
```

22. Add four lines (two horizontal, above and below, and two vertical to the left and to the right) to give the effect of a box around the five tiles and the two buttons.

23. Add a text box under the tiles. This will be the indication of the selected client. Make it approximately half the width of the report. Set the font to Arial Black 10 point and the expression for the text box value to

```
=IIF(IsNothing(Parameters!SelectedElement.Value) = True,
"Please click on a reseller", Parameters!SelectedElement.Value)
```

24. Add a table under the tiles and add a column (there should be a total of four columns). Apply the dataset `ListOutput`.

25. Add the following fields to the four columns, from left to right:

    a. `Make`

    b. `Model`

    c. `Color`

    d. `SalePrice`

26. Set the title row to have a grey background and centered text in bold. Add a border to the top and bottom of the table in the table properties.

There you have your tiled report. When displayed, the user can scroll through the client list and click or tap on any one of them to display the vehicles that they have sold to this customer for a given year.

How It Works

This report is in two parts. The top part is a list of clients that can be scrolled to display all the UK-based resellers. This list is tweaked, visually, to show five clients at a time in a horizontal structure called a tile display. Clicking a tile containing a client name refreshes the report to display the sales for the selected client in the table under the tile set.

As is the case with all the visual tricks in this chapter, a set of hidden parameters is used to convey all the information required to show the tiles and the selected client. These parameters are

- `ReportingYear`: This contains the year for the metrics.
- `BatchRow`: This indicates which row of the pivoted table of clients will be displayed in the tile set.
- `SelectedElement`: This parameter contains the name of a selected client (if there is one). This is used not only to filter the data used in the table that appears below the tile set, but also to change the appearance of the selected tile so that the user can see which client has been selected. This parameter is also displayed in a text box above the table. This is because it is possible to scroll through the client list while leaving a non-visible client selected. If you prefer, you can tweak the code to set the `@SelectedElement` parameter to nothing when scrolling through the list of clients.

Once the code and parameters are in place, it is the `Action` property of the text boxes in the tile set, or the scroll buttons to either side of the tiles, that takes over. As you can see in steps 19 and 21, this code simply tests that there is a previous or following row in the table that feeds data into the text boxes that make up the tile set, and returns the data if it exists. The code for the text boxes in the tile set (for the font, border, and colored bar) checks to see if the value of the selected element (in the text box that displays the client) matches the contents of the text box contents. If it does, then these visual elements are changed to draw attention to the selection.

As a final tweak, the length of the client name is checked, and if it is too long to be displayed comfortably in the text box, it is truncated and ellipses added.

■ **Note** In this example, the verification of the selected element uses a text box. It could just as easily compare the value of the parameter containing a selected element.

Adding a Carousel to Filter Data

Another form of interactive data selection is a carousel. This technique is similar to using tiles to subset data, and uses many of the same techniques to filter the data that is displayed. One major difference is that the selected element is always visible because it appears in the center of the carousel.

A carousel adds a sense of perspective to show the selected element as well as a defined number of elements both before and after the current selection. You can scroll forward and backwards through the elements that make up the carousel using the buttons at the right and left of the carousel, or you can click any visible element to position it in the center of the carousel. The chart will then update to display the chart data for the central element. This way the central element of the carousel is the selected element.

The trick to this visualization is to add an image as the background to each text box in the table. The design of the images gives the impression of perspective to the carousel. A report with a carousel appears to the user like the one shown in Figure 9-5.

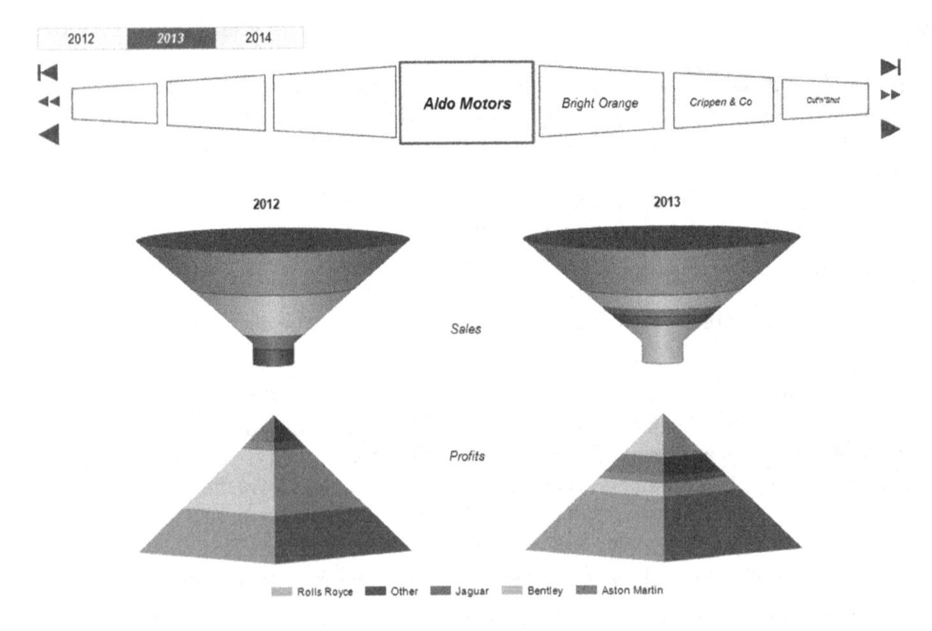

Figure 9-5. *A report using a carousel to select data*

As an added extra, this particular report compares sales and profits for the selected year with those for the previous year for the chosen reseller. This report extends the charting techniques that you saw previously in this book. Specifically, it shows how to create multiple chart areas as sub-charts inside a single chart.

The Source Data

This report is definitely quite complex. It requires the following four datasets to work properly:

- `pr_TabletCarouselCarousel`: This stored procedure fills the carousel with the list of clients (or clients, or resellers if you prefer).
- `pr_TabletCarouselMaxElements`: This stored procedure calculates the maximum number of records returned to the carousel. This number is used to prevent the carousel scrolling into infinity (or beyond).
- `pr_TabletCarouselClient`: This stored procedure returns the name of the client that is at the center of the carousel. This is then used as a parameter to select the data for the chart.
- `pr_TabletCarouselChart`: This stored procedure returns the data required by the chart.

The four code snippets are

```
DECLARE @ReportingYear INT = 2013
DECLARE @BatchRow INT = 5
DECLARE @ClientName VARHAR(100) = 'Aldo Motors'

-- Code.pr_TabletCarouselCarousel

IF OBJECT_ID('tempdb..#Tmp_Dealers') IS NOT NULL DROP TABLE tempdb..#Tmp_Dealers

CREATE TABLE #Tmp_Dealers (ID INT IDENTITY(1,1), ClientName NVARCHAR(100) COLLATE
DATABASE_DEFAULT)

INSERT INTO #Tmp_Dealers (ClientName)
VALUES (NULL), (NULL), (NULL)

INSERT INTO #Tmp_Dealers (ClientName)

SELECT DISTINCT
ClientName
FROM    Reports.CarSalesData
WHERE   CountryName = 'United Kingdom'
        AND YEAR(InvoiceDate) = @ReportingYear
ORDER BY ClientName

IF OBJECT_ID('tempdb..#Tmp_Output') IS NOT NULL DROP TABLE tempdb..#Tmp_Output
```

```
CREATE TABLE #Tmp_Output
(
 FirstElement VARCHAR(100) COLLATE DATABASE_DEFAULT
,SecondElement VARCHAR(100) COLLATE DATABASE_DEFAULT
,ThirdElement VARCHAR(100) COLLATE DATABASE_DEFAULT
,FourthElement VARCHAR(100) COLLATE DATABASE_DEFAULT
,FifthElement VARCHAR(100) COLLATE DATABASE_DEFAULT
,SixthElement VARCHAR(100) COLLATE DATABASE_DEFAULT
,SeventhElement VARCHAR(100) COLLATE DATABASE_DEFAULT
)

INSERT INTO #Tmp_Output (FirstElement,SecondElement ,ThirdElement ,FourthElement
,FifthElement, SixthElement, SeventhElement)

SELECT [1] AS FirstElement, [2] AS SecondElement, [3] AS ThirdElement, [4] AS FourthElement,
[5] AS FifthElement, [6] AS SixthElement, [7] AS SeventhElement
FROM
    (
     SELECT TOP (7) ROW_NUMBER() OVER (ORDER BY ID) AS SEQ, ClientName FROM #Tmp_Dealers
WHERE ID >= @BatchRow
        ) PT
PIVOT (MAX(ClientName) FOR SEQ IN ([1],[2],[3],[4],[5],[6],[7])) AS PT

SELECT * FROM #Tmp_Output

-- Code.pr_TabletCarouselMaxElements

SELECT COUNT(DISTINCT ClientName) AS MaxValue
FROM   Reports.CarSalesData
WHERE  CountryName = 'United Kingdom'
       AND YEAR(InvoiceDate) = @ReportingYear

-- Code.pr_TabletCarouselClient

SELECT
ClientName
FROM
(
SELECT DISTINCT
ROW_NUMBER() OVER (ORDER BY ClientName) AS RowNo
,ClientName
FROM   Reports.CarSalesData
WHERE  CountryName = 'United Kingdom'
       AND YEAR(InvoiceDate) = @ReportingYear
GROUP BY ClientName
) A
WHERE RowNo = @BatchRow

-- Code.pr_TabletCarouselChart
```

```
IF OBJECT_ID('tempdb..#Tmp_Output') IS NOT NULL DROP TABLE tempdb..#Tmp_Output

CREATE TABLE #Tmp_Output (Make NVARCHAR(50) COLLATE DATABASE_DEFAULT, CurrentYearSales
NUMERIC(18,2), PreviousYearSales NUMERIC(18,2), CurrentYearProfits NUMERIC(18,2),
PreviousYearProfits NUMERIC(18,2))

INSERT INTO #Tmp_Output (Make)

SELECT
CASE
WHEN Make IN ('Aston Martin','Rolls Royce','Bentley','Jaguar') THEN Make
WHEN Make IN ('Triumph','MG') THEN 'Classic Brit'
ELSE 'Other'
END
FROM        Reports.CarSalesData
WHERE       ReportingYear IN (@ReportingYear, @ReportingYear - 1)
            AND ClientName = @ClientName
GROUP BY
CASE
WHEN Make IN ('Aston Martin','Rolls Royce','Bentley','Jaguar') THEN Make
WHEN Make IN ('Triumph','MG') THEN 'Classic Brit'
ELSE 'Other'
END

-- Current Year Sales
;
WITH CurrSales_CTE
AS
(
SELECT
CASE
WHEN Make IN ('Aston Martin','Rolls Royce','Bentley','Jaguar') THEN Make
WHEN Make IN ('Triumph','MG') THEN 'Classic Brit'
ELSE 'Other'
END AS Make
,SUM(SalePrice) AS SalePrice
FROM        Reports.CarSalesData
WHERE       ReportingYear = @ReportingYear
            AND ClientName = @ClientName
GROUP BY
CASE
WHEN Make IN ('Aston Martin','Rolls Royce','Bentley','Jaguar') THEN Make
WHEN Make IN ('Triumph','MG') THEN 'Classic Brit'
ELSE 'Other'
END
)

UPDATE      Tmp

SET         Tmp.CurrentYearSales = CTE.SalePrice
```

```
FROM        #Tmp_Output Tmp
            INNER JOIN CurrSales_CTE CTE
                   ON Tmp.Make = CTE.Make

-- Previous Year Sales
;
WITH PrevSales_CTE
AS
(
SELECT
CASE
WHEN Make IN ('Aston Martin','Rolls Royce','Bentley','Jaguar') THEN Make
WHEN Make IN ('Triumph','MG') THEN 'Classic Brit'
ELSE 'Other'
END AS Make
,SUM(SalePrice) AS SalePrice
FROM       Reports.CarSalesData
WHERE      ReportingYear = @ReportingYear - 1
           AND ClientName = @ClientName
GROUP BY
CASE
WHEN Make IN ('Aston Martin','Rolls Royce','Bentley','Jaguar') THEN Make
WHEN Make IN ('Triumph','MG') THEN 'Classic Brit'
ELSE 'Other'
END
)

UPDATE      Tmp

SET         Tmp.PreviousYearSales = CTE.SalePrice

FROM        #Tmp_Output Tmp
            INNER JOIN PrevSales_CTE CTE
            ON Tmp.Make = CTE.Make

-- Current Year Profit
;
WITH CurrProfit_CTE
AS
(
SELECT
CASE
WHEN Make IN ('Aston Martin','Rolls Royce','Bentley','Jaguar') THEN Make
WHEN Make IN ('Triumph','MG') THEN 'Classic Brit'
ELSE 'Other'
END AS Make
,SUM(SalePrice) - SUM(CostPrice) - SUM(TotalDiscount) - SUM(DeliveryCharge) -
SUM(SpareParts) - SUM(LaborCost) AS Profit
FROM       Reports.CarSalesData
WHERE      ReportingYear = @ReportingYear
           AND ClientName = @ClientName
```

```
GROUP BY
CASE
WHEN Make IN ('Aston Martin','Rolls Royce','Bentley','Jaguar') THEN Make
WHEN Make IN ('Triumph','MG') THEN 'Classic Brit'
ELSE 'Other'
END
)

UPDATE      Tmp
SET         Tmp.CurrentYearProfits = CTE.Profit
FROM        #Tmp_Output Tmp
            INNER JOIN CurrProfit_CTE CTE
            ON Tmp.Make = CTE.Make

-- Previous Year Profit
;
WITH PrevProfit_CTE
AS
(
SELECT
CASE
WHEN Make IN ('Aston Martin','Rolls Royce','Bentley','Jaguar') THEN Make
WHEN Make IN ('Triumph','MG') THEN 'Classic Brit'
ELSE 'Other'
END AS Make
,SUM(SalePrice) - SUM(CostPrice) - SUM(TotalDiscount) - SUM(DeliveryCharge) -
SUM(SpareParts) - SUM(LaborCost) AS Profit
FROM        Reports.CarSalesData
WHERE       ReportingYear = @ReportingYear - 1
            AND ClientName = @ClientName
GROUP BY
CASE
WHEN Make IN ('Aston Martin','Rolls Royce','Bentley','Jaguar') THEN Make
WHEN Make IN ('Triumph','MG') THEN 'Classic Brit'
ELSE 'Other'
END
)

UPDATE      Tmp
SET         Tmp.PreviousYearProfits = CTE.Profit
FROM        #Tmp_Output Tmp
            INNER JOIN PrevProfit_CTE CTE
            ON Tmp.Make = CTE.Make

UPDATE      #Tmp_Output SET CurrentYearSales = 0 WHERE CurrentYearSales IS NULL
UPDATE      #Tmp_Output SET PreviousYearSales = 0 WHERE PreviousYearSales IS NULL
UPDATE      #Tmp_Output SET CurrentYearProfits = 0 WHERE CurrentYearProfits IS NULL
UPDATE      #Tmp_Output SET PreviousYearProfits = 0 WHERE PreviousYearProfits IS NULL

SELECT * FROM #Tmp_Output
```

The output from these four stored procedures is shown in Figure 9-6.

Code.pr_TabletCarouselCarousel

	FirstElement	SecondElement	ThirdElement	FourthElement	FifthElement	SixthElement	SeventhElement
1	Bright Orange	Crippen & Co	Cut'n'Shut	Honest John	Impressive Wheels	Luxury Rentals	Olde Englande

Code.pr_TabletCarouselChart

	Make	CurrentYearSales	PreviousYearSales	CurrentYearProfits	PreviousYearProfits
1	Aston Martin	288500.00	120000.00	159910.00	42350.00
2	Bentley	85250.00	264000.00	-11287.00	99986.00
3	Classic Brit	22500.00	NULL	13075.00	NULL
4	Jaguar	164250.00	39500.00	7038.00	-1436.00
5	Other	45000.00	NULL	26150.00	NULL
6	Rolls Royce	308500.00	89000.00	164989.00	-635.00

Code.pr_TabletCarouselMaxElements

	MaxValue
1	11

Code.pr_TabletCarouselClient

	ClientName
1	Honest John

Figure 9-6. *The output from the four stored procedures creating the carousel and its chart*

How the Code Works

The initial, and main, procedure underlying this visualization is `Code.pr_TabletCarouselCarousel`. This piece of code starts by creating a temporary table (`#Tmp_Dealers`) into which it places all the clients that have recorded sales for the chosen year for the UK. The choice of the UK is currently hard-coded. This temporary table, however, has three empty rows added before the data. This is because the carousel essentially starts with the fourth element in the list, the central box in the visualization. Then a second temporary table, `#Tmp_Output`, is filled with seven pivoted records (beginning with the one that has to be the leftmost box in the carousel). The starting record is a variable returned from SSRS.

The other three procedures are much simpler. `pr_TabletCarouselMaxElements` echoes a part of the previous code to count the total possible number of records that can appear in a carousel for a specific year. This figure will be used by the interface to prevent empty recordsets being displayed. `Code.pr_TabletCarouselClient` takes the variable returned from Reporting Services that indicates where the selection of records begins and returns the corresponding client name. Finally, `Code.pr_TabletCarouselChart` takes the client name and calculates the data required to display the four charts. This is little different from many of the code snippets that you have already seen that aggregate sales or profit figures for a specified year and client. The only trick here is that, as you need to compare clients over two years, the client list is generated for either of the two years (in case a client appears in one but not the other) and placed in a temporary table first. Subsequently, any required metrics are calculated as CTEs and the temporary table is updated. Finally, any NULL values are set to zero to ensure that the shared legend (read from the chart of current year sales) will contain *all* the clients because any NULL clients would not otherwise be included by SSRS.

Building the Report

Assuming that these stored procedures are clear, you can proceed to build the visualization. I have to warn you that this particular report is quite complex and can take a while to produce. Not only is the carousel complex to create, but the chart (which is in fact a single chart using four separate chart areas) requires a certain attention to detail. However, I feel that the result is worth the effort, and I hope that you do, too.

1. Follow steps 1 through 4 of the preceding example. Name the copy of the file `Tablet_Carousel.rdl`. This way you will have a base report containing the year selector.
2. Add the following datasets (this will add all the necessary parameters automatically):
   a. `Carousel`, using the stored procedure Code.pr_TabletCarouselCarousel
   b. `Maximumelements`, using the stored procedure `Code.pr_TabletCarouselMaxElements`
   c. `TabletCarouselChart`, using the stored procedure `Code.pr_TabletCarouselChart`
   d. `TabletCarouselClient`, using the stored procedure `Code.pr_TabletCarouselClient`
3. Add the following images from the folder `C:\BIWithSSRS\Images`:
   a. `CarouselMain.png`: This is the central box in the carousel.
   b. `CarouselLeft3.png`: This is the leftmost box in the carousel.
   c. `CarouselLeft2.png`: This is the second box from the left in the carousel.
   d. `CarouselLeft1.png`: This is the third box from the left in the carousel.
   e. `CarouselRight1.png`: This is the first box to the right of the central box in the carousel.
   f. `CarouselRight2.png`: This is the second box to the right of the central box in the carousel.
   g. `CarouselRight3.png`: This is the rightmost box in the carousel.
   h. `LeftOne.png`: This is the scroll button to slide the contents of the carousel one box to the left.
   i. `LeftAll.png`: This is the scroll button that jumps to the start of the list in the carousel.
   j. `LeftSet.png`: This is the scroll button that scrolls seven element leftwards (in effect filling the carousel with a complete set of elements).
   k. `RightOne.png`: This is the scroll button to slide the contents of the carousel one box to the right.
   l. `RightAll.png`: This is the scroll button that jumps to the end of the list in the carousel.
   m. `RightSet.png`: This is the scroll button that scrolls seven element rightwards (in effect filling the carousel with a complete new set of elements)

4. Add seven text boxes to the report. Make them all 0.9 inches high and approximately 1.2 inches wide (you will be resizing them later). Place them in a row under the year selector.

5. Name the text boxes TxtFirstElement, TxtSecondElement, etc. up to TxtSeventhElement (from left to right).

6. Add the images `CarouselLeft3`, `CarouselLeft2`, `CarouselLeft1`, `CarouselMain`, `CarouselRight1`, `CarouselRight2`, and `CarouselRight3` (from left to right) as the background images to the text boxes. Set the `BackgroundRepeat` property for each text box to Clip.

7. Adjust the width of each text box individually so that the width matches the image. You will find that the further a text box is from the center, the less wide it is. Ensure that the (tiny) space between each text box is identical. You may want to select the text boxes and use Format ➤ Horizontal Spacing ➤ Make Equal to get them perfectly positioned. Tweak the height of the text boxes if this is necessary to display the full image. Finally, align all the text boxes vertically (Format ➤ Align ➤ Tops). Remember that this is purely to obtain a visual effect.

8. Set the fonts for the text boxes as follows:

    a. TxtFirstElement: Arial 6 point italic

    b. TxtSecondElement: Arial 8 point italic

    c. TxtThirdElement: Arial 10 point italic

    d. TxtFourthElement: Arial 11 point bold

    e. TxtFifthElement: Arial 10 point italic

    f. TxtSixthElement: Arial 8 point italic

    g. TxtSeventhElement: Arial 6 point italic

9. Add six images to the report; anywhere under the carousel will do for the moment. Make them all 0.25 inches square. Add the six images `LeftOne`, `LeftSet`, `LeftAll`, `RightOne`, `RightSet`, and `RightAll`. Set the images' `Sizing` property to FitProportional.

10. Place the images to the left and right of the carousel as shown in Figure 9-5.

11. Set the expression for the leftmost text box (TxtFirstElement) to the following: `=First(Fields!FirstElement.Value, "Carousel")`

12. Repeat this operation for all the remaining text boxes. Modify the expression so that the second text box (TxtSecondElement) refers to the field SecondElement, and so on for all the text boxes.

13. Right-click the leftmost text box (TxtFirstElement) and select Text Box Properties from the context menu. Click Action on the left and set the following:

    a. Action: Go to report.

    b. Report: `=Globals!ReportName`.

14. Add two parameters:

    a. Name: ReportingYear - Value: `=Parameters!ReportingYear.Value`.

    b. Name: BatchRow - Value:

```
=IIF(Parameters!BatchRow.Value <= Sum(Fields!MaxValue.Value,
"MaximumElements") AND Not IsNothing(First(Fields!FirstElement.Value,
"Carousel"))
,Parameters!BatchRow.Value - 3
,Parameters!BatchRow.Value
)
```

15. Click OK to confirm your modifications.

16. Do exactly the same for the following text boxes; the expression for the BatchRow parameter will vary only slightly as shown below:

Text Box	Expression

TxtSecondElement	`=IIF(Parameters!BatchRow.Value <= Sum(Fields!MaxValue.Value, "MaximumElements") AND Not IsNothing(First(Fields!SecondElement.Value, "Carousel"))`
	`,Parameters!BatchRow.Value - 2`
	`,Parameters!BatchRow.Value`
	`)`

TxtThirdElement	`=IIF(Parameters!BatchRow.Value <= Sum(Fields!MaxValue.Value, "MaximumElements") AND Not IsNothing(First(Fields!ThirdElement.Value, "Carousel"))`
	`,Parameters!BatchRow.Value - 1`
	`,Parameters!BatchRow.Value`
	`)`

TxtFifthElement	`=IIF(Sum(Fields!MaxValue.Value, "MaximumElements") <= Parameters!BatchRow.Value`
	`,Parameters!BatchRow.Value`
	`,Parameters!BatchRow.Value + 1`
	`)`

TxtSixthElement	`=IIF(Sum(Fields!MaxValue.Value, "MaximumElements") <= Parameters!BatchRow.Value`
	`,Parameters!BatchRow.Value`
	`,Parameters!BatchRow.Value + 2`
	`)`

TxtSeventhElement	`=IIF(Sum(Fields!MaxValue.Value, "MaximumElements") <= Parameters!BatchRow.Value`
	`,Parameters!BatchRow.Value`
	`,Parameters!BatchRow.Value + 3`
	`)`

17. Set the Action properties for the six images to the left and right of the carousel exactly as you did for the leftmost text box (TxtFirstElement) in steps 13 and 14. The expression for the BatchRow parameter needs to be set as shown:

Image	Expression
LeftOne	=IIF(Parameters!BatchRow.Value -1 < 1, 1, Parameters!BatchRow.Value - 1)
LeftSet	=IIF(Parameters!BatchRow.Value -7 < 1, 1, Parameters!BatchRow.Value - 7)
LeftAll	=1
RightOne	=IIF(Sum(Fields!MaxValue.Value, "MaximumElements") <= Parameters!BatchRow.Value ,Parameters!BatchRow.Value ,Parameters!BatchRow.Value + 1)
RightSet	=IIF(Parameters!BatchRow.Value + 7 <= Sum(Fields!MaxValue.Value, "MaximumElements") ,Parameters!BatchRow.Value +7 ,Parameters!BatchRow.Value + (Sum(Fields!MaxValue.Value, "MaximumElements") - Parameters!BatchRow.Value))
RightAll	=Sum(Fields!MaxValue.Value, "MaximumElements")

18. Now for the chart. First, you need to add a chart to the report area under the carousel. Make it as wide as the carousel and around seven times taller. Make it a 3-D funnel chart and set it to use the dataset TabletCarouselChart.

19. Click inside the chart to display the Chart Data pane and add the following ∑ values: CurrentYearDales, PreviousYearSales, CurrentYearProfits, and PreviousYearProfits. Set the Category Groups to Make.

20. Right-click the funnel chart and select Series Properties. Choose Sum(PreviousYearSales) as the Value field. Select Make as the Category field in the Series Data pane. Confirm your modifications with OK.

21. Select the chart area for the chart by clicking close to the chart funnel but outside it. Rename the chart area to PrevYearSales.

22. Right-click inside the chart, but outside the existing chart area and legend. Select Add New Chart Area. Don't worry about the unprepossessing gray box that appears, and rename the new chart area PrevYearProfits.

23. In the Chart Data window, click the pop-up menu triangle to the right of the PreviousYearProfits series and select Series Properties from the context menu. Set the Category field to Make. Click Axes and Chart Area on the left, and select PrevYearProfits as the chart area to use. A funnel chart will now be displayed.

24. Repeat steps 22 and 23 twice to add two more chart areas. Name the one on the top right CurrYearSales, and the one on the bottom right CurrYearProfits.

25. In the Chart Data window, click the pop-up menu triangle to the right of the CurrentYearProfits series and select Series Properties from the context menu. Set the `Category` field to Make. Click on Axes and Chart Area on the left, and select CurrYearSales as the chart area to use. A funnel chart will now be displayed.

26. In the Chart Data window, click the pop-up menu triangle to the right of the CurrentYearProfits series and select Series Properties from the context menu. Set the `Category` field to Make. Click on Axes and Chart Area on the left, and select CurrYearProfits as the chart area to use. A funnel chart will now be displayed.

27. Right-click the legend and select Legend Properties from the context menu. Set the legend position to bottom center, and select wide table from the Layout pop-up menu. Confirm your changes.

28. Right-click the top left funnel and select Series Properties from the context menu. Click Legend on the left, and check the Do not show this series in a legend check box.

29. Do the same for the two bottom charts as you did in the previous step. The legend will now only display one make of car for the four charts.

30. Right-click the bottom left chart and select Change Chart Type from the context menu. Choose 3-D pyramid as the type of chart.

31. Do exactly the same for the bottom right chart as you did in the previous step.

32. In the Properties window, set the `CustomPosition` property as follows for the four charts:

    a. Enabled: True

    b. Height: 90

    c. Left: 10

    d. Top: 10

    e. Width: 80

33. Click the chart title, and in the Properties window, set the following properties:

Section	Property	Value
Caption	(Expression)	`=Cstr(Parameters!ReportingYear.Value - 1)`
CustomPosition	Enabled	True
	Height	5
	Left	10
	Top	0
	Width	30

34. Click inside the chart itself, and outside any of the chart areas, and select Add New Title from the context menu. Click the new chart title, and in the Properties Window, set the following properties:

Section	Property	Value
Caption	(Expression)	`=Parameters!ReportingYear.Value`
CustomPosition	Enabled	True
	Height	5
	Left	60
	Top	0
	Width	30

35. Click inside the chart itself, and outside any of the chart areas, and select Add New Title (again) from the context menu. Click the new chart title, and in the Properties window, set the following properties:

Section	Property	Value
Caption		Sales
CustomPosition	Enabled	True
	Height	5
	Left	0
	Top	30
	Width	100

36. Add a final new title, as you did in the previous step, and set the same properties, except that the Caption should be "Profits" and the CustomPosition Top property must be 60.

37. Click inside the chart itself, and outside any of the chart areas, and in the Properties Window, set the BorderStyle to None.

Well, that took a little while, but the effect will hopefully please your users. They can now flip from client to client singly, in sets, or directly leaping from start to end of the list and see the comparative sales and profits for the selected year and the previous year presented in an original way.

How It Works

This visualization is essentially in two parts:

- The *carousel* that lets you scroll through the list of clients
- The *chart* that displays metrics for the client in the center of the carousel

The Carousel

The carousel itself is a list of clients. Three principal tricks have been applied:

- The carousel is set to display the first client when first run.
- The recordset begins with three empty records. This means that the fourth record is the one that starts in the central box of the carousel. Consequently, this sleight of hand must be taken into account when scrolling through the carousel.
- The carousel is a pivoted dataset based on the initial selection of clients.

Virtually everything else is a question of shaping the appearance of the data in the carousel, hence the perspective effect produced by the background images and the text sizes that decrease the further the element is from the center of the carousel. Once the appearance has been defined, everything else is a question of handling the way the data in the carousel can be scrolled. In effect, it all boils down to setting the record that begins the set of seven records that is pivoted and displayed.

As you can see from the various expressions used to scroll through the data set, there are a few checks that have been added to the expressions. These ensure that scrolling cannot go beyond the limits of the available data.

The Chart

The chart is, perhaps surprisingly, a single chart. This is something that you have not seen in previous charts. While it can seem a little complex, it has the advantage of being easy to resize proportionally-something that is much harder when charts are separate, or contained in tables. It is also much easier to apply a single legend to chart collections like this.

When building a chart like this, you need to ensure that each chart has a data series that can be used by the individual chart element, as is the case with the dataset `TabletCarouselChart`. Then the trick is to add a separate chart area for each chart, and apply the required series to the chart area. Once this is done, everything else is position and aesthetics. There are a few points to note, however:

- When you set the custom position for titles, this is relative to the entire chart.
- When you set the inner plot position for the charts, this is relative to the individual chart area.

■ **Note** You can also specify the exact position of each chart area by setting the Custom Position for the chart area. However, you will have to do this for all the chart areas if you do it for one of them.

The buttons that have been added to the carousel try to cover all the ways that a user might want to scroll through a recordset: singly, in groups, or directly to the start or end. Although this may be overkill, at least you have examples of paging and scrolling code that you can reuse, totally or partially, in your own reports.

Of course, the chart that you added here is not the only visualization that you can use with a carousel-like selector. You can use tables, gauges, other chart types-indeed anything that you want. What matters is to give your users a more intuitive and visually appealing way of selecting a parameter in Reporting Services.

Adding Paged Recordsets

SSRS lets you flip forwards and backwards through larger datasets "out of the box." However, you need to display the toolbar for this, and the visual appeal is limited, to say the least. It is relatively easy in SSRS to enhance a table so that you can see instantly where you are in the recordset, and move through the recordset in a much more intuitive way. To show a way of doing this, let's produce a paged report that looks like the one in Figure 9-7.

2012 2013 2014 **United Kingdom**

Jan Feb Mar Apr May Jun Jul Aug Sep Oct Nov Dec

Make	Model	Color	Invoice Date	Sale Price
Aston Martin	DBS	Blue	01/01/2014	122,750.00
Jaguar	XJ6	Green	01/01/2014	42,250.00
Bentley	Continental	Blue	01/02/2014	42,250.00
Rolls Royce	Silver Ghost	Green	02/02/2014	90,750.00
Bentley	Arnage	Canary Yellow	01/03/2014	46,750.00
Rolls Royce	Silver Ghost	Blue	03/03/2014	91,750.00
Bentley	Azure	Red	01/04/2014	46,750.00

Rows 1 to 7 (of 27)

Figure 9-7. *A report with paged recordsets*

In a break with tradition, I think that it will also help to see the design view for this report. Consequently, you can see this in Figure 9-8.

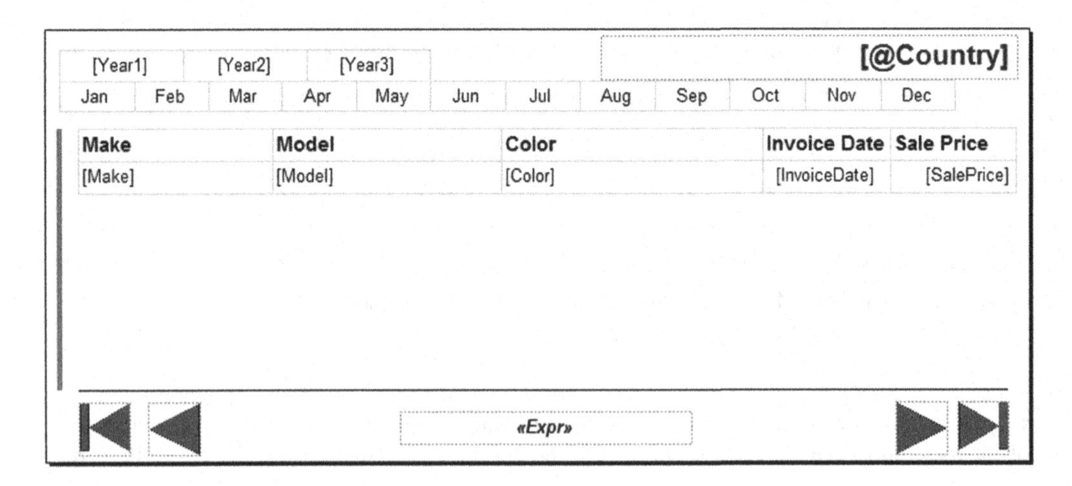

Figure 9-8. *The design view of a report with paged recordsets*

The Source Data

The source data for this output is a simple set of records beginning at a defined starting point in the record set. There is also a separate procedure to calculate the maximum number of records. This is used to prevent the interface spiraling off into infinity when paging the data.

```
-- Code.pr_BatchedSet

DECLARE @ReportingYear INT = 2014
DECLARE @ReportingMonth INT = 12
DECLARE @Country VARCHAR(50) = 'United Kingdom'
DECLARE @BatchStart INT = 1
DECLARE @SetNumber INT = 7

SELECT TOP (@SetNumber) Make, Model, InvoiceDate, Color, SalePrice
FROM
  (
   SELECT
              ROW_NUMBER() OVER (ORDER BY InvoiceDate) AS ID
             ,Make
             ,Model
             ,InvoiceDate
             ,Color
             ,SalePrice

   FROM      Reports.CarSalesData
   WHERE     ReportingYear = @ReportingYear
             AND ReportingMonth <= @ReportingMonth
             AND CountryName = @Country
) S
WHERE     ID > @BatchStart
ORDER BY  InvoiceDate

-- Code.pr_BatchLimit

SELECT COUNT(*) AS BatchMax
FROM

     (
            SELECT   Make ,Model, InvoiceDate, Color, SalePrice
            FROM     Reports.CarSalesData
            WHERE    ReportingYear = @ReportingYear
                     AND ReportingMonth <= @ReportingMonth
                     AND CountryName = @Country
      ) S
```

The output from these two procedures is shown in Figure 9-9.

Code.pr_BatchLimit

	Make	Model	InvoiceDate	Color	SalePrice
1	Triumph	TR4	2013-01-02 00:00:00.000	Black	22500.00
2	Triumph	TR5	2013-02-02 00:00:00.000	Dark Purple	25250.00
3	MGB	GT	2013-02-02 00:00:00.000	Green	39500.00
4	Jaguar	XJ6	2013-02-02 00:00:00.000	British Racing Green	39500.00
5	Bentley	Continental	2013-02-02 00:00:00.000	Green	110000.00
6	Jaguar	XK	2013-02-02 00:00:00.000	Dark Purple	44000.00
7	Aston Martin	DB4	2013-02-02 00:00:00.000	British Racing Green	110000.00
8	Rolls Royce	Wraith	2013-02-02 00:00:00.000	Silver	178500.00
9	Rolls Royce	Silver Ghost	2013-03-02 00:00:00.000	Green	110000.00
10	Aston Martin	DB9	2013-03-02 00:00:00.000	Black	178500.00

Code.pr_BatchLimit

	BatchMax
1	27

Figure 9-9. *The output for the two stored procedures in a paged recordset*

How the Code Works

The main procedure, `Code.pr_BatchedSet`, lists the sales according to the parameters passed into the procedure and a hard-coded country parameter. Then the output from this inner query is subset by selecting a specified number of records beginning at a defined record. Both the number of records to return and the starting record are passed in as parameters to make the code more reusable.

Building the Report

With the stored procedures in place, you can move on to creating this report.

1. Make a copy of the `.rdl` file named `__DateSelector`. Name the copy `Tablet_BatchedRecordset.rdl`, and open the copy.
2. Add the following datasets:
    a. `BatchSet`, using the stored procedure `Code.pr_BatchedSet`
    b. `BatchLimit`, using the stored procedure `Code.pr_BatchLimit`

3. Ensure that you have the following parameters; add them if not. Set them all to be hidden, and set their properties as follows:

Parameter	Property	Value
BatchStart	Data Type	Integer
	Allow null value	True
	Default Values	Specify values: 0
SetNumber	Data Type	Integer
	Default Values	Specify values: 7
Country	Data Type	Text
	Allow blank value	Checked
	Allow null value	Checked
	Default Values	Specify values: United Kingdom

4. Add the following images from the folder `C:\BIWithSSRS\Images`:
    a. `LeftAll.png`: This is the button to scroll back to the start of the record set.
    b. `RightAll.png`: This is the button to scroll forward to the end of the record set.
    c. `LeftOne.png`: This is the button to scroll forward to the next set of records.
    d. `RightOne.png`: This is the button to scroll backward to the previous set of records.
5. Add a table to the report. Apply the dataset `BatchSet`.
6. Add two more columns to the report. Add the fields `Make`, `Model`, `Color`, `InvoiceData`, and `SalePrice` to the detail row, in this order.
7. Add a vertical line (that you set to be 2.8 inches high) to the left of the table.
8. Set the line's `Hidden` property to True.
9. Align the tops of the line and the table by Ctrl-clicking both of them and selecting Format ➤ Align ➤ Tops.
10. Add a horizontal line just under the bottom of the vertical line. Make it as wide as the table.
11. Add four images to the report. Make each one 0.4 inches square (approximately). Add the four embedded images that you imported into the report in step 3, one to each of the image elements. Set their `Sizing` property to FitProportional.
12. Place the images as shown in Figure 9-8. Tweak the size of the images if you need to display the contents.

13. Add a text box at the bottom center of the report. Set the font to Arial Bold 10 point italic. Set the expression for this text box to the following:

```
="Rows " & Parameters!BatchStart.Value + 1 & " to " &
Iif(Sum(Fields!BatchMax.Value, "BatchLimit") <=
(Parameters!BatchStart.Value + Parameters!SetNumber.Value)
, Sum(Fields!BatchMax.Value, "BatchLimit")
, (Parameters!BatchStart.Value + Parameters!SetNumber.Value)) &
" (of " & Sum(Fields!BatchMax.Value, "BatchLimit") & ")"
```

14. Add a text box at the top right of the report. Set the font to Arial Bold 16 point blue. Set the expression for this text box to the following:

```
=Parameters!Country.Value
```

15. Set the Action property for each of the image buttons at the bottom of the report to the following:

Option	Value
Enable as an action	Go to report
Specify a report	`[&ReportName]`
Parameters	`Name: ReportingYear - Value: [@ReportingYear]`
	`Name: ReportingMonth - Value: [@ReportingMonth]`
	`Name: Country - Value: [@Country]`

16. For each of the image buttons, add the parameter `BatchStart` in the `Action` property with the following values for the given buttons:

Button Image	Value
LeftAll	0
RightAll	`=Cint(Sum(Fields!BatchMax.Value, "BatchLimit")) - Parameters!SetNumber.Value`
RightOne	`=IIF(` `(Parameters!BatchStart.Value + Parameters!SetNumber.Value) < Sum(Fields!BatchMax.Value, "BatchLimit")` `,Parameters!BatchStart.Value + Parameters!SetNumber.Value` `,Parameters!BatchStart.Value` `)`
LeftOne	`=IIF(Parameters!BatchStart.Value - Parameters!SetNumber.Value < 0, 0, Parameters!BatchStart.Value - Parameters!SetNumber.Value)`

That is all that you have to do. You can now visualize the report and click the scroll buttons to page through the report. As an added bonus, you can see the value of the `Country` parameter on the top right, as well as a record counter at the bottom of the report.

How It Works

As I mentioned when explaining the code, this is simply a way of paging data in a report. The heavy lifting is done by the stored procedure that returns a set of records for display. The interface merely handles how the necessary parameters are passed back to the stored procedure.

Two parameters are hard-coded with initial values in the report. These are the `Country` parameter (that selects the country for the sales) and the `SetNumber` parameter that sets the number of records displayed. Either or both of these could be made interactive using, say, a pop-up menu, as you saw in the last chapter.

Specifically, the interface maintains the state of the `BatchStart` parameter that sets the initial record in the set that is returned to the report. All this is done using the `Action` property of the image buttons, which also contain some checks and balances to prevent empty or invalid recordsets being displayed. These buttons work like this (I am using the image names to indicate which button is described):

- The *PrevList* button checks that the `BatchStart` parameter will not descend below zero if it is decremented, and if not, it decreases the `BatchStart` parameter by the value of the `SetNumber` parameter.
- The *NextList* button ensures that incrementing the `BatchStart` parameter will not exceed the number of records available, and if not, it adds the value of the `SetNumber` parameter to the `BatchStart` parameter. If the maximum value would be exceeded, then the value remains unchanged.
- The *StartList* button is the easiest of the lot; it resets the `BatchStart` parameter to zero.
- The *EndList* button calculates the value of the maximum available records minus the `SetNumber` parameter and uses this to display the final set of records.

So this is uniquely an interface trick, but one that users will probably prefer to paging through recordsets using the SSRS toolbar, especially on tablet devices.

The hidden line exists solely to ensure that the "footer" elements of the table remain in the same place whatever the number of records in the table. Without this, the buttons can move up if the final set of records contains fewer than seven records.

Conclusion

This chapter showed a few of the many ways that you can enhance the SSRS interface and give your users a more intuitive and friendly way of interacting with your reports. You saw how to page through data sets in several ways, from simple tables via tiles to carousels. All this is done using Reporting Services parameters and the `Action` properties of images and textboxes.

These examples are far from the limits of what is possible when you apply a little ingenuity and invention to the SSRS canvas. I hope that they will not only prove useful as they are, but will also inspire you to push the SSRS interface to new heights as you develop even more interesting and-dare I say-cooler reports.

CHAPTER 10

BI for SSRS on Tablets and Smartphones

"Give me Mobile BI" is the cry that has been coming from the executive suite for some time now. Admittedly, until SQL Server 2012 SP1 was released, all a Microsoft Reporting Services specialist could do in answer to this request was to shuffle their feet and look sheepish while they tried to implement a third-party add-on.

Now, however, the landscape has changed. Thanks to SQL Server 2012 SP1 (and naturally SQL Server 2014) you can output reports to a host of mobile devices, including iPads and iPhones as well as Android and Windows phones and tablets. Moreover, nothing has fundamentally changed as far as SSRS is concerned. You still develop reports as you did before. All you have to do is ensure that your reports are designed for the output device you will be using. In the case of both tablets and smartphones, this means the following:

- Design your output as a function of the size of the phone or tablet's screen.
- Take account of the height-to-width aspect ratio of the output device's screen.
- Tweak your reports to be used with the device held either vertically or horizontally for optimum viewing.
- Do not force the same report to appear on a multitude of output devices. Be prepared to start by building "widgets" that display the data, and then reuse the widgets in possibly several different reports, where each report is tailored to the size and aspect (height to width) ratio of each device.
- Attempt to use shared datasets so that your initial effort can be reused more easily.

So you are likely to be looking at a minimum of redesigning, and possibly rewriting, a good few reports. However, this could be the case with any suite of reports that have to be reworked for mobile output, whatever the tool used to create them. When all is said and done, the constraints come from the output device. Good mobile reports are the ones that have been designed with the specific mobile device in mind.

In this chapter, you will build on some of the visualizations that you saw in previous chapters. More specifically, you will look at further techniques that you can use to build gauges. This will extend the knowledge that you acquired in Chapter 3.

Designing Mobile Reports

There is definitely an art to designing tablet and smartphone reports. In any case, there is no one way of doing things, and most people will disagree on the best approach anyway. However, there are probably a few core guidelines that you need to take into account when designing reports for tablets and smartphones. These include the following:

- *Don't overload the screen.* It is too easy to think "the user only has to zoom in." The result is that you create a report that is hard to read. The user will probably want to view their data immediately and clearly, without having to use pinch movements to read the data, be it a chart, gauge, or table.
- *Hide the Report Toolbar and mask parameters using a custom interface.* Users of mobile devices are accustomed to slick apps with state of the art interfaces. While you can never get to the highest levels of swish user interfaces using Reporting Services, you can at least make the interaction smoother.
- *Develop a clear interface hierarchy.* You will inevitably need a couple of reports on a tablet, and possibly several on a smartphone, to do the work of one report designed for a large laptop screen. So accept this, and be prepared to break up existing reports into separate reports, and to drill through from report to report.
- *Learn about firewalls.* You will have to take corporate firewalls into account when preparing to deploy mobile BI using SSRS. So it is a worthwhile investment to make friends with the IT people who deal with this, or learn about it if the buck stops with you.

Delivering Mobile Reports

At the start of this chapter I mentioned that mobile BI has only become practical since SQL Server 2012 SP1. The other main point to remember is that if you are intending to view reports on mobile devices, you can only access them using the Web Service Report Viewer. You cannot use the Report Manager to view or browse reports. Indeed, if you try to use the Report Manager all you will see is a discouragingly blank screen.

Using Report Viewer is not a handicap in any way when it comes to displaying BI reports. However, report navigation may well require some tweaking. There are a couple of main reasons for this:

- Do you really want to show your users a report navigation screen that looks like it stems from the 1990s, and one that has never even heard of interactive interfaces? Your users will swipe to another app in microseconds!
- The URL that returns a report will be almost impossible to memorize.

An example of the first of these two limitations is shown in Figure 10-1.

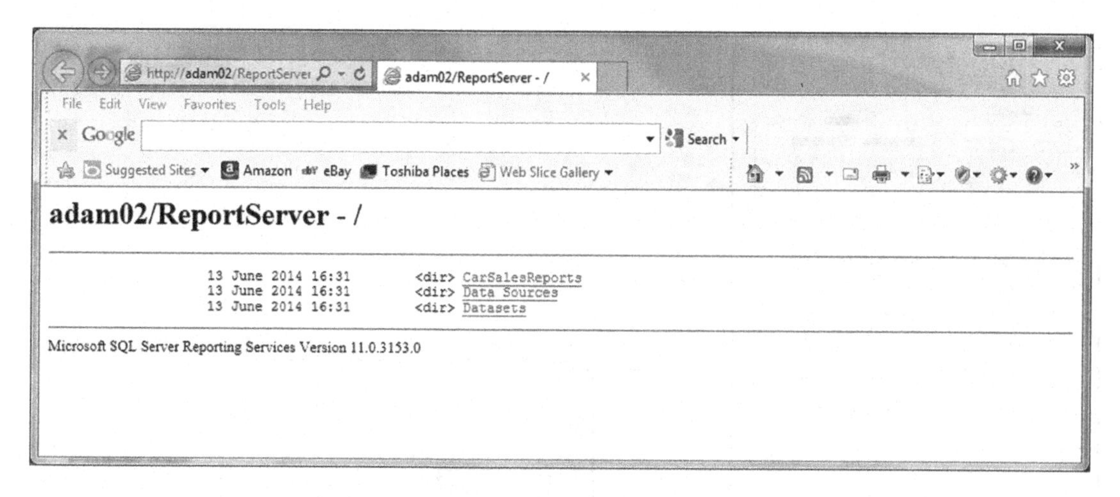

Figure 10-1. *The standard report web services interface*

As to the URLs, well, you will find one or two of those further on in this chapter. So be prepared to wrap your mobile BI reports in a navigation hierarchy, as described at the end of this chapter. It can require a little extra effort, but it is a major step on the road to ensuring that business users buy into your mobile BI strategy.

Tablet Reports

I am prepared to bet that tablets are the most frequently-used output platform for mobile business intelligence. They combine portability, ease of use, and a practical screen size, and can become an extremely efficient medium for BI using SSRS.

This is not to say, however, that the transition from laptop or PC to tablet will always be instantaneous. You will almost certainly need to adapt reports to tablet display for at least some of the following of reasons:

- The screen is smaller than most laptops, and despite often extremely high resolution, cannot physically show all that a desktop monitor is capable of displaying.
- Even if you can fit all of a report that was designed for a large desktop screen onto a tablet, users soon get tired of zooming in and out to make the data readable.
- Interactivity will require the application of many of the "revamping" techniques introduced in Chapter 8.

Let's now look at one of the principal techniques that you can use when creating business intelligence reports for tablets.

Multi-Page Reports

A classic way to make large report easier to use is to separate different elements on separate "pages." Obviously, in Reporting Services, pages are a display, not a design concept. Yet you can use paging effectively when creating or adapting reports for tablet devices. As an example, think back to the dashboard Dashboard_Monthly that you assembled in Chapter 7. A dashboard like this that was designed for a large high-resolution monitor would be unreadable if it were squeezed onto a 9-inch tablet screen. Yet if you separate out its component elements and adjust the layout a little, you end up with a presentation that is both appealing and easy to read, as shown in Figure 10-2.

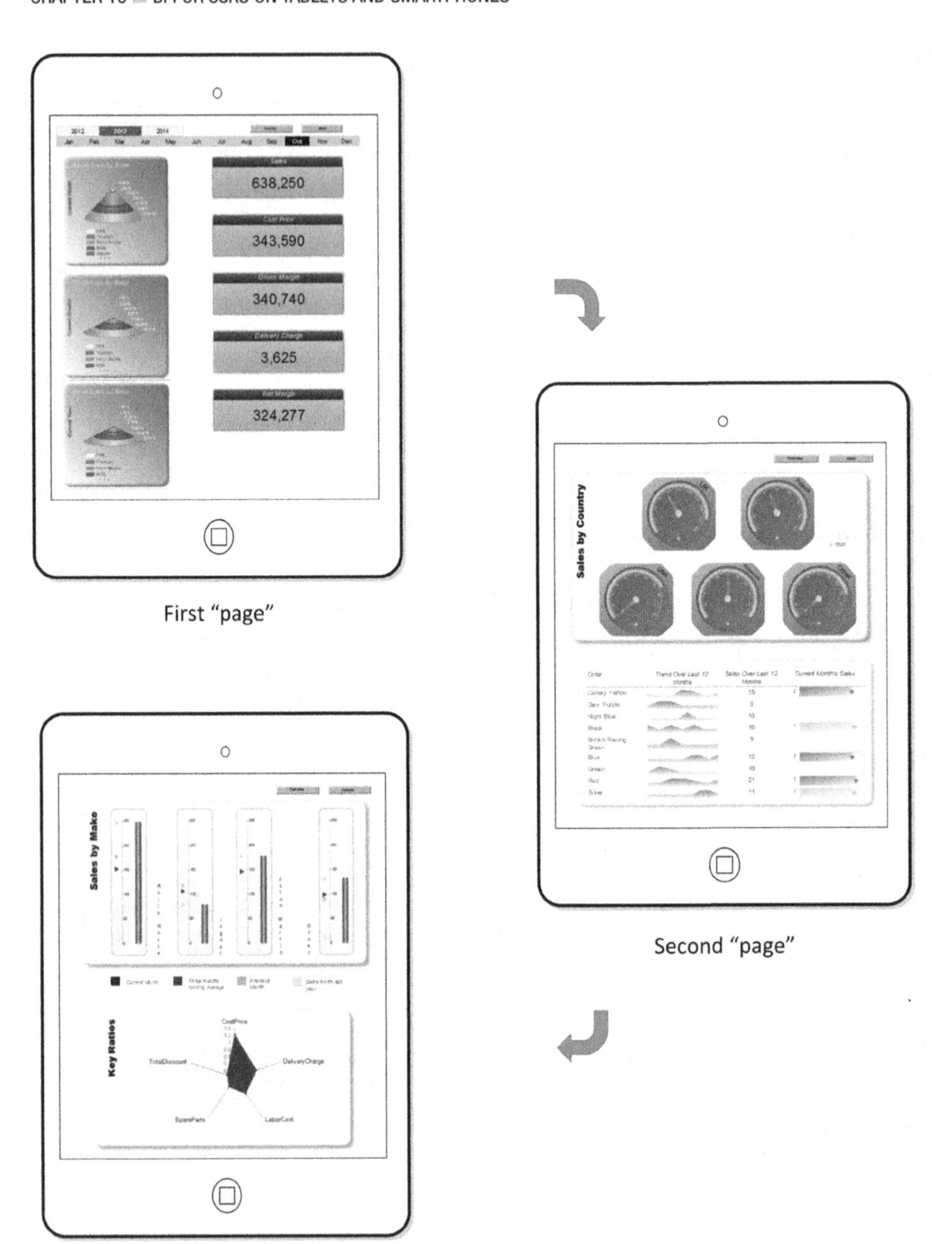

***Figure 10-2.** A dashboard broken down into separate screens for mobile BI*

This is still a single SSRS report, but it uses page breaks to separate the elements into three parts. Then you add buttons that jump to bookmarks inside the report to make flipping from page to page easier. To make the point, Figure 10-3 shows the design view of the report. As you can see, the three "pages" are nothing more than a vertical report layout.

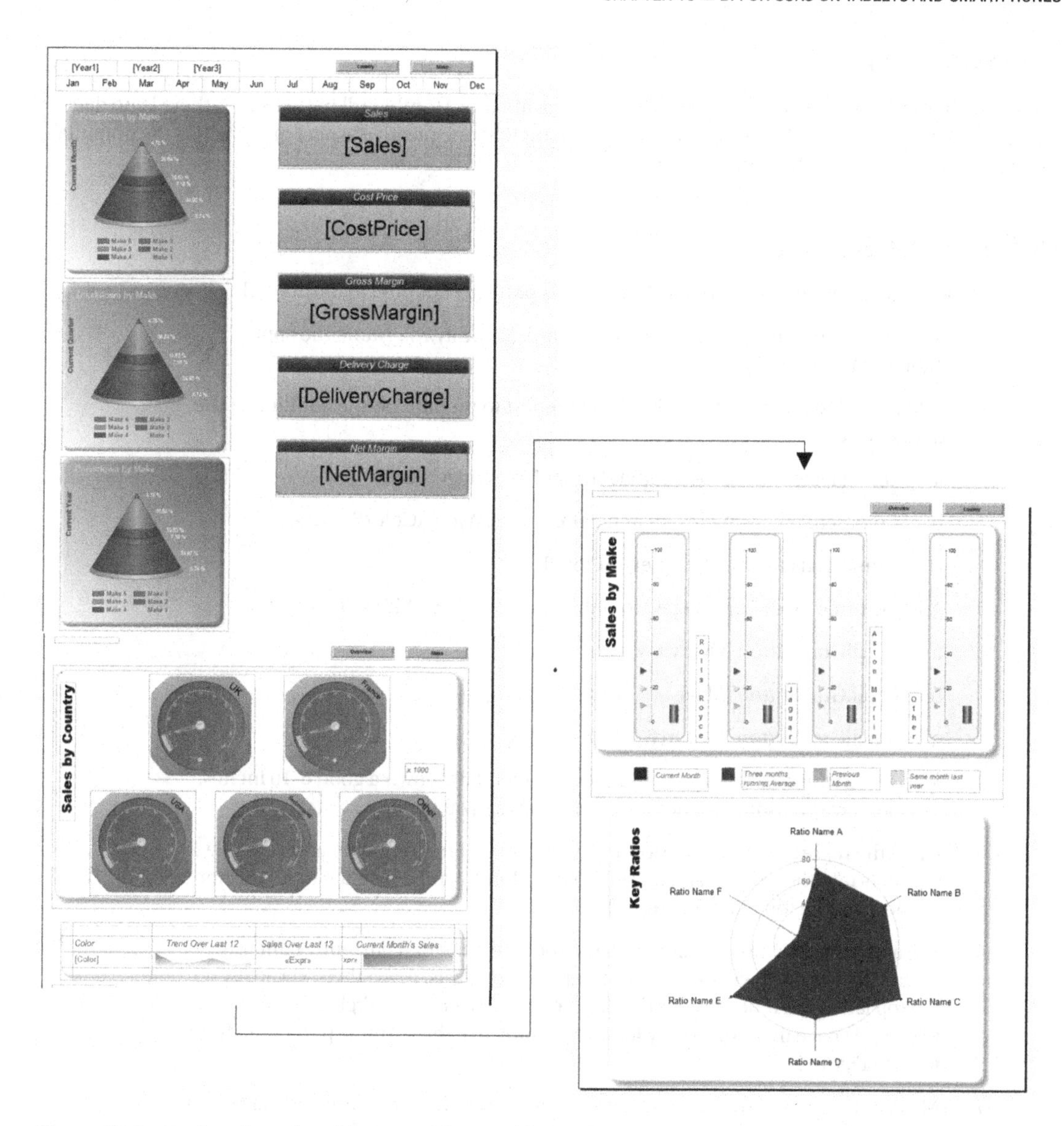

***Figure 10-3.** Design view of a tablet report designed for multi-page display*

The Source Data

Assuming that you have already built the dashboard Dashboard_Monthly.rdl in Chapter 7, there is no data needed; you have it already. This is to underline the point that one positive aspect of SSRS BI is the potential reusability of the various objects that you create.

Building the Report

Let's see how to adapt the source dashboard so that it is perfectly adapted to dashboard delivery.

1. Make a copy of the report named Dashboard_Monthly.rdl. Name the copy Tablet_Dashboard_Monthly.rdl.
2. Add the following datasets to the copied report to prepare for the addition of the date selector:
   a. YearSelector, using the shared dataset DateWidgetYear
   b. ColorScheme, using the shared dataset DateWidgetColorScheme
   c. Dummy, using the SQL query SELECT
3. Add the three following images from the directory C:\BIWithSSRS\Images:
   a. SmallGrayButton_Overview
   b. SmallGrayButton_Make
   c. SmallGrayButton_Country
4. Select everything in the report and move all elements down a good 6-10 inches. This will give you some room to tweak the existing objects.
5. Open the report __DateSelector.rdl that you created in Chapter 9, and copy the contents into the report Tablet_Dashboard_Monthly.rdl. Place the date selector elements at the top left of the report.
6. Cut and paste the three pyramid charts from the table that currently contains them. This way you will have three independent charts. Place them vertically on the top left of the report under the date selector. Make them slightly smaller, and align and space them until they look something like the leftmost part of the first page in Figure 10-3.
7. Make four copies of the table containing the figures that were originally at the top of the dashboard. Delete all the columns but one, leaving a different metric each time, until you have five separate tables, each containing *one* of the sales metrics from the table. Place them vertically on the top right of the report under the date selector. Align and space them until they look something like the rightmost part of the first page in Figure 10-3.
8. Leaving about an inch of clear space, place the gauges for the sales by country and the table of color sales under each other and under the elements that you rearranged in steps 6 and 7. They should look like the center page in Figure 10-3.

9. Drag both remaining visualizations out of the rectangle that contains them, and then delete the container rectangle. Again, leaving about an inch of clear space, place the two elements under each other and under the elements that you rearranged in step 8. They should look like the third page in Figure 10-3.

10. Add two image elements at the top right of the report just above the month list. Set the left-hand one to use the image SmallGrayButton_Country and the right-hand one to use the image SmallGrayButton_Make.

11. Set the images to a size of around 1 inch by 0.2 inches.

12. Copy the images twice. Place one copy above and at the right of the second group of elements (gauges for the sales by country and the table of color sales), and one copy above and at the right of the third group of elements (gauges for the sales by make and the chart of key ratios).

13. For the second set of images, set the left-hand one to use the image SmallGrayButton_Overview and the right-hand one to use the image SmallGrayButton_Make.

14. For the third set of images, set the left-hand one to use the image SmallGrayButton_Overview and the right-hand one to use the image SmallGrayButton_Country. Refer back to Figure 10-3 (the design view) if you need to see exactly how to set these items in the report.

15. Add a rectangle just under the bottom pyramid chart. Set the following properties:

    a. Hidden: False

    b. BackgroundColor: No Color

    c. BorderStyle: None

    d. PageBreak > PageLocation: Start

    e. Bookmark: MiddlePage

16. Copy this rectangle and place the copy just below the table of color sales. Set its bookmark property to BottomPage.

17. Ensure that the tops of the sets of buttons are always just below the *bottom* of the rectangles.

18. Select the tablix containing the years in the year selector, and using the Properties window, set its bookmark property to TopPage.

19. Right-click the top left image button (it should display Country) and select Image Properties from the context menu. Select Action on the left and then Go to bookmark as the action to enable.

20. Enter MiddlePage as the bookmark, and click OK.

21. Do the same for all the image buttons, using the following bookmarks per button:

    a. Country: MiddlePage

    b. Make: BottomPage

    c. Overview: TopPage

You can now preview the report. When you click the buttons you should jump to the next part of the report. You may need to tweak the height of the sections of the report to match the display size of the actual mobile device that you are using.

How It Works

This report uses a tried-and-tested technique of using the vertical parts of a report as separate pages when displayed. A rectangle (which, although technically not hidden, is not visible because it has no border or fill) is used to force the page breaks. The same rectangles serve to act as bookmarks, except for the top of the page, where using the year selector is more appropriate. Then a series of images (here they are buttons) are used to provide the action, which jumps to the appropriate bookmark if the image is clicked or tapped.

Quite possibly the hardest part when rejuggling dashboards like this is deciding how best to reuse, and how to group, the existing visualizations. In reality, you might find yourself re-tweaking an original dashboard more than you did here. So do not be afraid to remove objects or add other elements if it suits the purpose of the report.

You do not need to use images for the buttons that allow users to jump around a report. Any SSRS object that triggers an action will do. However, tweaking a handful of images so that they contain the required text takes only a few minutes, and it certainly looks professional. Also, these buttons could become a standard across all your reports and consequently familiar to all your users.

Creating Tabbed Reports

Sometimes you need to present information in a series of separate areas that are nonetheless part of a whole. In these cases, scrolling down through a report (or jumping to a different part of the report) is distracting, and moving to a completely different report can confuse the user.

One classic, yet effective, way to overcome these issues is to design reports that break down the available information into separate sections. You then make these sections available to the user as "tabs" on the report. This avoids an unnecessarily complex navigation path through a set of reports, and lets the user focus on a specific area of information. These reports group different elements in separate sections (or "tabs") where one click on the tab displays the chosen subset of data. This is a bit like having an Excel file with multiple worksheets, only in an SSRS report.

A tabbed report is a single report consisting of two or more sections. These sections are laid out vertically in SSRS, one above the other. The trick is only to make visible the elements that make up one section at a time when viewing the report.

This approach is a little different from most of the techniques you have seen so far in this chapter. Up until now you have been filtering data in charts or tables. What you will be doing now is making report elements visible or invisible as an interface technique.

In this example, the report will have three tabs as visual indicators at the top of the report. The sections of the report are implemented as three sets of elements (charts, tables, text boxes, etc.), one above the other in the actual report design. The key trick is to handle the Hidden property of each element so that it is set by clicking on the appropriate tab. The final report appears to the user like one of the tabs shown in Figure 10-4.

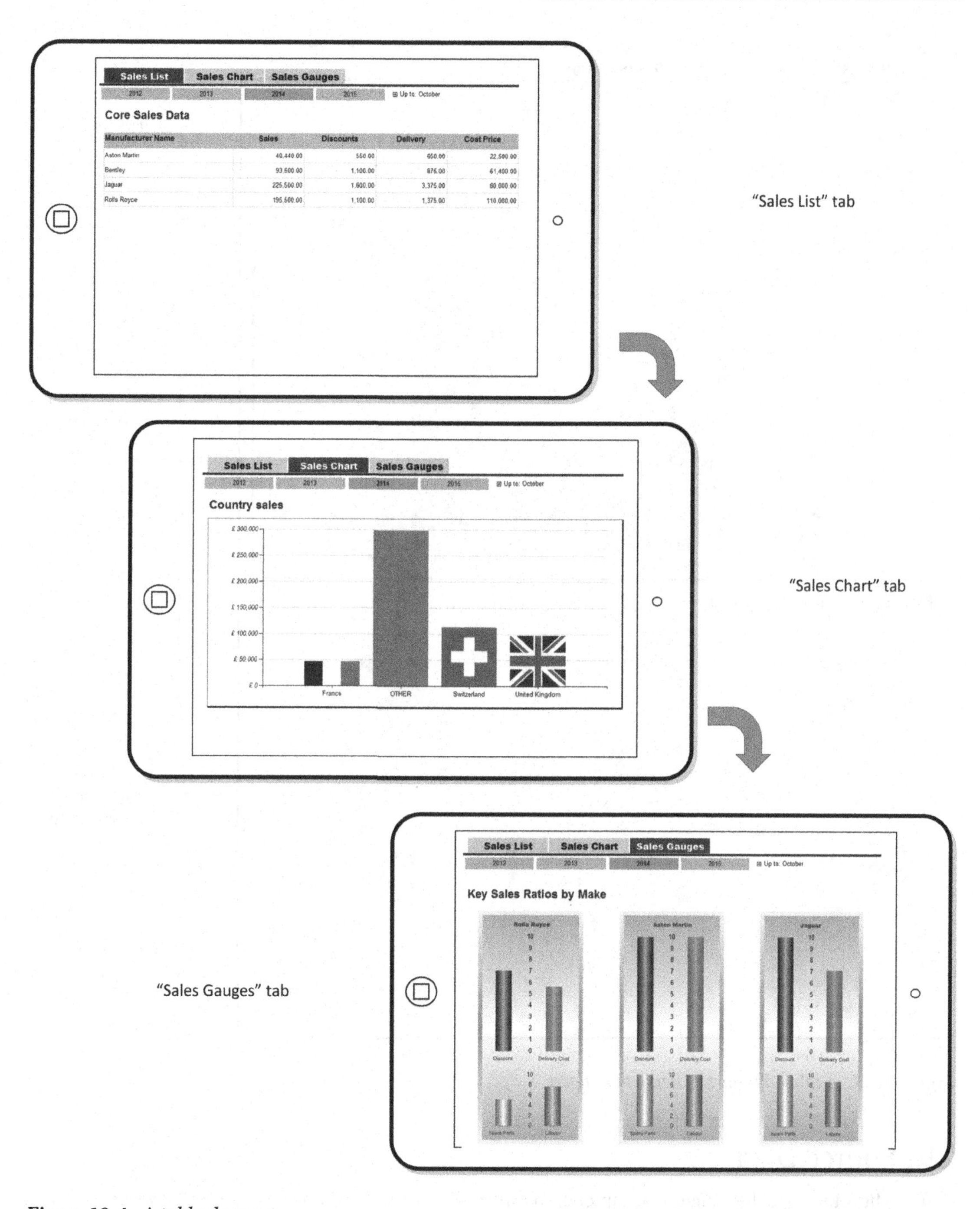

__Figure 10-4.__ A tabbed report

However, the report is somewhat different in Report Designer, shown in Figure 10-5. Here, as you can see, the three tabbed screens are, in effect, a single report.

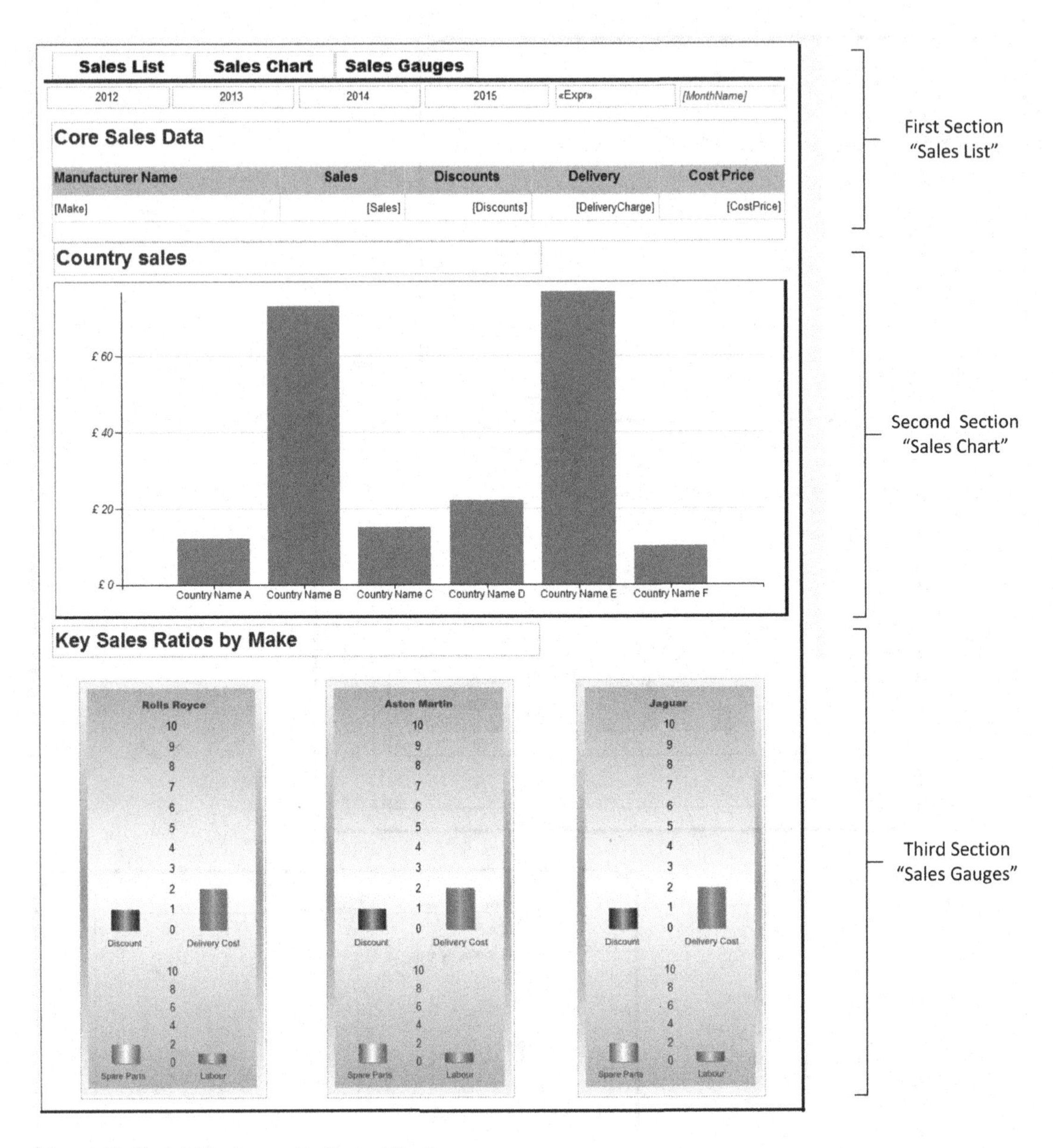

Figure 10-5. *A tabbed report in Report Designer*

The Source Data

You saw how to create the gauges in Chapter 4, so I will not repeat the code for them here. However, the code for the initial table and the column chart is as follows:

```
DECLARE @ReportingYear INT = 2014
DECLARE @ReportingMonth INT = 10

-- Code.pr_TabletTabbedReportCountrySales
```

```
SELECT
CASE
WHEN CountryName IN ('France','United Kingdom','Switzerland', 'United States', 'Spain') THEN
CountryName
ELSE 'OTHER'
END AS CountryName
,SUM(SalePrice) AS Sales
FROM        Reports.CarSalesData
WHERE       ReportingYear = @ReportingYear
            AND ReportingMonth = @ReportingMonth
GROUP BY    CountryName

-- Code.pr_TabletTabbedReportSalesList
SELECT
 Make
,SUM(SalePrice) AS Sales
,SUM(TotalDiscount) AS Discounts
,SUM(DeliveryCharge) AS DeliveryCharge
,SUM(CostPrice) AS CostPrice
FROM        Reports.CarSalesData
WHERE       ReportingYear = @ReportingYear
            AND ReportingMonth = @ReportingMonth
GROUP BY    Make

-- Code.pr_TabletTabbedReportRatioGauges

IF OBJECT_ID('tempdb..#Tmp_GaugeOutput') IS NOT NULL DROP TABLE #Tmp_GaugeOutput

CREATE TABLE #Tmp_GaugeOutput
(
ManufacturerName NVARCHAR(80) COLLATE DATABASE_DEFAULT
,Sales NUMERIC(18,6)
,Discount NUMERIC(18,6)
,DeliveryCharge NUMERIC(18,6)
,SpareParts NUMERIC(18,6)
,LabourCost NUMERIC(18,6)
,DiscountRatio NUMERIC(18,6)
,DeliveryChargeRatio NUMERIC(18,6)
,SparePartsRatio NUMERIC(18,6)
,LabourCostRatio NUMERIC(18,6)
)

INSERT INTO #Tmp_GaugeOutput (ManufacturerName, Sales, Discount, DeliveryCharge, SpareParts,
LabourCost)

SELECT
Make
,SUM(SalePrice) AS Sales
,SUM(TotalDiscount) AS Discount
,SUM(DeliveryCharge) AS DeliveryCharge
,SUM(SpareParts) AS SpareParts
,SUM(LaborCost) AS LabourCost
```

```
FROM          Reports.CarSalesData
WHERE         ReportingYear = @ReportingYear
              AND ReportingMonth = @ReportingMonth
GROUP BY      Make

UPDATE      #Tmp_GaugeOutput
SET         DiscountRatio =
                                          CASE
                                          WHEN Discount IS NULL or Discount = 0 THEN 0
                                          ELSE (Discount / Sales) * 100
                                          END
           ,DeliveryChargeRatio =
                                          CASE
                                          WHEN DeliveryCharge IS NULL or DeliveryCharge = 0
                                          THEN 0
                                          ELSE (DeliveryCharge / Sales) * 100
                                          END
           ,SparePartsRatio =
                                          CASE
                                          WHEN SpareParts IS NULL or SpareParts = 0 THEN 0
                                          ELSE (SpareParts / Sales) * 100
                                          END
           ,LabourCostRatio =
                                          CASE
                                          WHEN LabourCost IS NULL or LabourCost = 0 THEN 0
                                          ELSE (LabourCost / Sales) * 100
                                          END

SELECT * FROM #Tmp_GaugeOutput
```

How the Code Works

These code snippets return a few key metrics aggregated by make or country for a specific year and month. Running this code returns the data shown in Figure 10-6.

Code.pr_TabletTabbedReportSalesList

	Make	Sales	Discounts	DeliveryCharge	CostPrice
1	Aston Martin	40440.00	550.00	650.00	22500.00
2	Bentley	93500.00	1100.00	875.00	61400.00
3	Jaguar	225500.00	1600.00	3375.00	60000.00
4	Rolls Royce	195500.00	1100.00	1375.00	110000.00

Code.pr_TabletTabbedReportCountrySales

	CountryName	Sales
1	France	46750.00
2	Switzerland	112750.00
3	United Kingdom	97750.00
4	OTHER	297690.00

Code.pr_TabletTabbedReportRatioGauges

	ManufacturerName	Sales	Discount	DeliveryCharge	SpareParts	LabourCost	DiscountRatio	DeliveryChargeRatio	SparePartsRatio	LabourCostRatio
1	Aston Martin	40440.000000	550.000000	650.000000	895.000000	1250.000000	13.600396	16.073195	22.131553	30.909990
2	Bentley	93500.000000	1100.000000	875.000000	3900.000000	1974.000000	11.764706	9.358289	41.711230	21.112299
3	Jaguar	225500.000000	1600.000000	3375.000000	5140.000000	1974.000000	7.095344	14.966741	22.793792	8.753880
4	Rolls Royce	195500.000000	1100.000000	1375.000000	1000.000000	1500.000000	5.626598	7.033248	5.115090	7.672634

***Figure 10-6.** The output for the table in the tabbed report*

Building the Report

This report is nothing more than a series of objects whose visibility is controlled by the action properties of the text boxes that are used to give a "tabbed" appearance. As I want to concentrate on the way that objects are made to appear and disappear, I will be somewhat succinct about how to create the objects (table, chart, and gauges) themselves, as they are essentially variations on a theme of visualizations that you have already seen in previous chapters.

1. Make a copy of the .rdl file named __DateSelector. Name the copy Tablet_TabbedReport.rdl, and open the copy.
2. Remove the month elements as described for the tiled report earlier in this chapter. Leave the ReportingMonth parameter, however.
3. Add the following three parameters, all of which are Boolean and hidden:
    a. TabSalesList - default value: False
    b. TabSalesChart - default value: True
    c. TabSalesGauge - default value: True
4. Add the following datasets:
    a. CountrySales, using the stored procedure Code.pr_TabletTabbedReportCountrySales
    b. SalesList, using the stored procedure Code.pr_TabletTabbedReportSalesList
    c. RatioGauges, using the stored procedure Code.pr_TabletTabbedReportRatioGauges
    d. MonthList, using the shared dataset ReportingFullMonth
5. Embed the following seven images into the report (all .png files from the directory C:\BIWithSSRS\Images that you have downloaded from the Apress web site): EuropeFlag, GermanFlag, USAFlag, GBFlag, SpainFlag, FranceFlag, and SwissFlag.
6. Copy the pop-up menu text box and table that you created in Chapter 8. You can see this type of visualization in Figure 8-5. Set the table for the "menu" to use the dataset MonthList.
7. Add three text boxes in a row above the year selector. Format them to Arial Black 14 point centered. Enter the following texts (in this order): Sales List, Sales Chart, and Sales Gauges. Name the three text boxes (also in this order): TxtTabSalesList, TxtTabSalesChart, and TxtTabSalesGauges.
8. Add a line under the text boxes. Set the LineColor to dark blue and the LineWidth to 3 points. This is the "tabbed" header that you can see in Figure 10-4.
9. Create a table and delete the second (detail) row. Delete all but one column. Add two more rows and set the dataset to SalesList. In the Properties window, name this table TabSalesList.
10. Create a second table of five columns and drag it into the second row of the table that you created previously. As the nested table will inherit the outer table's dataset, you can set the detail row to use the fields Make, Sales, Discounts, DeliveryCharge, and CostPrice (in this order).

11. Add the text Core Sales Data to the top row of the outer table. Format the table as you see fit, possibly using Figure 9-11 as an example.

12. Add a text box under the table and name it TxtChartTitles. Set the font to Arial blue 18 point.

13. Under this title, add a column chart that you name Chart. Set it to use the dataset CountrySales. Set the ∑ Values to use the Sales field and the category groups to use the CountryName field.

14. Click any column in the chart and set it to use an embedded background image where the Value for the image is set in the Properties window using the following expression:

```
=Switch (
        Fields!CountryName.Value= "United Kingdom", "GBFlag"
        ,Fields!CountryName.Value= "France", "FranceFlag"
        ,Fields!CountryName.Value= "UnitedStates", "USAFlag"
        ,Fields!CountryName.Value= "Germany", "GermanFlag"
        ,Fields!CountryName.Value= "Spain", "SpainFlag"
        ,Fields!CountryName.Value= "Switzerland", "SwissFlag"
        ,Fields!CountryName.Value= "OTHER", "EuropeFlag"
        )
```

15. Format this chart as you want, possibly using Figure 6-13 from Chapter 6 as a model.

16. Add a text box under the chart and name it TxtGaugeTitles. Set the font to Arial blue 18 point.

17. Create a gauge similar to the one that you created in Chapter 3 for Figure 3-3. However, you must add a second scale by right-clicking the gauge and selecting Add Scale.

18. The first scale needs the following properties setting in the Linear Scale Properties dialog:

Parameter	Property	Value
Layout	Position in gauge (percent)	50
	Start margin (percent)	40
	End margin (percent)	11
	Scale bar width (percent)	0
Labels	Placement relative to scale	Cross
Major Tick Marks	Hide major tick marks	Checked
Minor Tick Marks	Hide minor tick marks	Checked

19. The second scale needs the following properties setting in the Linear Scale Properties dialog:

Parameter	Property	Value
Layout	Position in gauge (percent)	50
	Start margin (percent)	8
	End margin (percent)	70
	Scale bar width (percent)	0
Labels	Placement relative to scale	Cross
Major Tick Marks	Hide major tick marks	Checked
Minor Tick Marks	Hide minor tick marks	Checked

20. In this case, however, use the dataset RatioGauges and add a total of *four* bar pointers, *two for each scale.* You do this by right-clicking inside the gauge and selecting Add Pointer For and selecting the relevant scale for each pointer that you add. These pointers must use (clockwise from top left) the following fields: DeliveryChargeRatio, DiscountRatio, LaborCostRatio, SparePartsRatio.

21. Set all the pointers to 15 percent width. The two left-hand pointers should have their distance from the scale set to 33 percent. The two right-hand pointers should have their distance from the scale set to 15 percent.

22. Add titles under each pointer. See Chapter 3 for techniques on how to do this. Add a title at the top of the gauge and enter the text Rolls Royce.

23. Filter the gauge on the make Rolls Royce.

24. Make two copies of the gauge and place them side by side with the first gauge. Name the second gauge GaugeAstonMartin and filter it on Make = Aston Martin. Name the third gauge GaugeJaguar and filter it on Make = Jaguar.

    Now that you have all the elements in place, you can add the final touch: managing the visibility of the various parts of the report.

25. Select the table named TabSalesList and in the Properties window, set its Hidden property to `=Parameters!TabSalesList.Value`.

26. Select the text box title for the chart as well as the chart, and set their Hidden property to `=Parameters!TabSalesChart.Value`.

27. Select the three gauges and the text box title for the gauges and set their Hidden property to `=Parameters!TabSalesGauge.Value`.

28. Click the TxtTabSalesList text box, and set the following properties:

Property	Value

BackgroundColor	`=Switch(`
	`Parameters!TabSalesList.Value=False,"Blue"`
	`,1=1`
	`,"LightBlue"`
	`)`

Color	`=Switch(`
	`Parameters!TabSalesList.Value=False,"White"`
	`,1=1`
	`,"Black"`
	`)`

29. Set the same properties for the two other text boxes (TxtTabSalesChart and TxtTabGauges), only change the parameter reference in the expression to the appropriate parameter. So, for the text box TxtTabSalesChart, you will see Parameters!TabSalesChart.Value and the text box TxtTabSalesGauge you should use Parameters!TabSalesGauge.Value.

30. Right-click the text box TxtTabSalesList and select Text Box Properties from the context menu. Click Action on the left, and set the following options:

Option	Parameter	Value
Enable as action		Go to report
Specify a report		`[&ReportName]`
Use these parameters	ReportingYear	`[@ReportingYear]`
	TabSalesList	False
	TabSalesChart	True
	TabSalesGauge	True
	ReportingMonth	`[@ReportingMonth]`

31. The dialog should look like Figure 10-7.

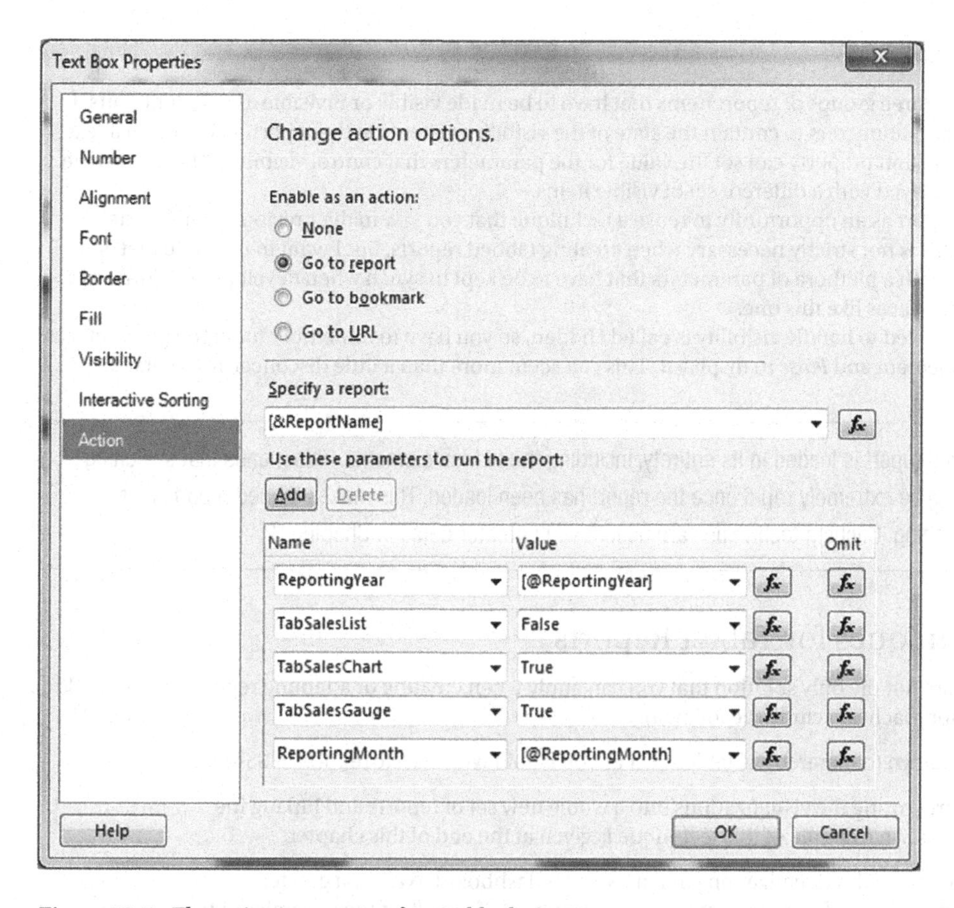

Figure 10-7. *The action parameters for a tabbed report*

32. Do exactly the same for the text box TxtTabSalesChart, only alter the two following parameters:
    a. TabSalesList: True
    b. TabSalesChart: False
    c. TabSalesGauge: True
33. Do exactly the same for the text box TxtTabSalesGauge, only alter the two following parameters:
    a. TabSalesList: True
    b. TabSalesChart: True
    c. TabSalesGauge: False

That's it. You have defined a report where two out of three will parts of the report always be hidden, giving the impression that the user is flipping from tab to tab when a text box at the top of the report is clicked or tapped.

How It Works

Because there are three groups of report items that have to be made visible or invisible as coherent units, I find it easier to use parameters to contain the state of the visibility of items in the report. Then, when a text box is clicked, its action property can set the value for the parameters that control visibility. This causes the report to be redisplayed with a different set of visible items.

I used this report as an opportunity to reuse a technique that you saw in the previous chapter: the pop-up menu. This is not strictly necessary when creating tabbed reports, but I want to illustrate that you can soon end up with a plethora of parameters that have to be kept in synch when developing reports with more complex interfaces like this one.

The property used to handle visibility is called Hidden, so you have to remember to set the parameter to *True* to hide an element and *False* to display it. This can seem more than a little disconcerting at first.

■ **Note** A tabbed report is loaded in its entirety, including the hidden elements. This means that switching from tab to tab can be extremely rapid once the report has been loaded. This makes tabbed reports ideal candidates for caching. Caching techniques are described in greater detail in Chapter 12.

Other Techniques for Tablet Reports

Tabbed reports are not the only solution that you can apply when creating or adapting reports for handheld devices. Other approaches include the following:

- Drilldown to hierarchical tables, just as you would with classic reports in SSRS.
- Restructuring the visualizations into a whole new set of reports and linking the reports. An example of this technique is given at the end of this chapter.
- Simplifying the visualizations that make up a dashboard to allow a greater concentration of widgets in a smaller space. Most, if not all, of the examples in this chapter do this in some way.
- Removing dashboard elements to declutter the report.

Smartphone Reports

There is probably one word that defines successful reports for smartphones: *simplicity*. As in most areas, simplicity in report design is often harder than complexity. While we all have our opinions about design, here are a few tips:

- Isolate truly key elements. Remove anything not essential from the report.
- Create single-focus screens. Ideally the information should be "bite-sized."
- Deliver key data higher in the sequence of reports. Give less detail at higher levels and reserve granularity for further reports lower in the hierarchy.
- Limit the number of metrics you deliver on each screen.
- Consider providing data or titles as tooltips. It can save screen real estate.
- Share datasets. This encourages widget reuse.

Let's see how these principles can be applied in practice. I'll cover using gauges and charts to display metrics, and then provide a few tips on delivering text-based metrics on a smartphone.

Multiple Gauges

A user looking at their phone for business intelligence will be a busy person and their attention span will be limited. After all, many other distractions are jostling for their time and focus. So you need to give them the information they want as simply and effectively as possible. Gauges can be an ideal way to achieve this objective.

Inevitably you will have to limit the information that can be displayed efficiently. This will depend on the size of the screen you are targeting, so there are no definite limits. In this example, I will use six gauges to show car sales for the current month. Moreover, I will deliberately break down the display into

- Five specific makes of cars sold
- One gauge for all the others

This could require tweaking the source data to suit the desired output, but this is all too often what you end up doing when designing BI for smartphones. This final output will look like Figure 10-8. In this particular visualization, there is no facility for selecting the year and month; the current year and month are displayed automatically.

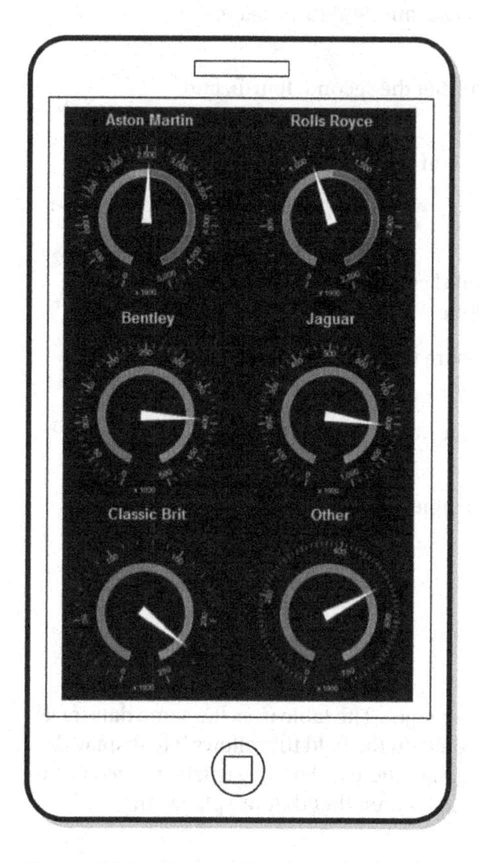

Figure 10-8. *Gauges showing sales for the current month*

The Source Data

You can get the figures for sales for the current month using the following SQL (Code.pr_SmartPhoneCarSalesGauges). You saw this in Chapter 3, so I will refer you to back to this chapter rather than show it all again here.

Building the Gauges

Now it is time to put the pieces together and assemble the gauges for the smartphone display. You will be building on some of the experience gained in Chapter 3, specifically the report named _CarSalesGauge.rdl.

1. Make a copy of the report named _CarSalesGauge.rdl. Name the copy SmartPhone_CarSalesGauges.rdl.
2. Right-click the ReportingYear and ReportingMonth parameters (in turn) and set the parameter to hidden.
3. Increase the size of the report so that it is three times wider and five times taller than the gauge. Place the gauge on the right of the report. Resize the gauge so that it is 1.5 inches square.
4. Add a table to the left of the report. Delete the detail row. Make the table three columns by six rows, adding the required number of columns and rows. Set the dataset to SmartPhoneCarSalesGauges.
5. Set the first, third, and fifth rows to be 0.25 inches tall. Set the second, fourth, and sixth rows to be 1.5 inches tall.
6. Set the first and third columns to be 1.5 inches wide, and the second column to be 0.25 inches wide.
7. Set all the cell backgrounds to black.
8. Add the titles shown in Figure 10-8 to the first and third columns of the first, third, and fifth rows. Set the font to Arial 10 point white.
9. Drag the gauge into the left-hand cell of row two. Ensure that the filter is set to Aston Martin.
10. Copy the gauge into the cells of the left-hand and right-hand columns of rows two, four, and six.
11. Filter each of the gauges to apply only the make that appears in the title above the gauge.
12. Resize the report to fit to the size of the table.

How It Works

This report simply uses a table as a placeholder to contain the six gauges. The table uses the same dataset as the gauges, even though it does not use it. Each gauge is then filtered on the field that allows it to display data for a single record in the output recordset. The big advantage of gauges here is that they can be resized easily, and will even resize if you adjust the height or width of the row and column that they are placed in.

Slider Gauges

A type of presentation that is extremely well suited to smartphones is the lateral gauge. These gauges have been popularized by certain web sites, and are sometimes called "slider" gauges. An example is shown in Figure 10-9.

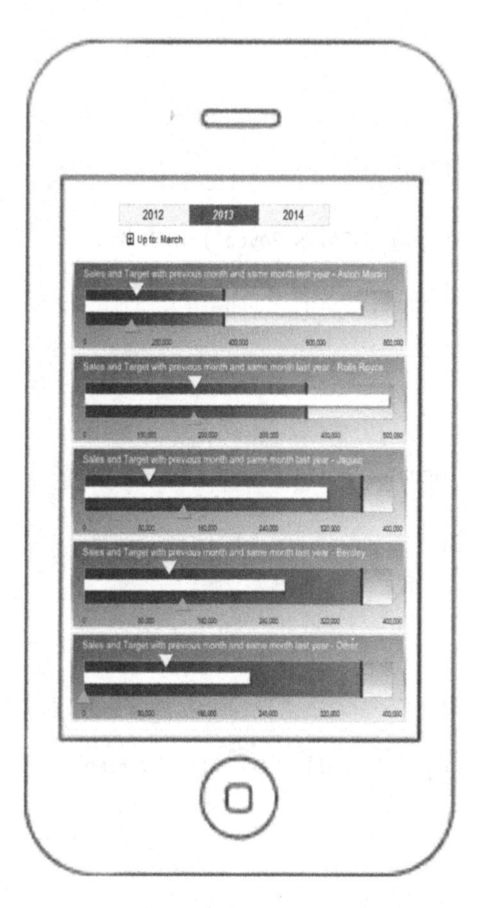

Figure 10-9. *A slider gauge*

The Source Data

The code needed to produce this mobile report is as follows:

```
DECLARE @ReportingYear INT = 2014
DECLARE @ReportingMonth INT = 10

IF OBJECT_ID('Tempdb..#Tmp_Output') IS NOT NULL DROP TABLE Tempdb..#Tmp_Output

CREATE TABLE #Tmp_Output
(
Make NVARCHAR(80) COLLATE DATABASE_DEFAULT
,Sales NUMERIC(18,6)
```

```
,SalesBudget NUMERIC(18,6)
,PreviousYear NUMERIC(18,6)
,PreviousMonth NUMERIC(18,6)
,ScaleMax INT
)

INSERT INTO #Tmp_Output
(
Make
,Sales
)

SELECT      CASE
                WHEN Make IN ('Aston Martin','Bentley','Jaguar','Rolls Royce') THEN Make
                ELSE 'Other'
            END AS Make
            ,SUM(SalePrice)
FROM        Reports.CarSalesData
WHERE       ReportingYear = @ReportingYear
            AND ReportingMonth <= @ReportingMonth
GROUP BY    CASE
                WHEN Make IN ('Aston Martin','Bentley','Jaguar','Rolls Royce') THEN Make
                ELSE 'Other'
            END

-- Previous Year Sales
;
WITH SalesPrev_CTE
AS
(
SELECT       CASE
                WHEN Make IN ('Aston Martin','Bentley','Jaguar','Rolls Royce') THEN Make
                ELSE 'Other'
             END AS Make
             ,SUM(SalePrice) AS Sales
FROM         Reports.CarSalesData
WHERE        ReportingYear = @ReportingYear - 1
             AND ReportingMonth <= @ReportingMonth
GROUP BY      CASE
                WHEN Make IN ('Aston Martin','Bentley','Jaguar','Rolls Royce') THEN Make
                ELSE 'Other'
              END
)

UPDATE        Tmp
SET           Tmp.PreviousYear = CTE.Sales
```

```
FROM         #Tmp_Output Tmp
             INNER JOIN SalesPrev_CTE CTE
             ON Tmp.Make = CTE.Make
;
WITH Budget_CTE
AS
(
SELECT        SUM(BudgetValue) AS BudgetValue
              ,BudgetDetail
FROM          Reference.Budget
WHERE         BudgetElement = 'Sales'
              AND Year = @ReportingYear
              AND Month <= @ReportingMonth

GROUP BY      BudgetDetail
)

UPDATE       Tmp
SET          Tmp.SalesBudget = CTE.BudgetValue

FROM         #Tmp_Output Tmp
             INNER JOIN Budget_CTE CTE
             ON Tmp.Make = CTE.BudgetDetail

 -- Previous month sales

 ;
 WITH PreviousMonthSales_CTE
 AS
 (
SELECT
SUM(SalePrice) AS Sales
,CASE
     WHEN Make IN ('Aston Martin','Bentley','Jaguar','Rolls Royce') THEN Make
     ELSE 'Other'
 END AS Make
FROM       Reports.CarSalesData
WHERE      InvoiceDate >= DATEADD(mm, -1 ,CONVERT(DATE, CAST(@ReportingYear AS CHAR(4))
           + RIGHT('0' + CAST(@ReportingMonth AS VARCHAR(2)),2) + '01'))
           AND InvoiceDate <= DATEADD(dd, -1 ,CONVERT(DATE, CAST(@ReportingYear AS CHAR(4))
           + RIGHT('0' + CAST(@ReportingMonth AS VARCHAR(2)),2) + '01'))
GROUP BY
 CASE
     WHEN Make IN ('Aston Martin','Bentley','Jaguar','Rolls Royce') THEN Make
     ELSE 'Other'
 END
 )

UPDATE          Tmp
SET             Tmp.PreviousMonth = CTE.Sales
FROM          #Tmp_Output Tmp
               INNER JOIN PreviousMonthSales_CTE CTE
               ON Tmp.Make = CTE.Make
```

```
-- Scale maximum

UPDATE      #Tmp_Output
SET         ScaleMax =
                         CASE
                         WHEN Sales >= SalesBudget THEN (SELECT Code.fn_ScaleDecile (Sales))
                         ELSE (SELECT Code.fn_ScaleDecile (SalesBudget))
                         END
-- Output

SELECT      *
FROM      #Tmp_Output
```

Running this code gives the results shown in Figure 10-10 (for March 2013):

	Make	Sales	SalesBudget	PreviousYear	PreviousMonth	ScaleMax
1	Aston Martin	716300.000000	360000.000000	120000.000000	132750.000000	800000
2	Bentley	261250.000000	360000.000000	127500.000000	110000.000000	400000
3	Jaguar	315250.000000	360000.000000	128500.000000	83500.000000	400000
4	Other	215750.000000	360000.000000	NULL	106000.000000	400000
5	Rolls Royce	493000.000000	360000.000000	177000.000000	178500.000000	500000

Figure 10-10. *The data for slider display*

How the Code Works

This code first creates a temporary table, and then it adds the sales data for the current year up to the selected month, sales up to the same month for the previous year, the budget data for the current year, and the sales for the preceding month.

This code block also includes the maximum value for the scale inside the table where the core data is held. This is because the maximum value can change for each make of car.

Building the Report

So, with the data in place, here is how to build the report.

1. Make a copy of the report _DateSelector and name the copy SmartPhone_SalesAndTargetWithPreviousMonthAndPreviousPeriod.rdl.
2. Add the following two datasets:
   a. MonthList, based on the shared dataset ReportingFullMonth
   b. MonthlyCarSalesWithTargetPreviousMonthAndPreviousYear, based on the stored procedure Code.pr_MonthlyCarSalesWithTargetPreviousMonthAndPreviousYear
3. Delete the table containing the months.
4. Copy the text box and table that act as a pop-up menu from the report Tablet_TabbedReport.rdl that you created earlier in this chapter, and place them under the year selector.

5. Add a gauge to the report and choose Horizontal as the gauge type. Apply the dataset MonthlyCarSalesWithTargetPreviousMonthAndPreviousYear.

6. Add three new pointers and one range so that there are four pointers and two ranges in total. Set the pointers to use the following fields (the easiest way is to click the gauge so that the Gauge Data pane appears):

   a. LinearPointer1: PreviousMonth

   b. LinearPointer2: SalesBudget

   c. LinearPointer3: Sales

   d. LinearPointer4: PreviousYear

7. Add the following tooltips in the Properties window for three of the four pointers:

   a. LinearPointer1: ="Last Month"

   b. LinearPointer3: ="This Month"

   c. LinearPointer4: ="Same Month Last Year"

8. Set the four pointer properties as follows (right-click each one individually and select Linear Pointer Properties from the context menu):

Pointer	Option	Parameter	Value
LinearPointer1	Pointer options	Pointer type	Marker
		Marker style	Triangle
		Placement relative to scale	Inside
		Distance from scale (percent)	17
		Width (percent)	21
		Length (percent)	15
	Pointer fill	Fill style	Solid
		Color	Yellow
LinearPointer2	Pointer options	Pointer type	Marker
		Marker style	Rectangle
		Placement relative to scale	Cross
		Distance from scale (percent)	5
		Width (percent)	3
		Length (percent)	40
	Pointer fill	Fill style	Solid
		Color	Blue

(*continued*)

Pointer	Option	Parameter	Value
LinearPointer3	Pointer options	Pointer type	Bar
		Bar start	Scale start
		Placement relative to scale	Cross
		Distance from scale (percent)	6
		Width (percent)	13
	Pointer fill	Fill style	Solid
		Color	White
LinearPointer4	Pointer options	Pointer type	Marker
		Marker style	Triangle
		Placement relative to scale	Outside
		Distance from scale (percent)	6
		Width (percent)	21
		Length (percent)	15
	Pointer fill	Fill style	Solid
		Color	Orange

9. Set the following properties in the Linear Range Scale Properties dialog for the two ranges:

Range	Option	Parameter	Value
Range 1	General	Start range at scale value	0
		End range at scale value	`=Fields!SalesBudget.Value`
		Placement relative to scale	Cross
		Distance from scale (percent)	5
		Start width (percent)	40
		End width (percent)	40
	Fill	Fill style	Gradient
		Color	Blue
		Secondary color	Cornflower blue
		Gradient style	Left Right
	Border	Line style	Solid
		Line color	Silver

(continued)

Range	Option	Parameter	Value
Range 2	General	Start range at scale value	=Fields!SalesBudget.Value
		End range at scale value	=Fields!ScaleMax.Value
		Placement relative to scale	Cross
		Distance from scale (percent)	5
		Start width (percent)	40
		End width (percent)	40
	Fill	Fill style	Gradient
		Color	Lime green
		Secondary color	White
		Gradient style	Top Bottom
	Border	Line style	Solid
		Line color	Silver

10. Right-click the scale, select Scale Properties, and set the following properties:

Section	Property	Value
General	Minimum	0
	Maximum (expression)	=Fields!ScaleMax.Value
Layout	Position in gauge (percent)	57
	Start margin (percent)	4
	End margin (percent)	4
	Scale bar width (percent)	1
Labels	Placement (relative to scale)	Outside
	Distance from scale	28
Font	Font	Arial
	Size	8 point
	Color	Dim gray
Number	Category	Number
	Use 1000 separator	Checked
Major Tick Marks	Hide major tick marks	Checked
Minor Tick Marks	Hide minor tick marks	Checked

11. Add a table to the report area and delete all but one column and the header row. Set the remaining cell to be the width of the gauge and approximately 0.8 inches high.

12. Drag the gauge into the table cell.

13. Right-click the gauge and set the following Gauge Panel Properties. This will filter the gauge data so that only the data for the current record is displayed:

Section	Property	Value
Filters	Expression	Make
	Operator	=
	Value	Make
Fill	Background color of the gauge panel	White

14. Right-click the gauge and set the following Gauge Properties:

Section	Property	Value
Back Fill	Fill style	Gradient
	Color	Dim gray
	Secondary Color	White smoke
	GradientStyle	Diagonal left
Frame	Style	Simple
	Width (percent)	4.5
Frame Fill	Fill style	Gradient
	Color	White smoke
	Secondary Color	Dim gray
	GradientStyle	Horizontal center
Frame Border	Line style	None

How It Works

This gauge uses three different pointer styles as well as pointer placement to add a lot of information to a single gauge. The "main" pointer is the central bar (which contains the data for sales to the selected month in the year) while the budget is a bar across the gauge that lets the user see how sales relate to budget. Because they are ancillary data, the pointers for last month's sales and the sales up until the same month for the previous year are shown as small triangular pointers above and below the main pointer.

The budget is then reflected in the two ranges to indicate more clearly how sales relate to budget. Finally, tooltips are added for all the pointers so that the user can see which pointer represents which value.

■ **Note** Every time a new user sees this gauge, they ask how they can move the "slider" pointers and change the data. You have been warned!

Text-Based Metrics

I would never advise creating BI visualizations for smartphones that rely on lots of text. Users will simply not read reams of prose on a tiny mobile device. Instead, think of some of the apps that you currently use. They probably have simple screens and large buttons. Smartphone BI should emulate this approach. As an example, consider the list of sales by make in Figure 10-11. You will note that this example also gives the user the possibility to select the year and month using some of the techniques for interactive selection described earlier in this chapter.

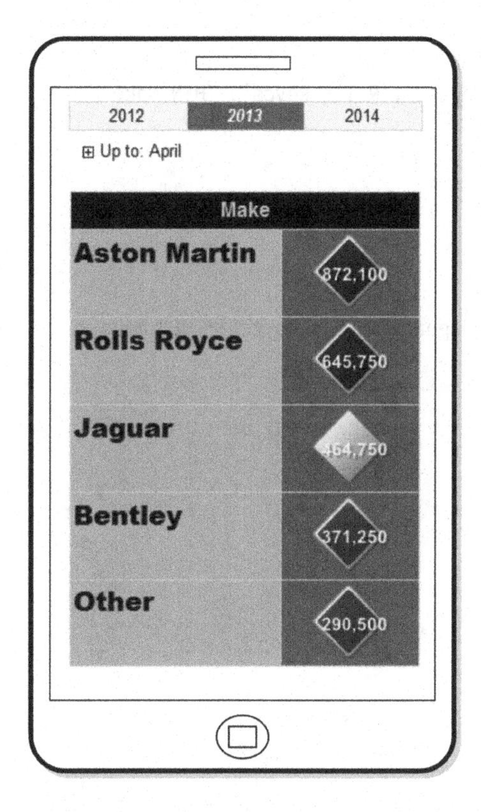

Figure 10-11. *Sales by make for smartphone display*

The Source Data

The code needed to deliver this visualization is not overly complex, and is as follows:

```
DECLARE @ReportingYear INT = 2014
DECLARE @ReportingMonth INT = 10

IF OBJECT_ID('Tempdb..#Tmp_Output') IS NOT NULL DROP TABLE Tempdb..#Tmp_Output

CREATE TABLE #Tmp_Output
(
Make NVARCHAR(80) COLLATE DATABASE_DEFAULT
,Sales NUMERIC(18,6)
```

```
,SalesBudget NUMERIC(18,6)
,SalesStatus TINYINT
)

INSERT INTO #Tmp_Output
(
Make
,Sales
)

SELECT    CASE
            WHEN Make IN ('Aston Martin','Bentley','Jaguar','Rolls Royce') THEN Make
            ELSE 'Other'
          END AS Make
          ,SUM(SalePrice)
FROM      Reports.CarSalesData
WHERE     ReportingYear = @ReportingYear
          AND ReportingMonth <= @ReportingMonth
GROUP BY  CASE
             WHEN Make IN ('Aston Martin','Bentley','Jaguar','Rolls Royce') THEN Make
             ELSE 'Other'
          END
;
WITH Budget_CTE
AS
(
SELECT      SUM(BudgetValue) AS BudgetValue
             ,BudgetDetail
FROM        Reference.Budget
WHERE       BudgetElement = 'Sales'
            AND Year = @ReportingYear
            AND Month <= @ReportingMonth
GROUP BY    BudgetDetail
)

UPDATE      Tmp
SET         Tmp.SalesBudget = CTE.BudgetValue
FROM        #Tmp_Output Tmp
            INNER JOIN Budget_CTE CTE
            ON Tmp.Make = CTE.BudgetDetail

-- Set Sales Status

UPDATE      #Tmp_Output
SET SalesStatus =
                 CASE
                 WHEN Sales < (SalesBudget * 0.9) THEN 1
                 WHEN Sales >= (SalesBudget * 0.9)  AND Sales <= (SalesBudget * 1.1) THEN 2
```

```
                WHEN Sales > (SalesBudget * 1.1) THEN 3
                ELSE 0
                END
-- Output

SELECT   *
FROM     #Tmp_Output
```

Running this code produces the table shown in Figure 10-12.

	Make	Sales	SalesBudget	SalesStatus
1	Aston Martin	872100.000000	480000.000000	3
2	Bentley	371250.000000	480000.000000	1
3	Jaguar	464750.000000	480000.000000	0
4	Other	290500.000000	480000.000000	1
5	Rolls Royce	645750.000000	480000.000000	3

Figure 10-12. *The data for sales by make for the current year*

How the Code Works

Using the approach that you have probably become used to by now in this book, a temporary table is used to hold the sales data for a hard-coded set of vehicle makes. Then, using a CTE, the budget data for these makes is added, and a status flag is calculated on a scale of 1 through 3.

Building the Display

Let's assemble the table to display this key data on your smartphone.

1. Make a copy of the report _DateSelector and name the copy SmartPhone_CarSalesByCountryWithFlagDials.rdl. Delete the table containing the months.
2. Add the following two datasets:
    a. MonthList, based on the shared dataset ReportingFullMonth
    b. MonthlyCarSalesWithTargetAndStatus, based on the stored procedure Code.pr_MonthlyCarSalesWithTargetAndStatus
3. Copy the text box and table that act as a pop-up menu from the report SmartPhone_CarSalesByCountryWithFlagDials.rdl that you created earlier in this chapter, and place them under the year selector.
4. Add a table to the report and delete the third column. Apply the dataset MonthlyCarSalesWithTargetAndStatus.
5. Set the left column to 1.75 inches wide and the right column to 1.25 inches wide (approximately). Set the details row to be about 1/2 inch high. Merge the cells on the top row.

6. Add the text Make to the top row and center it. Set the text box background to black and the font to light gray Arial 12 point bold.

7. Add the field Make to the left. Set the text box background to silver and the font to black Arial 16 point bold.

8. Set the text box background of the right column to dim gray.

9. Drag an indicator from the toolbox into the right-hand column cell and select 3 signs as the indicator type.

10. Click the indicator (twice if necessary) and when the Gauge Data pane appears, select SalesStatus as the ∑ Value.

11. Right-click the indicator and select Indicator Properties from the context menu. Set the Value and States measurement unit property to numeric.

12. Add an indicator state so that there are a total of four. Set all to use the diamond icon. Set the indicator properties as follows:

Indicator	Color	Start	End
First	No Color	0	0
Second	Maroon	1	1
Third	Green	2	2
Fourth	Dark Blue	3	3

13. Right-click the indicator and select Add Label from the context menu. Then right-click the label and select Label Properties from the context menu. Set the following properties:

Section	Property	Value
General	Text (Expression)	`=Microsoft.VisualBasic.Strings.Format(Fields!Sales.Value, "#,#")`
	Text alignment	Center
	Vertical alignment	Middle
	Top (percent)	30
	Left (percent)	10
	Width (percent)	90
	Height (percent)	50
Font	Font	Arial
	Style	Bold
	Color	Yellow

14. Right-click the table and select Tablix properties from the context menu. Select Sorting on the left and click Add. Select Sales as the column to sort by and Z to A as the sort order.

How It Works

This visualization uses a table structure to show the records from the source data where the data is not only directly input into a table cell (the Make) but is also part of the indicator, as a text. However, when text is added to an indicator you must use an expression to display it and some simple Visual Basic to format the numbers. One advantage of a table here is that you can resize the indicator simply by adjusting the height and width of the column that contains it; the text will resize automatically as well.

Because I am demonstrating smartphone delivery, the color scheme is deliberately brash. You may prefer more moderate pastel shades. Just be aware that this medium has little respect for subtlety or discretion.

Multiple Charts

Some, but not all, chart types can be extremely effective when used on smartphones. I do not want to dismiss any of the more "traditional" chart types as unsuitable, so the only comment I will make is that it's best to deliver a single chart per screen if the charts are complex or dense line or bar charts.

If you need to deliver multiple charts, consider using a chart like the one from the tabbed report at the start of this chapter. While it can be a little laborious to configure multiple chart areas in this way, it does make resizing the whole chart to suit a specific mobile device extremely easy.

Alternatively, if you need multiple charts to make comparisons easier for the user, consider a trellis chart structure. You saw an example of this at the end of Chapter 4 if you need to refer back to refresh your memory.

Smartphone and Tablet Report Hierarchy

You may find that the way you present the hierarchy of reports for smartphone users is not the same way you create a reporting suite for tablet devices. This is because smartphone users are (probably) more used to simpler, more bite-sized chunks of information. So consider the following tips:

- Breaking down reports into sub-elements, with bookmark links inside the report, as you saw at the start of this chapter.
- A menu-style access to a report hierarchy, using buttons rather than lists (as you saw for tablets above) to navigate down and up a hierarchy.
- Use graphically consistent buttons in all reports.
- Using visualizations that are "tappable" and that become part of the navigation when possible.

As you have seen the first of these solutions already, let's take a quick look at a simple implementation of the second. This is fairly "classic" stuff, but it is worth showing how it can be applied to some of the tablet-based reports that you have developed so far in this book.

Access to a Report Hierarchy

To hide the undeniably horrendous Web Services interface, all you have to do is to set up a home page for your suite of reports and send this URL to your users. They can then add it to the favorites menu in their browser, and use it as the starting point for accessing the set of tablet-oriented BI that you have lovingly crafted. A simple example is shown in Figure 10-13.

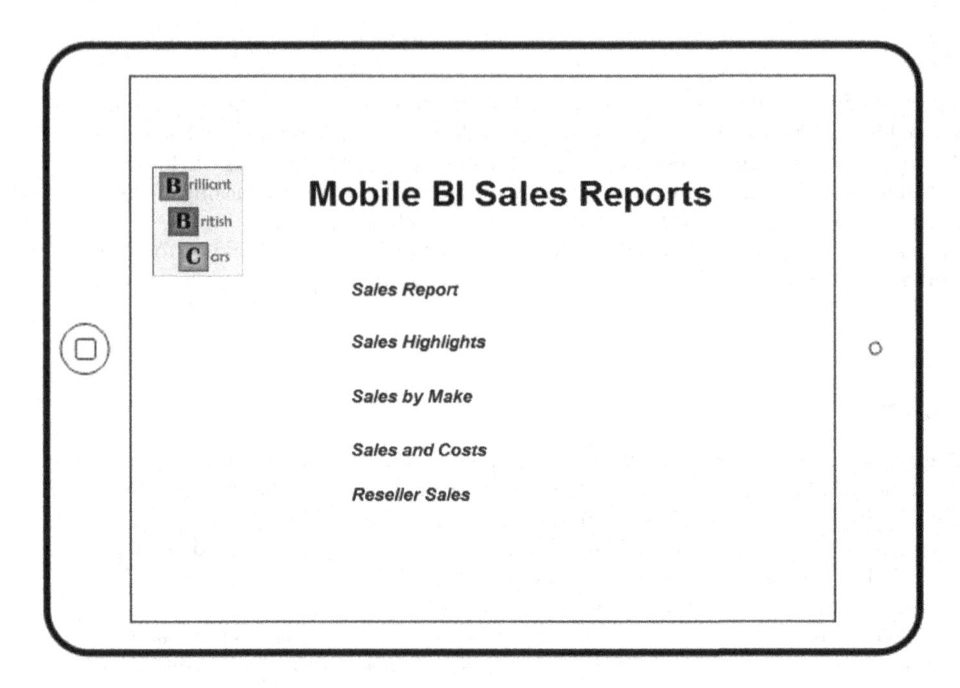

Figure 10-13. *Accessing reports through a custom interface*

I am presuming here that you have deployed your reports to a SQL Server Reporting Services Instance, and that you have structured the reports into a set of subfolders. In this specific instance, I have set up a virtual folder named CarSalesReports, with a subfolder named Tablet that contains all the reports in the sample solution that begin with Tablet_.

Here is how to create this entry page.

1. Create a new, blank report and save it under the name Tablet_Reportsmenu.rdl.
2. Add the image CarsLogo.png.
3. Drag this image to the top left of the report.
4. Add a suitably formatted title such as "Mobile BI Sales Reports."
5. Add a text box. Enter the text Sales Report, and set the text to Arial 12 point bold in blue.
6. Right-click the text box and select Text Box Properties from the context menu.
7. Click Action on the left.
8. Select Go to URL and enter the following URL:

```
http://localhost/ReportServer/Pages/ReportViewer.aspx?%2fCarSalesRep
orts%2fTablet%2fTablet_TabbedReport&rs:Command=Render
```

9. Confirm with OK.

10. Add four new text boxes, all formatted like the first one. Set the texts and URLs to the following:

    a. Sales Highlights: `http://localhost/ReportServer/Pages/ReportViewer.aspx?%2fCarSalesReports%2fTablet%2fTablet_SimpleSlicer&rs:Command=Render`

    b. Sales by Make: `http://localhost/ReportServer/Pages/ReportViewer.aspx?%2fCarSalesReports%2fTablet%2fTablet_ChartHighlight&rs:Command=Render`

    c. Sales and Costs: `http://localhost/ReportServer/Pages/ReportViewer.aspx?%2fCarSalesReports%2fTablet%2fTablet_SlicerHighlight&rs:Command=Render`

    d. Reseller Sales: `http://localhost/ReportServer/Pages/ReportViewer.aspx?%2fCarSalesReports%2fTablet%2fTablet_Tile&rs:Command=Render`

11. Deploy the report (to the Tablet subdirectoy of the CarSalesReports virtual directory) and run it using the following URL:

```
http://localhost/ReportServer/Pages/ReportViewer.aspx?%2fCarSalesRep
orts%2fTablet%2fTablet_ReportsMenu&rs:Command=Render
```

You can now display the home page for your reporting suite, and then drill down into any report with a simple tap on the tablet.

How It Works

As you can see, you are using the Web Services URL for every report and studiously avoiding the Report Manager interface. This is to ensure that tablets (and that means iPads too) can display the report. However, all that your users see is a structured drill-down to the reports that they want, so hopefully the few minutes of work to set up a menu-driven access to your reporting suite will be worth the effort.

Structuring the Report Hierarchy

Now that you have seen the principles of creating an interface to replace the Report Manager, you should be able to extend this to as many sublevels as you need. One extra point is that you might find it useful to add a small text box, using the principles that you just saw, to each of the subreports. This text box uses the URL of the home menu in its action property. This way you can flip back from any report at a sublevel to the home page or to another sublevel. If you are thinking "yes, but I have the browser's back button for that," remember that your users could be using the postback techniques that you have been using in this and previous chapters. If so, the back button will only display the same report, but using the previous set of parameters and not return to a higher level in the report hierarchy.

This technique can also be used to break down existing reports into multiple separate reports. In the case of both the tabbed report and the bookmark report that you saw earlier, you could "cut" a report into separate .rdl files and use action properties to switch to another report.

Conclusion

This chapter showed you some of the ways you can deliver business intelligence to tablets and smartphones using SSRS. You saw that you frequently have to tailor the way information is delivered to a tablet or phone. If you have an existing report, you may have to adapt it for optimum effect on a mobile device. In any case, there are a series of methods that you can apply to save space and use the available screen real estate to greatest effect. These can involve creating tabbed reports, breaking reports down into separate reports, or using bookmarks to flip around inside a report.

Where smartphones are concerned, you probably need to think in terms of simplicity first. In this chapter, you applied the "less is more" principle and consequently you removed any element that added clutter to a screen. This way you key BI metrics are visible, instantly and clearly, for your users.

CHAPTER 11

Standardizing BI Report Suites

So far in this book you have seen myriad ways to deliver eye-catching business intelligence to corporate users. A successful BI project can-indeed, possibly should-be a victim of its own success. That can mean users requesting more and more reports, dashboards, and output to different devices. If this happens, it is easy to get swamped by the amount of work that must be done. So you need to know some of the approaches that can save time when building BI report suites.

The time-saving techniques that you will look at in this chapter include

- Planning for reuse
- Using shared datasets
- Copying multiple elements between reports
- Using database images and image datasets
- Creating "styles" in reports to ensure homogeneity across reports

These techniques will use the same sample database that you have used so far in this book, so I hope that you have already downloaded it from the Apress web site.

Planning for Reuse

If you have spent some time developing a report, and you subsequently find yourself needing all or part of it in another report, your first reaction should probably be to think "how can I make this reusable?" Equally, if you are seeing demand grow for output to various handheld devices, or reports that share common elements, you should probably think in terms of doing some or all of the following:

- Build the "components" that make up dashboards and tablet reports, and then assemble them as required by user demand (as you have done in previous chapters).
- Create "template" reports that contain common data sources and datasets (as you have done in previous chapters).
- Design stored procedures carefully, so that they can either be reused in different reports or can serve as the basis for new stored procedures.
- Think in terms of shared datasets (more on this later).
- Build parameter lists as stored procedures, especially if you are using cascading parameters.

Of course, each project's requirements will be different, and some solutions will not apply to every environment. Nonetheless, bearing these ideas in mind can help you to save time and energy.

Shared Datasets

A technique that you have been applying throughout this book is the centralized definition of source data as shared datasets. To reiterate, a shared dataset is a query (either as a code snippet or as a stored procedure) that is defined like any report-specific dataset. However, it can be applied to *any* report simply by selecting it from the list of shared datasets in a report project.

Moreover, once a shared dataset has been added to a report, you can apply filters and/or calculated fields just as you can to a standard dataset. You can even apply the same shared dataset several times (providing that you give each instance a different name). This can allow you, for instance, to have a project-wide dataset that you then filter on multiple different criteria inside a report.

A shared dataset must be based on a shared data source, as you have done with the shared datasets used as examples in earlier chapters of this book. Moreover, a shared dataset can be cached (and scheduled) to accelerate report delivery, but more on that in the final chapter.

■ **Note** As you have probably noticed, Visual Studio adds a visual indicator (a small arrow in the bottom left corner) to any shared datasets that are inside the datasets folder of the Report Data window.

There are, inevitably, a few limitations to shared datasets. Essentially, they are fairly rigidly defined when created, and cannot be altered significantly when applied to a report. More specifically,

- A shared dataset's data source cannot be changed in the report.
- A shared dataset's query cannot be changed in the report.
- The fields in a shared dataset cannot be altered, but you can add calculated fields.

Generally, I recommend the use of shared datasets when creating BI reporting suites. Defining a clear set of core data is a good discipline when planning a collection of reports. Forcing yourself to specify the required data and then breaking this down into separate, reusable, shared datasets can be a profitable part of the initial analysis.

■ **Note** I advise against creating very wide generic shared datasets; these can slow down report delivery because they send quantities of unneeded data through the network. Efficient shared datasets are those that deliver just the right amount of data.

Adding Multiple Data Sources, Datasets, or Image Files

You may find that you have a component file that contains multiple datasets (and possibly multiple data sources too). Another eventuality is that a report that you are building requires dozens of images that you have already added to a previous report. Adding them one by one (especially to multiple destination files) would be extremely laborious. Consequently, you would like to copy them from the existing report into the new one. Fortunately, there is an easy way to copy multiple image files, data sources, or datasets between reports. This involves using the XML code that constitutes the SSRS `.rdl` file. What you have to do is to open the source code, locate the XML that you want to transfer, and copy it into the destination file.

Copying Images Between Reports

Here is an example of how to copy several images between two reports. As a source for the images that you require, you will be using the file `_ColorSalesWithBorder.rdl`. This report contains eight images that you need in another report. Manually reinserting these images would be slow, painful, and could lead to errors. Here is how to copy them all at once.

1. Right-click the file containing the images (or other reference elements) that you want to copy into another report. In this case, it is `_ColorSalesWithBorder.rdl`.
2. Select View Code. The file will open as XML.
3. Click Edit ➤ Find and Replace ➤ Quick Find. Enter EmbeddedImages as the text to find and click Find.
4. Select the EmbeddedImage tags, including the opening and closing tags, for the collection of images that you want to transfer, then click Copy. You will see something like the XML in Figure 11-1.


```
<EmbeddedImages>
  <EmbeddedImage Name="EmbossedBorder1_Bottom">
    <MIMEType>image/jpeg</MIMEType>
    <ImageData>/9j/4AAQSkZJRgABAQEAYABgAAD/2wBDAAIBAQIBAQICAgICAgICAwUDAwMDAwYEBAMFBwYHBwcGBwcICQsJCAgKCAcHC
  </EmbeddedImage>
  <EmbeddedImage Name="EmbossedBorder1_BottomLeft">
    <MIMEType>image/jpeg</MIMEType>
    <ImageData>/9j/4AAQSkZJRgABAQEAYABgAAD/2wBDAAIBAQIBAQICAgICAgICAwUDAwMDAwYEBAMFBwYHBwcGBwcICQsJCAgKCAcHC
B//Z</ImageData>
  </EmbeddedImage>
  <EmbeddedImage Name="EmbossedBorder1_BottomRight">
    <MIMEType>image/jpeg</MIMEType>
    <ImageData>/9j/4AAQSkZJRgABAQEAYABgAAD/2wBDAAIBAQIBAQICAgICAgICAwUDAwMDAwYEBAMFBwYHBwcGBwcICQsJCAgKCAcHC
KKAP/9k=</ImageData>
  </EmbeddedImage>
  <EmbeddedImage Name="EmbossedBorder1_Left">
    <MIMEType>image/jpeg</MIMEType>
    <ImageData>/9j/4AAQSkZJRgABAQEAYABgAAD/2wBDAAIBAQIBAQICAgICAgICAwUDAwMDAwYEBAMFBwYHBwcGBwcICQsJCAgKCAcHC
  </EmbeddedImage>
  <EmbeddedImage Name="EmbossedBorder1_Right">
    <MIMEType>image/jpeg</MIMEType>
    <ImageData>/9j/4AAQSkZJRgABAQEAYABgAAD/2wBDAAIBAQIBAQICAgICAgICAwUDAwMDAwYEBAMFBwYHBwcGBwcICQsJCAgKCAcHC
  </EmbeddedImage>
  <EmbeddedImage Name="EmbossedBorder1_Top">
    <MIMEType>image/jpeg</MIMEType>
    <ImageData>/9j/4AAQSkZJRgABAQEAYABgAAD/2wBDAAIBAQIBAQICAgICAgICAwUDAwMDAwYEBAMFBwYHBwcGBwcICQsJCAgKCAcHC
  </EmbeddedImage>
  <EmbeddedImage Name="EmbossedBorder1_TopLeft">
    <MIMEType>image/jpeg</MIMEType>
    <ImageData>/9j/4AAQSkZJRgABAQEAYABgAAD/2wBDAAIBAQIBAQICAgICAgICAwUDAwMDAwYEBAMFBwYHBwcGBwcICQsJCAgKCAcHC
  </EmbeddedImage>
  <EmbeddedImage Name="EmbossedBorder1_TopRight">
    <MIMEType>image/jpeg</MIMEType>
    <ImageData>/9j/4AAQSkZJRgABAQEAYABgAAD/2wBDAAIBAQIBAQICAgICAgICAwUDAwMDAwYEBAMFBwYHBwcGBwcICQsJCAgKCAcHC
  </EmbeddedImage>
</EmbeddedImages>
```


Figure 11-1. *The XML for embedded images*

5. View the code for the "destination" report, and search for the EmbeddedImage tags. Paste the XML that you copied between the `<EmbeddedImage> ... </EmbeddedImage>` tags.
6. Save the code and close the file.
7. Reopen the file normally. You will see the images that you just copied if you expand the `Images` folder in the Report Data window.

Hints and Tips

There are a couple of points to note about this technique.

- If your destination report (the dashboard, in your example) does not contain the `<EmbeddedImage> ... </EmbeddedImage>` tags (because there are no images in the report already), you can add them manually. These tags must only be nested in the `<Report>..</Report>` tags, and not inside any other tag.
- If you need to copy multiple images that are *not* part of a contiguous bloc, as they are in this example, you could consider copying and pasting the required image tags into a Notepad file before pasting them all together into the destination file.

Copying Datasets Between Reports

If you are copying data sources between reports, the procedure is largely identical to the one you just saw for images. The only difference is that you need to look for the `<DataSources>...</DataSources>` Xml tags. An example is

```
<DataSources>
  <DataSource Name="CarSales_Reports">
    <DataSourceReference>CarSales_Reports</DataSourceReference>
    <rd:SecurityType>None</rd:SecurityType>
    <rd:DataSourceID>41799262-6788-4b94-8049-3c91f78c06a7</rd:DataSourceID>
  </DataSource>
</DataSources>
```

Once you have found the data source(s) that you want, copy them into the destination report. If your destination report does not already have any data sources, you will need to include the outer `<DataSources>...</DataSources>` XML tags. Otherwise, you can add the data source itself (the snippet enclosed by the `<DataSource>...</DataSource>` XML tags) inside the `<DataSources>...</DataSources>` tags, but *not* inside any existing the `<DataSource>...</DataSource>` tags.

Copying Datasets Between Reports

Datasets follow the same principles, only you need to look for the `<DataSets>...</DataSets>` tags. An example is

```
<DataSets>
  <DataSet Name="CurrentMonth">
    <SharedDataSet>
      <SharedDataSetReference>CurrentMonth</SharedDataSetReference>
    </SharedDataSet>
```

```
    <Fields>
      <Field Name="CurrentMonth">
        <DataField>CurrentMonth</DataField>
        <rd:TypeName>System.Int32</rd:TypeName>
      </Field>
    </Fields>
  </DataSet>
</DataSets>
```

Once again, if your destination report does not already have any datasets, you will need to include the outer <DataSets>...</DataSets> XML tags. Otherwise, you can add the data source itself (the snippet enclosed by the <DataSet>...</DataSet> XML tags) inside the <DataSets>...</DataSets> tags, but *not* inside any existing the <DataSet>...</DataSet> tags.

Images in BI Projects

In this book, up until now you have only used embedded images in the reports that you have created. This approach is a great solution for a few images, for images that never change, or for images that are only used in a single report. However, if you have hundreds of images, or if you have images that need to be added to (and probably updated in) dozens-or hundreds-of reports, then using database-based images is probably a more efficient solution.

Using images from a database is not difficult, but it does require a little planning and organization. Firstly, you need a clean table structure in which to store your images, and it helps to have all the image metadata that you could require as well. Then, as you will soon see, you need ways of selecting images from a table of images and delivering them to the report. This section explains all of the basic techniques you will use to centralize and distribute images in SSRS.

Loading and Importing Database Images

If you are aiming to store report images in a SQL Server table, you need to be able to load and unload binary files to and from a database. This is not difficult, but to save you having to dig out the code, an example follows. Firstly, you need a table to store the images. The code for this is

```
CREATE TABLE Data.ReportImages
(
        ID int IDENTITY(1,1) NOT NULL,
        FileName nvarchar(50) NULL,
        FileExtension nvarchar(5) NULL,
        FileType nvarchar(5) NULL,
        ImageDescription nvarchar(max) NOT NULL,
        BinaryData varbinary(max) NULL,
        DateLoaded datetime NULL
)
```

Loading a single image requires a short code snippet like the following one:

```
INSERT INTO Data.ReportImages
(FileName, FileExtension, FileType, ImageDescription, BinaryData)
SELECT 'UK_Round', '.png', 'PNG', 'UK Flag'
, * FROM OPENROWSET(BULK N'C:\BIWithSSRS\Images\UnionJack.png', SINGLE_BLOB) img
```

You do not have to add or use fields such as `FileName` and `FileExtension` if you do not want to, but they can be useful in certain circumstances. Indeed, you might want to add fields that allow you to categorize your images, or even list the reports they are used in. You might also want to add fields to store the image height and width, or possibly categorize the image resolution (high, medium, or low).

■ **Note** If you need further information on images in databases, or have a requirement to load multiple images at once, export or update images, or if you simply prefer to use SQL Server Integration Services to load your image files into a database, these techniques are explained in my book *SQL Server 2012 Data Integration Recipes* (Apress 2012).

Using Images from a SQL Server Table

I will assume that you have created the table `Data.ReportImages` using the DDL given above, and that you have loaded the following images from the directory `C:\BIWithSSRS\Images` (assuming that you have downloaded the sample data from the Apress web site):

- `UnionJack`
- `UK_Round`
- `SwissFlag_Round`
- `FrenchFlag_Round`
- `GermanFlag_Round`

Once you have loaded the image files, running a simple SELECT query against the table `Data.ReportImages` should return the data shown in Figure 11-2.

	ID	FileName	FileExtension	FileType	ImageDescription	BinaryData	DateLoaded
1	1	UnionJack	.png	PNG	UK Flag	0x89504E470D0A1A0A0000000D49484452000000011E000000...	2014-09-07 09:48:26.213
2	2	UK_Round	.jpg	JPG	UK Flag Round	0xFFD8FFE000104A46494600010100000100010000FFDB008...	2014-09-19 17:07:08.440
3	3	SwissFlag_Round	.jpg	JPG	UK Flag Round	0xFFD8FFE000104A46494600010100000100010000FFDB008...	2014-09-19 17:08:06.127
4	4	FrenchFlag_Round	.jpg	JPG	UK Flag Round	0xFFD8FFE000104A46494600010100000100010000FFDB008...	2014-09-19 17:08:20.830
5	5	GermanFlag_Round	.jpg	JPG	UK Flag Round	0xFFD8FFE000104A46494600010100000100010000FFDB008...	2014-09-19 17:08:32.427

***Figure 11-2.** The image table with a few files loaded*

You will now see how to use these images in

- Image items
- Text boxes
- Table backgrounds
- Gauge backgrounds

Adding a Dataset for Images to a Report

As any images that you want to use are in a database, you will need a dataset to deliver them from the database to the report. However, you will need a way to define which image is going to be used from all those in the table. There are really only two solutions:

- Set up a dataset for each image so that a single image is returned.
- Pivot a group of images and select the appropriate image to apply from the fields that are available in the pivot set.

Let's see an example of each.

Multiple Image Datasets

If your report only contains a handful of images at the most, then you are probably best served by doing the following:

- Creating a shared dataset that returns the image data.
- Creating multiple datasets based on this shared dataset where each derived dataset has a filter applied to return only the required image.

To make this clearer, here is an example.

1. Create a stored procedure named `Code.pr_Images` using the following code:

```
CREATE PROCEDURE Code.pr_Images

AS

  SELECT
  BinaryData AS ImageData
  ,FileType
  ,FileName

  FROM Data.ReportImages
```

2. Create a new, shared dataset named `ReportImages` using the data source `CarSales_Reports`. Set it to use the stored procedure `Code.pr_Images`.
3. Create a new, blank report and add the shared data source `CarSales_Reports`.
4. Add a new dataset using the shared dataset named `ReportImages`. Name the first instance of this shared dataset UKFlagImage.
5. Click Filters on the left.

6. Click Add.
7. Set the following filter attributes:
    a. Expression: FileName
    b. Operator: =
    c. Value: UK_Round
8. Click OK.

Once your first dataset has been set up, you can create as many similar datasets as you have images that you want to use in your report. The only trick is to use a suitable name for the dataset, and filter each dataset to use the appropriate record from the image table. You will see how to apply this image in a couple of pages.

Pivoted Image Datasets

As you can probably imagine, having a dataset per image can rapidly become unmanageable when reports have more than a few images. An alternative solution is to deliver images to reports using a single dataset. This dataset consists of a single record where each *column* in the dataset contains an image. In other words, you are simply pivoting the source table.

As a simple example, take a look at the following code snippet, which returns a chosen subset of images from the table `Data.ReportImages`:

```
CREATE PROCEDURE Code.pr_ImageFlags

AS

SELECT
 (SELECT BinaryData FROM Data.ReportImages WHERE FileName = 'UK_Round') AS UK_Round
,(SELECT BinaryData FROM Data.ReportImages WHERE FileName = 'SwissFlag_Round')
  AS SwissFlag_Round
,(SELECT BinaryData FROM Data.ReportImages WHERE FileName = 'FrenchFlag_Round')
  AS FrenchFlag_Round
```

The output from this 'sproc looks something like Figure 11-3.

	UK_Round	SwissFlag_Round	FrenchFlag_Round	GermanFlag_Round
1	0xFFD8FFE000104A4...	0xFFD8FFE000104A464946...	0xFFD8FFE000104A464...	0xFFD8FFE000104A464...

Figure 11-3. *Pivoted image output*

As it is the column name that specifies which image is contained in the binary data, you effectively have a useable set of images to apply in a report. To use this pivoted source of database images, follow these steps.

1. Create a report and add the dataset `UKFlagImage` that you created previously.
2. Create a new, shared dataset named `ReportMultiImages` using the data source `CarSales_Reports`. Set it to use the stored procedure `Code.pr_ImageFlags`.
3. Create a new, blank report and add the shared data source `CarSales_Reports`.
4. Add a new dataset using the shared dataset named `ReportImages`. Name this shared dataset `ImageFlags`.

■ **Note** You do not *have* to use a shared dataset, but as this chapter is all about reusability I want to stick to the principles that you should be applying across the board.

Hints and Tips

There is one tip that you may find useful when using multiple images from a database in SSRS.

- It can be a good idea to prepare several sets of pivoted images (grouping them thematically, for instance) rather than trying to create a single massive image set containing every image in your table of images. This also makes applying the images easier, as you will not be scrolling through vast lists of images.

Database Images in Image Items

I realize that many readers will find this illustration a little obvious, but in the interest of completeness, here is how to use a database image in an image item.

1. Drag an Image item from the toolbox onto the report area. The image properties dialog will appear.
2. Select Database as the image source from the pop-up.
3. Select `=First(Fields!ImageData.Value, "UKFlagImage")` as the field to use.
4. Select Image/png as the MIME type.

Be warned that you will not see the image until you preview or deploy and display the image. I realize that this can be frustrating at first.

■ **Note** If you are using the pivoted image dataset (`ImageFlags`) in step 3, then you must be sure to use the appropriate *dataset* name as well as the *column* name that maps to the image that you want to see displayed.

Database Images in Text Boxes

To ensure that you can use database images in all possible circumstances (while accepting that this may be a bit simplistic for some readers), here is how to add a database-based image as the background to a text box.

1. Create a report and add the dataset `UKFlagImage` that you created previously.
2. Add a text box to the report where you want to use an image as a text box background.
3. Display the Properties window.

4. Expand the BackgroundImage category and set the following elements:
    a. Source: Database
    b. Value: `=First(Fields!ImageData.Value, "UKFlagImage")`
    c. MIMEType: image/png

Once again, you will only see the image when the report is previewed, or deployed and viewed.

Database Images in Table Backgrounds

Adding an image from a database to a table as the table background is virtually identical to the way that you added an image to a text box. For the sake of completeness, here is how to do it.

1. Create a report and add the dataset `UKFlagImage` that you created previously.
2. Select the table where you want to use an image as a text box background.
3. Display the Properties window.
4. Expand the BackgroundImage category and set the following elements:
    a. Source: Database
    b. Value: `=First(Fields!ImageData.Value, "UKFlagImage")`
    c. MIMEType: image/png

Once again, you will only see the image when the report is previewed, or deployed and viewed.

■ **Note** An image sourced from a database is an image like any other. So you will still have to set any relevant properties such as the `Sizing` property for image items and the `BackgroundRepeat` for text boxes.

Database Images in Gauge Backgrounds

Using database images for a gauge is only a slight variation on the techniques that you have seen so far.

1. Select the gauge (and not the gauge panel) where you want to add an image.
2. Expand the BackFrame category, then the FrameImage category, and set the following elements:
    a. Source: Database
    b. Value: `=First(Fields!ImageData.Value, "UKFlagImage")`
    c. MIMEType: image/png

Once again, you will only see the image when the report is previewed, or deployed and viewed.

■ **Note** Remember that if you have an image database, you may want first to create a shared dataset to query the image table, and secondly to add this dataset to any reports that you are using as templates (or models) for future reports. In the example solution used in this book, you would add a shared dataset like this to the reports `__BaseReport.rdl` and `__DateSelector.rdl`, for instance.

Centralized Style Information

As many developers have noted over the years, SSRS does not have any form of built-in style sheets. This means, in practice, that you can expend considerable effort doing the following:

- Creating objects (tables, text elements, chart attributes) that look the same across dozens of reports.
- Updating objects in dozens of reports when the corporate style gurus decree a minor change to an aspect of a report.

There is a (fairly) simple workaround to this problem-a problem that affects all SSRS reporting, and not just BI reports, incidentally-and it is to record style attributes in a central database table. These attributes are then applied from the database table. Although this approach can add a little time to the initial development of a report, this investment is rapidly recovered because

- Standards are defined.
- Presentation attributes are applied across an entire reporting suite.
- Stylistic and typographical errors are minimized.
- Any update is made once at a central point, and then the reports that refer to the style are updated automatically the next time they are displayed.

It has to be said that applying styles from a database table is a little laborious; the approach is not as simple or harmonious as the use of style sheets in Microsoft Word, for example. However, I can only repeat that if this approach is implemented early on in a project, it is truly worth the effort.

As an example, let's suppose that you have analyzed a series of presentational attributes that you will be applying to a set of BI reports. These attributes cover the following aspects of a report:

- Header text
- Subheader text
- Table text (header, subheader, body text, and totals)
- Chart titles
- Chart axis titles (X and Y)

Clearly, there could be much, much more but this is surely enough to give you an idea of what can (and must) be done.

These elements are stored in a table with the following DDL:

```
CREATE TABLE Reference.SSRS_Styles(
     ID int IDENTITY(1,1) NOT NULL,
     StyleFamily varchar(50) NULL,
     HEADER_Font_Colour varchar(50) NULL,
```

```
HEADER_Font_Size varchar(50) NULL,
HEADER_Font_Family varchar(50) NULL,
HEADER_Font_Style varchar(50) NULL,
HEADER_Font_Weight varchar(50) NULL,
SUBHEAD_Font_Colour varchar(50) NULL,
SUBHEAD_Font_Size varchar(50) NULL,
SUBHEAD_Font_Family varchar(50) NULL,
SUBHEAD_Font_Style varchar(50) NULL,
SUBHEAD_Font_Weight varchar(50) NULL,
TABLE_MAINTITLE_Font_Colour varchar(50) NULL,
TABLE_MAINTITLE_Font_Size varchar(50) NULL,
TABLE_MAINTITLE_Font_Family varchar(50) NULL,
TABLE_MAINTITLE_Font_Style varchar(50) NULL,
TABLE_MAINTITLE_Font_Weight varchar(50) NULL,
TABLE_MAINTITLE_Background_Colour varchar(50) NULL,
TABLE_SUBTITLE_Font_Colour varchar(50) NULL,
TABLE_SUBTITLE_Font_Size varchar(50) NULL,
TABLE_SUBTITLE_Font_Family varchar(50) NULL,
TABLE_SUBTITLE_Font_Style varchar(50) NULL,
TABLE_SUBTITLE_Font_Weight varchar(50) NULL,
TABLE_SUBTITLE_Background_Colour varchar(50) NULL,
TABLE_TOTAL_Font_Colour varchar(50) NULL,
TABLE_TOTAL_Font_Size varchar(50) NULL,
TABLE_TOTAL_Font_Family varchar(50) NULL,
TABLE_TOTAL_Font_Style varchar(50) NULL,
TABLE_TOTAL_Font_Weight varchar(50) NULL,
TABLE_TOTAL_Background_Colour varchar(50) NULL,
TABLE_STANDARD_Font_Colour varchar(50) NULL,
TABLE_STANDARD_Font_Size varchar(50) NULL,
TABLE_STANDARD_Font_Family varchar(50) NULL,
TABLE_STANDARD_Font_Style varchar(50) NULL,
TABLE_STANDARD_Font_Weight varchar(50) NULL,
TABLE_STANDARD_Background_Colour varchar(50) NULL,
CHART_MAINTITLE_Font_Colour varchar(50) NULL,
CHART_MAINTITLE_Font_Size varchar(50) NULL,
CHART_MAINTITLE_Font_Family varchar(50) NULL,
CHART_MAINTITLE_Font_Style varchar(50) NULL,
CHART_MAINTITLE_Font_Weight varchar(50) NULL,
CHART_MAINTITLE_Background_Colour varchar(50) NULL,
CHART_SUBTITLE_Font_Colour varchar(50) NULL,
CHART_SUBTITLE_Font_Size varchar(50) NULL,
CHART_SUBTITLE_Font_Family varchar(50) NULL,
CHART_SUBTITLE_Font_Style varchar(50) NULL,
CHART_SUBTITLE_Font_Weight varchar(50) NULL,
CHART_SUBTITLE_Background_Colour varchar(50) NULL,
CHART_Y_AXIS_Font_Colour varchar(50) NULL,
CHART_Y_AXIS_Font_Size varchar(50) NULL,
CHART_Y_AXIS_Font_Family varchar(50) NULL,
CHART_Y_AXIS_Font_Style varchar(50) NULL,
CHART_Y_AXIS_Font_Weight varchar(50) NULL,
CHART_X_AXIS_Font_Colour varchar(50) NULL,
```

```
    CHART_X_AXIS_Font_Size varchar(50) NULL,
    CHART_X_AXIS_Font_Family varchar(50) NULL,
    CHART_X_AXIS_Font_Style varchar(50) NULL,
    CHART_X_AXIS_Font_Weight varchar(50) NULL,
 CONSTRAINT PK_SSRS_Styles PRIMARY KEY CLUSTERED
(
    ID ASC
)WITH (PAD_INDEX = OFF, STATISTICS_NORECOMPUTE = OFF, IGNORE_DUP_KEY = OFF,
ALLOW_ROW_LOCKS = ON, ALLOW_PAGE_LOCKS = ON) ON PRIMARY
) ON PRIMARY
```

As you can see, the table reflects the some of the various properties that you can set in the Properties window for text boxes (including inside tables) and charts. Indeed, you *must* create a separate field in the style table for each property element that you wish to set in this way.

A small snippet of what some of the data in this table will look like is shown in Figure 11-4.

	ID	StyleFamily	HEADER_Font_Colour	HEADER_Font_Size	HEADER_Font_Family
1	1	Normal	Black	9pt	Arial
2	2	Special	Blue	18pt	Arial Black

Figure 11-4. *Data in a style table*

As you can see, this approach lets you set multiple "families" (or groups) of attributes identified by a reference, Normal or Special in this example. You will use this criterion to set the overall presentation of a set of visual attributes in a few minutes.

In keeping with the logic used so far in this book, here is the stored procedure that will be used to return data from the table to a report:

```
CREATE PROCEDURE Code.pr_StyleOutput
(
@StyleFamily VARCHAR(50)
)

AS

SELECT  *
FROM    Reference.SSRS_Styles
WHERE   StyleFamily = @StyleFamily
```

Applying Style Information from a Database Table

Once you have a table that contains style definitions for a report suite, you can start applying the style attributes, rather than hard-coding them as you have done up until now. Here is a short example to show how to format a text box.

1. Create a new, shared dataset named `StyleOutput` using the data source `CarSales_Reports`. Set it to use the stored procedure `Code.pr_ StyleOutput`.
2. Create a new, blank report and add the shared data source `CarSales_Reports`.

3. Add a new dataset using the shared dataset named `StyleOutput`. Name this shared dataset `StyleOutput`.
4. While the Dataset Properties dialog is open, click Parameters on the left.
5. Click Add.
6. Set (though it should appear by default) @StyleFamily as the parameter name and Special as the parameter value.
7. Add a text box and enter some text. Tradition dictates that this really should be Hello World...
8. Select the text box and display the Properties window.
9. Expand the `Font` property.
10. Set the `FontFamily` property to the expression `=First(Fields!HEADER_Font_Family.Value, "StyleOutput")`
11. Set the `FontSize` property to the expression `=First(Fields!HEADER_Font_Size.Value, "StyleOutput")`
12. Set the `Color` property to the expression `=First(Fields!HEADER_Font_Colour.Value, "StyleOutput")`

That is all that you have to do, at least for this text box. As was the case with images earlier, you will not see the results of your formatting until you either preview or deploy and view the report.

The advantages of this approach are probably not instantly obvious. So to make the point, double-click the dataset `StyleOutput` and alter the parameter value from Special to Normal, then preview the report. A completely different set of attributes will be applied, and the text box will no longer be in blue, Arial Black 18 point, but black Arial 9 point.

So, despite the fact that applying the properties takes a little longer, you now have a centralized repository for key visual attributes for a report. You can decide at the level of a report which set of styles is applied as well as ensure that a single change to a table element is reflected across a suite of reports.

Hints and Tips

Just to give you a few more ideas, consider these tips.

- If you are reusing widgets targeting different devices (for smartphones, tablets, or laptops, for instance), you can define a style set for each type of device. So, for instance, smartphones could have smaller text but more striking colors, laptops could use pastel shades, etc.
- Rather than using a single style table as I did here, you could have smaller (and more manageable) tables focusing on a type of object: one for tables, one for charts, and one for gauges, for instance.
- Create preformatted tables using the colors, fonts, and borders that you want to use across your reportset. All you then need to do is change the dataset and apply the fields in cells and expressions.

Conclusion

In this chapter, you saw a series of ideas that enable you to standardize presentation across a range of BI reports. You saw techniques that range from ways to copy essential objects such as dataset and images between reports, to ways of sharing commonly used items between reports. Then you saw how to store repeatedly-used images in a database and from there deliver them to reports, both individually and as sets of images. Finally, you saw that report properties can be centralized too in a database and used as the basis for standardized presentation.

So now it is time to move on to the last chapter and take a look at a few ways of delivering your BI dashboards and tablet visualizations to your users faster and more smoothly.

CHAPTER 12

Optimizing SSRS for Business Intelligence

So far in this book you have seen a multitude of ways of creating business intelligence reports, dashboards, and mobile output using SQL Server Reporting Services. Yet there is one vital part of the puzzle that needs to be discussed. It is how to deliver the data to your users as fast as possible. Few things will cause your users to complain-or worse, request rival products-than reports that take more than a few seconds to appear on their PC, tablet, or smartphone.

Consequently, it is fundamental to learn the variety of techniques that you can apply to speed up report delivery. They range from restructuring the source data to delivery options such as caching reports and datasets.

Optimizing BI delivery is, inevitably, a vast subject. It follows that all we have time for here is a whirlwind tour of some of the possibilities that you can adapt to deliver your business intelligence faster. As many of these techniques have been explained in considerable detail in other books and articles, my aim here is more to remind you of the available solutions rather than to provide an exhaustively detailed description of all that can be done.

This chapter will briefly discuss the following:

- Denormalized data sources
- Report-specific tables
- Indexed views
- Indexes (including filtered indexes and covering indexes)
- Columnstore indexes
- Caching reports and datasets
- In-memory data sources

Most of the techniques that you will see in this chapter can be used in many different contexts, and not just for BI. By this I mean that indexed views may be particularly useful when your source data is a copy of an OLTP data source, but they can also come in handy when you are creating a denormalized data source for reporting, too. Similarly, covering indexes can be used to great effect in a wide range of reporting scenarios, and not just when you are building a specific denormalized data set for a BI report suite.

When considering the range of options that you could implement to accelerate BI delivery to your users, it is important to remember that there is probably not one solution. In most cases, there are many techniques that you can apply conjointly to accelerate report delivery rather than focusing on a single option. Indeed, you may find that you are better off applying multiple tweaks to your BI process, as many of the ideas that you will find can be complementary rather than exclusive. It follows that a more holistic approach can pay real dividends when considering which techniques to apply to your reporting suite. So be prepared to try out (and eventually apply) several of the ideas that you will see in this chapter to your reports. After all, experimentation and some trial and error are all part of the process.

These ideas are not the only ways to speed up report delivery. Far from it. However, an in-depth analysis of all the potential tricks that exist to accelerate BI delivery is outside the scope of this short chapter. So, without more ado, onto some of the ways to make your BI not only smarter but faster.

Normalized Data Sources

A majority of the business intelligence data sources that I have seen over the years have essentially been normalized database tables. In many cases, they were copies of OLTP data from a source database. Often the copied data was a subset of the source data, given that only a partial set of the original data was required for BI delivery. In a few rare cases the BI hooked directly into the OLTP system (though this nearly always proved to be a bad idea).

So I am prepared to bet that you may well have to use SQL Server data (or data from any other relational database system) as your data source. While this is not a problem in any way, there are a few techniques and structural approaches that can be applied to make these data sources more efficient when it comes to delivering data to dashboards and scorecards. These techniques include

- Creating reporting tables to subset only the required data
- Applying the right indexes
- Judicious use of indexed views

While you may not need all or indeed any of these techniques, it is a good idea to appreciate how they can be used. You can then approach your BI project with a more overall vision, and build a platform that combines efficiency with clarity.

Reporting Tables

To begin with, let me say that I am using the term “reporting tables” very idiosyncratically; I mean tables that are specifically created as repositories for data that will be used for some or all of your BI reports. In some cases you may even find it productive to create tables that are specific to certain BI visualizations. Moreover, I am assuming that these tables will be truncated and loaded, or updated, regularly with every reporting cycle (such as at the end of a daily processing schedule, for instance). It follows that such tables are really “temporary” persisted tables with a reporting focus.

Often tables like this will be denormalized to some extent. By this I mean that the core dataset is based on a more traditional, normalized OLTP database, but (where possible) data from some of the peripheral lookup tables have been integrated into the more “major” tables. The objective in this case is to create a clearer set of fewer, flattened reporting tables, and thus simplify access to the data while at the same time making the data model more comprehensive. This approach is not without its downsides, notably that there is much repetition of data.

The advantages of such highly-targeted data sources are

- They can be specifically customized to optimize report delivery for a few reports that otherwise would take a while to render.
- They can contain a specific subset of data, and selection criteria, for a few frequently-used reports.
- They can be processed out of hours and specifically before users start viewing their reports.

The disadvantages of such data sources are

- They can require considerable preparation.
- They necessitate specific maintenance, especially if source data structures change.
- They have to be integrated into an existing processing cycle, and can extend processing time.

Nonetheless, there are circumstances where the advantages outweigh the drawbacks. So if you have reports that need a specific dataset, and these reports take time to render, consider creating data source tables specifically for these reports. And, of course, index them coherently to get the best out of your tables.

In a reporting scenario, reporting tables can have one big advantage. They can be updated at the end of a regular ETL cycle, say a daily or weekly process, and in many cases this only means adding new records, not updating or deleting existing records. This approach offers the following advantages:

- Adding the new records can be much, much faster than truncating and reloading entire tables.
- The data that is added is a small proportion of the entire dataset, so indexes on these tables can be updated at less frequent intervals, say weekly in the case of a daily table update.

As an extremely simple example, suppose that you want to create a reporting table based on the view Reports.CarSalesData that you have been using throughout this book. At its simplest, the code would be

```
IF OBJECT_ID ('Reports.CarSalesReporting') IS NOT NULL DROP TABLE Reports.CarSalesReporting

SELECT *
INTO   Reports.CarSalesReporting
FROM   Reports.CarSalesData
```

A table like this one can then become the basis for your suite of BI reports. This way the heavy-duty processing is carried out *once* at the end of the data load cycle.

Aggregated Reporting Tables

The table that you created in the previous example (Reports.CarSalesReporting) is nothing more than an agglomeration of all the fields that you need for all your reports. This is, however, probably not all that a reporting table can provide. To begin with, an OLTP system probably provides detailed records of every transaction. However, a BI system (equally probably) only needs data aggregated at a daily, or even monthly level. So one of your first considerations should be to aggregate the data in reporting tables to the lowest level of grain required by your reports.

For instance, you could begin by aggregating all the detail-level records in your OLTP system into a reporting table that summarizes data at the daily level. This could

- Reduce the number of records in the reporting table.
- Remove the need for endless repeated aggregations in every report query.

If you also have reports at higher level of grain, you could then use the daily aggregations table as a basis for weekly and monthly (even quarterly) tables. In my experience, the hard work from a processing viewpoint is at the level of daily aggregations. Re-aggregating days into weeks, months, and quarters can be extremely rapid. These tables can be processed (either in their totality or by adding fresh data only) as part of the daily/weekly/monthly processing cycle. In any case, the reports that use this data will be rendered much faster than if they were sourced in tables containing all the unaggregated data.

■ **Note** Denormalized reporting tables will probably need careful indexing, just as copies of OLTP tables will.

Indexes

As a seasoned SQL Server developer, you doubtless know the value of a well-defined indexing strategy. Well, efficient BI report delivery means using indexes effectively too. Nothing, of course, can replace a balanced analysis of your existing and missing indexes using the tools and Dynamic Management Views (DMVs) that are part of the SQL Server toolkit. However, as a starting point, you should probably consider indexing any fields that are used for

- Parameters
- Filters
- Grouping

Indeed, I always recommend listing all the fields that are used in the parameters of your reports and stored procedures because these are frequently a key baseline for the fields that you will need to index.

As you are probably aware, entire books have been written about indexing, detecting required indexes, and defining the indexes (and the types of index) to apply to a database. For more information on these topics, I recommend *Expert Performance Indexing for SQL Server 2012* by Jason Strate (Apress 2012).

Report Parameters

If you want a list of all the parameters used by your reports, you can query the Catalog table from the ReportServer database. This is much easier and faster than opening every report and looking at the report parameters individually. These parameters could be a valuable indication of the fields that you need to index in the underlying tables.

The code snippet that will list the parameters used in each report is the following:

```
WITH SSRS_CTE
AS
(
SELECT        Name
             ,CAST(Parameter AS XML) AS ParameterXML
FROM         ReportServer.dbo.Catalog
WHERE        Parameter IS NOT NULL
)

SELECT DISTINCT  Name, R.value('(.)','nvarchar(150)') AS SSRSParameter
FROM             SSRS_CTE
CROSS APPLY      ParameterXML.nodes('Parameters/Parameter/Name') as SSRS(R)
ORDER BY         Name, SSRSParameter
```

Running this code against the ReportServer database where the sample reports have been deployed will give something like the output shown in Figure 12-1 (although only the first few lines are shown).

	Name	SSRSParameter
1	__BaseReport	ReportingMonth
2	__BaseReport	ReportingYear
3	_BubbleChartMarginByAgeAndMake	ReportingMonth
4	_BubbleChartMarginByAgeAndMake	ReportingYear
5	_ColourSales	ReportingMonth
6	_ColourSales	ReportingYear
7	_CountryChartPercentageToTarget	ReportingMonth
8	_CountryChartPercentageToTarget	ReportingYear
9	_CountryHighlight	Country
10	_CountryHighlight	IsFrance
11	_CountryHighlight	IsGermany
12	_CountryHighlight	IsSpain
13	_CountryHighlight	IsSwitzerland
14	_CountryHighlight	IsUnitedKingdom
15	_CountryHighlight	IsUSA
16	_CountryHighlight	ReportingMonth
17	_CountryHighlight	ReportingYear
18	_DateWidget	ReportingMonth
19	_DateWidget	ReportingYear

Figure 12-1. *Report parameters*

Stored Procedure Parameters

As mentioned earlier, the parameters that you pass into the stored procedures that return data to your BI reports can indicate that the fields that they filter on might be valid candidates for indexes. So it is worth knowing that you can list all the parameters used in all the stored procedures in a database with the following code snippet:

```
SELECT DISTINCT  name
FROM             sys.parameters
ORDER BY         name
```

When applied to the CarSales_Reports database, you will see output like that shown in Figure 12-2. The output is not the whole list that is returned, but you should get the idea.

	name
1	
2	@BatchRow
3	@BatchStart
4	@ClientName
5	@Color
6	@Country
7	@definition
8	@diagramname
9	@InputMaxValue
10	@IsAstonMartin
11	@IsBentley
12	@IsJaguar
13	@IsRollsRoyce
14	@Make
15	@new_diagramname
16	@owner_id
17	@ReportingMonth
18	@ReportingYear

Figure 12-2. *Stored procedure parameters*

You will need to use your discretion and your knowledge of the reporting suite to decide if a parameter really indicates that a corresponding index is necessary on the source data table. Also, you will need to guess which field a parameter relates to and in which table. Moreover, there will be parameters that have nothing at all to with reports (@new_diagramname is an example). And there could be parameters that relate to data types that cannot or should not be indexed (such as the @IsBentley and @IsAstonMartin fields, amongst others, in this sample data). Nonetheless, a good look at the parameters used by your stored procedures can provide an invaluable starting point for a reporting suite indexing strategy.

Covering Indexes

Business intelligence visualizations often only use a few columns of data; think of the charts and gauges you built in previous chapters. However, they are often highly selective. So you will probably be able to isolate "core" fields that are used to filter the data that goes into gauges, tables, and charts. However, indexing a field that is used for parameterizing or filtering a visualization is not the end of the matter. Although you will certainly have accelerated the search for the relevant records, remember that an index could then have to use key lookups to return data not in the index from the data table itself-and slow the query down considerably.

The classic solution to this problem is to create covering indexes. These are indexes that search on specified fields, but also contain the data for other fields. The trick is to return the fields required elsewhere in your table, gauge, or chart.

As an example, consider the code snippet you used to create the gauges in Figure 10-11 in Chapter 10 (Sales by make for smartphone display).

```
declare @ReportingYear int = 2014
declare @ReportingMonth int = 10

SELECT    CASE
             WHEN Make IN ('Aston Martin','Bentley','Jaguar','Rolls Royce') THEN Make
             ELSE 'Other'
          END AS Make
          ,SUM(SalePrice)
FROM      Reports.CarSalesReporting WITH (INDEX(IX_CarSalesReporting_MakeAndSales))

WHERE     ReportingYear = @ReportingYear
          AND ReportingMonth <= @ReportingMonth
GROUP BY  CASE
              WHEN Make IN ('Aston Martin','Bentley','Jaguar','Rolls Royce') THEN Make
              ELSE 'Other'
          END
```

This code only needs four fields:

- Make
- ReportingYear
- ReportingMonth
- SalePrice

So a covering index that contains these four fields (either as part of the index or as included data) will mean that *no* key lookups will be required. The code for this index is

```
CREATE INDEX IX_CarSalesReporting_MakeAndSales ON Reports.CarSalesReporting (Make,
ReportingYear, ReportingMonth) INCLUDE (SalePrice)
```

If you create this index on the reporting table Reports.CarSalesReporting and then run the query in the code snippet above, you will see in the execution plan that only the index IX_CarSalesReporting_MakeAndSales is used to return the data.

Admittedly, creating potentially dozens of covering indexes can be time-consuming, but consider the following points:

- It is not difficult to go through the set of reporting stored procedures and run every separate SELECT statement and look at the execution plan.
- Once you have built the first few covering indexes, you will probably find that many of them are reused in multiple visualizations.
- A core set of covering indexes can frequently become that basis for a multitude of stored procedures if you think carefully about the included fields when you create the index so that you return the data required by several queries. Adding a judicious set of metrics, for instance, will not make an index over wide, but it can enable it to be used by dozens of visualizations. This way you will avoid creating a plethora of indexes, and consequently reduce the time spent reindexing the tables.
- This encourages you to use stored procedures rather than ad-hoc code in SSRS!

Filtered Indexes

Business intelligence is, by its very nature, selective. It follows that some BI displays will not need to look at an entire dataset. As an example, consider a set of gauges that only show data for the countries of continental Europe. You will not need any information for the United Kingdom or the United States. This is where a filtered index could be a useful solution.

To show this, here is the code that returns cost data for the European countries where Brilliant British Cars currently sells its prestigious vehicles:

```
declare @ReportingYear int = 2014
declare @ReportingMonth int = 10

SELECT    Make
         ,SUM(CostPrice) AS Cost
FROM     Reports.CarSalesReporting
WHERE    CountryName IN ('France', 'Germany', 'Spain', 'Switzerland')
         AND ReportingYear = @ReportingYear
         AND ReportingMonth <= @ReportingMonth
GROUP BY Make
```

This code snippet is not difficult to write or understand. Yet if these countries only make up a few percent of the sales, you are “wasting” a large percentage of the index (that you have presumably created) to return the data. Remember also that a filtered index can even contain an INCLUDE statement.

Here, then, is a quick filtered index to make this code even more efficient:

```
CREATE INDEX IX_CarSalesReporting_EuropeSales ON Reports.CarSalesReporting (CountryName,
ReportingYear, ReportingMonth)
INCLUDE (CostPrice)
WHERE    CountryName IN ('France', 'Germany', 'Spain', 'Switzerland')
```

The index will be smaller, and the data returned faster, because only the relevant information is used in the index.

As a final point, if you are fine-tuning your data preparation to this level, you might need to add a hint to the query that returns the data to force the query optimizer to use your specific index. In this case, the FROM clause in the stored procedure that returns the data will look like this:

```
FROM     Reports.CarSalesReporting  WITH (INDEX (IX_CarSalesReporting_EuropeSales))
```

Indexed Views

If you do not want to isolate reporting data in specific tables, you can apply another solution that has been around for nearly 15 years: indexed views. An indexed view will be stored just as a table with a clustered index is stored, and so the data will be persisted on disk and consequently (in most cases, anyway) accessed faster than it would be from multiple tables.

Indexed views have a series of limitations, and I advise you to consult the Microsoft documentation before you start adding them to your BI solution. The one major drawback is that they apply schemabinding; that is, once an indexed view is created, you cannot alter the definition of the source tables unless you drop the indexed view first. So you will inevitably end up creating indexed views once your source data structures are settled.

As a simple example, here is the code to create an indexed view that is a nearly exact copy of the Reports.CarSalesDataview that you used as the source for nearly all the tables, charts, and gauges that you built during the course of this book. The only difference is that I have added the InvoiceLineID field, as it is unique, and an indexed view must have a unique field (or a unique combination of fields) to use as the basis for a unique clustered index.

Once the view is created, the code first adds a unique clustered index,and then a few other indexes to enhance report delivery.

```
CREATE VIEW  Reports.CarSalesData_Indexed
WITH SCHEMABINDING
AS

SELECT
 InvoiceLineID
,Data.Invoices.InvoiceDate
,Data.Stock.Make COLLATE DATABASE_DEFAULT AS Make
,Data.Countries.CountryName COLLATE DATABASE_DEFAULT AS CountryName
,Data.Clients.IsDealer
,Data.InvoiceLines.SalePrice
,Data.Stock.CostPrice
,Data.Invoices.TotalDiscount
,Data.Invoices.DeliveryCharge
,Data.Stock.SpareParts
,Data.Stock.LaborCost
,Data.Clients.ClientName COLLATE DATABASE_DEFAULT AS ClientName
,Data.Stock.Model COLLATE DATABASE_DEFAULT AS Model
,Data.Colors.Color COLLATE DATABASE_DEFAULT AS Color
,YEAR(Data.Invoices.InvoiceDate) AS ReportingYear
,MONTH(Data.Invoices.InvoiceDate) AS ReportingMonth
,Data.Stock.Registration_Date
,Data.Stock.VehicleType COLLATE DATABASE_DEFAULT AS VehicleType
,Data.Invoices.InvoiceNumber COLLATE DATABASE_DEFAULT AS InvoiceNumber
,Data.Countries.CountryISOCode COLLATE DATABASE_DEFAULT AS CountryISOCode
,Data.Clients.OuterPostode COLLATE DATABASE_DEFAULT AS OuterPostode
,Data.Clients.Region COLLATE DATABASE_DEFAULT AS Region

FROM        Data.Invoices
            INNER JOIN Data.InvoiceLines
            ON Data.Invoices.InvoiceID = Data.InvoiceLines.InvoiceID
            INNER JOIN Data.Stock
            ON Data.InvoiceLines.StockID = Data.Stock.StockID
            INNER JOIN Data.Clients
            ON Data.Invoices.ClientID = Data.Clients.ClientID
            INNER JOIN Data.Countries
            ON Data.Clients.CountryID = Data.Countries.CountryID
            INNER JOIN Data.Colors
            ON Data.Stock.ColorID = Data.Colors.ColorID
```

```
GO

-- Now some indexes - beginning with the unique clustered indexes

CREATE UNIQUE CLUSTERED INDEX CX_CarSalesData_Indexed_InvoiceLineID ON Reports.CarSalesData_
Indexed (InvoiceLineID)

CREATE INDEX IX_CarSalesData_Indexed_Make ON Reports.CarSalesData_Indexed (Make)
CREATE INDEX IX_CarSalesData_Indexed_Color ON Reports.CarSalesData_Indexed (Color)
CREATE INDEX IX_CarSalesData_Indexed_Model ON Reports.CarSalesData_Indexed (Model)

GO
```

Even if an indexed view is updated as the source data changes, you will probably want to update its indexes at regular intervals. One quick way to do this is to use the following code snippet:

```
ALTER INDEX ALL ON Reports.CarSalesData_Indexed
REBUILD WITH (FILLFACTOR = 100, SORT_IN_TEMPDB = ON,
              STATISTICS_NORECOMPUTE = ON);
```

This indexed view can now be used for most of the visualizations you have seen in this book.

Partitioned Data Sources

Partitioning the data tables that supply report data is essentially a way of minimizing disk I/O. This can considerably reduce the time taken to return data to a report. Partitioning, at its simplest, consists of segregating data using a key field or fields so that a query will only read the necessary data partitions and thus avoid superfluous disk access.

Without going into all the potential techniques that are available to partition data, the two classic partitioning modes that you could find yourself using are

- Vertical partitioning
- Horizontal partitioning

Vertical Partitioning

Vertical partitioning consists of dividing a wide table up into several narrower tables that share a common unique ID that is repeated in each table. This approach lets you subset your data manually into tables that separate out less-frequently used fields. You can also group fields by report sets, so that some of the vertical tables map to certain reports.

This approach can require a lot of manual labor to get right. Only a close analysis of your reports will allow you to define which fields belong to a core table and which fields are more "secondary" in their use. However, this approach can bring considerable speed benefits (particularly for the reports that only use the core dataset).

Here is an example of one way of partitioning the table Reports.CarSalesReporting, where you first make a narrower "core" table of the data that is used more frequently, and then a second table that will not be used by many reports.

```
-- Main table, used for many reports

IF OBJECT_ID ('Reports.CarSales_Main') IS NOT NULL DROP TABLE Reports.CarSales_Main

SELECT
 InvoiceLineID
,Data.Invoices.InvoiceDate
,Data.Stock.Make COLLATE DATABASE_DEFAULT AS Make
,Data.Countries.CountryName COLLATE DATABASE_DEFAULT AS CountryName
,Data.Clients.IsDealer
,Data.InvoiceLines.SalePrice
,Data.Stock.CostPrice
,Data.Invoices.TotalDiscount
,Data.Invoices.DeliveryCharge
,Data.Stock.SpareParts
,Data.Stock.LaborCost
,Data.Clients.ClientName COLLATE DATABASE_DEFAULT AS ClientName
,Data.Stock.Model COLLATE DATABASE_DEFAULT AS Model
,Data.Colors.Color COLLATE DATABASE_DEFAULT AS Color
,YEAR(Data.Invoices.InvoiceDate) AS ReportingYear
,MONTH(Data.Invoices.InvoiceDate) AS ReportingMonth

INTO Reports.CarSales_Main

FROM  Data.Invoices
      INNER JOIN Data.InvoiceLines
      ON Data.Invoices.InvoiceID = Data.InvoiceLines.InvoiceID
      INNER JOIN Data.Stock
      ON Data.InvoiceLines.StockID = Data.Stock.StockID
      INNER JOIN Data.Clients
      ON Data.Invoices.ClientID = Data.Clients.ClientID
      INNER JOIN Data.Countries
      ON Data.Clients.CountryID = Data.Countries.CountryID
      INNER JOIN Data.Colors
      ON Data.Stock.ColorID = Data.Colors.ColorID

-- Secondary table, used in fewer reports

IF OBJECT_ID ('Reports.CarSales_Secondary') IS NOT NULL DROP TABLE Reports.CarSales_
Secondary

SELECT
 InvoiceLineID
,Data.Stock.Registration_Date
,Data.Stock.VehicleType COLLATE DATABASE_DEFAULT AS VehicleType
,Data.Invoices.InvoiceNumber COLLATE DATABASE_DEFAULT AS InvoiceNumber
,Data.Countries.CountryISOCode COLLATE DATABASE_DEFAULT AS CountryISOCode
,Data.Clients.OuterPostode COLLATE DATABASE_DEFAULT AS OuterPostode
,Data.Clients.Region COLLATE DATABASE_DEFAULT AS Region
```

```
INTO Reports.CarSales_Secondary

FROM  Data.Invoices
      INNER JOIN Data.InvoiceLines
      ON Data.Invoices.InvoiceID = Data.InvoiceLines.InvoiceID
      INNER JOIN Data.Stock
      ON Data.InvoiceLines.StockID = Data.Stock.StockID
      INNER JOIN Data.Clients
      ON Data.Invoices.ClientID = Data.Clients.ClientID
      INNER JOIN Data.Countries
      ON Data.Clients.CountryID = Data.Countries.CountryID
      INNER JOIN Data.Colors
      ON Data.Stock.ColorID = Data.Colors.ColorID
```

As you can see from the code, there is only one field that overlaps the two tables. This is the InvoiceLineID field, which is the unique identifier for each table. In a real-world application, this field would almost inevitably have to be a unique clustered index on both tables.

This approach will necessitate using the appropriate vertically partitioned table(s) in the report's underlying query. If only the "core" table is used, you will select data from that one table only. If you need data from both (or all) vertically partitioned tables, you will have to join the tables on the shared key column. You can, of course, create a view based on the joined tables to avoid having to rewrite the JOIN clauses in every stored procedure, and use the view in the stored procedures that feed into your report datasets.

Horizontal Partitioning

Horizontal partitioning is, fortunately, a process that is built into the Enterprise version of SQL Server. It is particularly useful in a BI context, as frequently reports look at data for a specific time period, and horizontal partitioning is frequently defined using time periods. Horizontal partitioning also has the advantage of being largely automatic once it has been set up. Although it can require a little initial effort, the efficiency gains can be considerable for BI reports.

Although you may already know how to set up partitioning in SQL Server, I prefer to give a simple example based on a denormalized table, because it is a classic requirement in BI scenarios. This example will partition by the data in the ReportingYear column of the table that you will create.

1. Create as many filegroups as you will have partitions for your table. The following code snippet creates three filegroups (ReportsTo2013, Reports2014, and Reports2015).

   ```
   ALTER DATABASE CarSales_Reports
   ADD FILEGROUP ReportsTo2013;
   GO
   ALTER DATABASE CarSales_Reports
   ADD FILEGROUP Reports2014;
   GO
   ALTER DATABASE CarSales_Reports
   ADD FILEGROUP Reports2015;
   GO
   ```

2. Create the files that will be used by each filegroup, and add the file to each of the filegroups, as shown in the following code snippet:

```
ALTER DATABASE CarSales_Reports
ADD FILE
(
    NAME = ReportsTo2013,
    FILENAME = 'C:\BIWithSSRS\Database\ReportsTo2013.ndf',
    SIZE = 1MB,
    MAXSIZE = 500MB,
    FILEGROWTH = 1MB
)
TO FILEGROUP ReportsTo2013;

ALTER DATABASE CarSales_Reports
ADD FILE
(
    NAME = Reports2014,
    FILENAME = 'C:\BIWithSSRS\Database\Reports2014.ndf',
    SIZE = 1MB,
    MAXSIZE = 500MB,
    FILEGROWTH = 1MB
)
TO FILEGROUP Reports2014;

ALTER DATABASE CarSales_Reports
ADD FILE
(
    NAME = Reports2015,
    FILENAME = 'C:\BIWithSSRS\Database\Reports2015.ndf',
    SIZE = 1MB,
    MAXSIZE = 500MB,
    FILEGROWTH = 1MB
)
TO FILEGROUP Reports2015;
GO
```

3. Create the partition function that will separate data out into the three partitions. Remember that this means specifying only two break points, as all other values will by default be sent to the third partition.

```
CREATE PARTITION FUNCTION BIReporting (int)
    AS RANGE LEFT FOR VALUES (2014, 2015) ;
GO
```

4. Now you can create the partition scheme that uses the partition function to distribute the records into the required partitions. The following code snippet will do this:

```
CREATE PARTITION SCHEME BIReportingPartition
    AS PARTITION BIReporting
    TO (ReportsTo2013, Reports2014, Reports2015) ;
GO
```

5. Now that the partition logic is in place, you can create a table to use it. The following DDL does just this, as it specifies the partition scheme to use and the column to use to separate the data:

```
CREATE TABLE Reports.CarSalesReporting_Partitioned
(
InvoiceDate datetime NULL,
Make nvarchar(50) NULL,
CountryName nvarchar(50) NULL,
IsDealer bit NULL,
SalePrice money NULL,
CostPrice numeric(18, 2) NULL,
TotalDiscount numeric(18, 2) NULL,
DeliveryCharge smallmoney NULL,
SpareParts numeric(18, 2) NULL,
LaborCost numeric(18, 2) NULL,
ClientName nvarchar(150) NULL,
Model nvarchar(50) NULL,
Color nvarchar(50) NULL,
ReportingYear int NULL,
ReportingMonth int NULL,
Registration_Date date NULL,
VehicleType nvarchar(50) NULL,
InvoiceNumber nvarchar(50) NOT NULL,
CountryISOCode nchar(3) NULL,
OuterPostode varchar(2) NULL,
Region varchar(50) NULL
)
ON BIReportingPartition (ReportingYear) ;
GO
```

6. Finally, you can load data into the partitioned table, just as you would for any non-partitioned table. For the sake of completeness, here is the code to do this:

```
INSERT INTO Reports.CarSalesReporting_Partitioned
(
InvoiceDate
,Make
,CountryName
,IsDealer
,SalePrice
,CostPrice
,TotalDiscount
,DeliveryCharge
,SpareParts
,LaborCost
,ClientName
,Model
,Color
```

```
,ReportingYear
,ReportingMonth
,Registration_Date
,VehicleType
,InvoiceNumber
,CountryISOCode
,OuterPostode
,Region
)

SELECT
InvoiceDate
,Make
,CountryName
,IsDealer
,SalePrice
,CostPrice
,TotalDiscount
,DeliveryCharge
,SpareParts
,LaborCost
,ClientName
,Model
,Color
,ReportingYear
,ReportingMonth
,Registration_Date
,VehicleType
,InvoiceNumber
,CountryISOCode
,OuterPostode
,Region

FROM    Reports.CarSalesData
```

You can now use the partitioned table as the basis for the stored procedures that feed into your dashboards and other BI reports. The advantage is that when you are returning data for a single year or a couple of years, only the partitions that contain the required data will be read, reducing disk I/O and response times.

This was, as you can see, an extremely simple example. In practice, your partitions could be more numerous, more complex, or both. You could even combine vertical and horizontal partitioning to achieve optimum response times. It will depend on the size and depth of your data.

■ **Note** When you create indexes on partitioned tables, it is best to create them on the same partition scheme as the table. This way SQL Server will ensure that the indexes, too, make best use of the horizontal partitioning.

Columnstore Indexes

If you are building business intelligence delivery with SQL Server-and as a reader of this book, I presume that you are-then there is one word that should have brought balm to your troubled spirit when SQL Server 2012 was released. That word is *columnstore*. As you probably know, columnstore indexes are a way of storing data column by column rather than row by row and then also compressing the data. The result is that the sort of wide and deep reporting tables that have been the basis for much BI delivery have finally become a valid and efficient way not only of grouping and preaggregating data, but also of delivering it.

The reasoning is simple. If your gauge only requires a handful of columns from a 200-column table, then a columnstore index will ensure that only those columns are read from disk. This cuts down massively on wasted I/O and can reduce the corresponding time taken to display the report.

A columnstore index is created using code like the following:

```
CREATE NONCLUSTERED COLUMNSTORE INDEX IX_CarSalesReporting_ColumnStore
ON Reports.CarSalesReporting
(
InvoiceDate
,Make
,CountryName
,IsDealer
,SalePrice
,CostPrice
,TotalDiscount
,DeliveryCharge
,SpareParts
,LaborCost
,ClientName
,Model
,Color
,ReportingYear
,ReportingMonth
)

GO
```

■ **Note** Note that I have deliberately not selected all the columns in the source table. This is to make the point that you *can* subset the data that you keep in a columnstore index, if you want to.

Once you have created a columnstore index, you can query your source table perfectly normally. SQL Server will, in my experience, make a valid decision as to whether to use the columnstore index or another possible index. You can, of course, force the query optimizer to use a columnstore index just as you would for a traditional index.

There are, inevitably, a few clouds to the silver lining (if you will excuse the metaphor). Specifically, it is worth knowing the following limitations when setting up columnstore indexes:

- A columnstore index can *only* include columns with the following data types: int, big int, small int, tiny int, money, smallmoney, bit, float, real, char(n), varchar(n), nchar(n), nvarchar(n), date, datetime, datetime2, small datetime, time, datetimeoffset (with precision <=2), decimal or numeric (with precision <= 18).
- Only one columnstore index per table is possible.
- A columnstore index cannot be created on an indexed view.
- A computed column cannot be part of a columnstore index.
- A columnstore index cannot have more than 1,024 columns.
- In SQL Server 2014, and if your existing table has existing indexes, you need to drop all indexes, except for the clustered index, before creating a clustered columnstore index.
- Columnstore indexes require, in some cases, a considerable amount of memory when they are created. If SQL Server cannot obtain sufficient memory grants, the columnstore index creation will fail.
- Columnstore indexes cannot be altered-only dropped and recreated.
- In SQL Server 2012, if you need to add data to the underlying table, then you have to drop the columnstore index first. In SQL Server 2014, this limitation no longer applies.

You may well find that few, if any, of these limitations really apply (or are any real problem) in a BI scenario. The most probable obstacle (for SQL Server 2012) is the requirement to drop and recreate the columnstore index when appending data to the underlying table. However, if the time taken to recreate the columnstore index once the table has been updated is not a problem, then the noticeable decrease in response times for dashboards and other BI reports will probably be worth the effort.

Dimensional Data Sources

So far in this chapter you have looked only at relational data sources, and specifically SQL Server. I realize that you could also be using SQL Server Analysis Services as a data provider, and that a few comments on how best to use this particular source of information are required.

As optimizing SSAS is a vast and specialized area, here are a few high-level reminders of the basic things you need to think about when delivering BI in SSRS from SSAS. They are as follows:

- Always take a look at any MDX that was generated by SSDT and see if it can be rewritten more efficiently. The MDX code that is created automatically when you generate reports is renowned for being verbose and slow. Frequently you only need to tweak it a little to get much more performant code and reports that rendered much faster.
- Once you have developed your report suite, it is probably a good time to apply the Usage Based Optimization Wizard to tune your SSAS cube. As you now have some real-world report queries, it is a good moment to let SSAS use them to tune the aggregations in your cube. If you can make the time, it may be worth building aggregations manually.

- Take a step back and consider some classic cube optimizations. This could include the following:
    - Partitioning the cube.
    - Optimizing measure group design to ensure that all related measures are contained in fewer measure groups.
    - Optimizing dimension design to match your hierarchies to the use of drill-down patterns in your reports.
- As you can now see which reports are using what queries, you can consider SSAS cache warming. This means taking a good look at any frequently-used MDX queries and converting them to generic versions that will ensure that regularly-used information is in the cache.

Dataset Caching

Shared datasets (though not datasets that are specific to a report) can be cached. Caching a dataset consists of running the query that a shared dataset is based on, and then storing the result set in the SSRS cache. A cached data set will be created for each parameter combination in the underlying query. Once a dataset is in the cache, any report that contains the shared dataset will use the cached data if the combination of parameters matches that of the cached data.

When you enable a shared dataset for caching, you specify for how long the data will be cached. This can be either a specified number of minutes, or until a scheduled cache expiration. This ensures that stale data is not used to deliver your BI.

Any shared dataset that you have created can be cached. By doing this, you

- Reduce the number of times that the dataset query runs against the external data source.
- Ensure a stable, unchanging set of data across a series of reports.
- Return data to the reports using cached datasets faster.

Remember that you need to cache a dataset for every possible combination of parameters in the dataset's source query because every resulting dataset is different. This can be a problem in some circumstances, but a potential solution is given in the "Parameters or Filters" section.

You also need to be aware that not every query is ideal for a cached dataset. Table 12-1 lists some of the cases when you should and should not consider caching a dataset.

Table 12-1. *The Case for and Against Caching Datasets*

Consider caching	Cases where caching may not be productive
The dataset query takes a while to run.	The underlying query executes swiftly.
There are many reports that use the dataset.	Few reports require this dataset.
The reports using the dataset are queried frequently.	The reports using the dataset are queried infrequently.
It is vital to return data to the report as fast as possible.	Users can reasonably wait a few more seconds to display the report.
The number of parameter combinations is small, or you can reasonably cache only a frequently used set of parameter combinations.	The required number of parameter combinations would be huge.

Moreover, there are a couple of cases where a cached dataset cannot be used. They are the following:

- The data source that the shared dataset is based on has Windows Integrated credentials or prompts for credentials.
- The shared dataset filter or the query contains an expression using a reference to the global collection user.

Anyway, assuming that you have decided that a shared dataset fits the criteria for caching, the next question is, how do you go about this?

Data Source Credentials

Firstly, you need to store credentials for any data sources that are used by shared datasets. This is fairly logical, as datasets may be set to be cached following a specific, unattended schedule, and so need to be able to be run without needing a user to input credentials.

1. Ensure that you have a valid SQL Server user using SQL Server security with sufficient privileges in the database(s) that contain the source data to execute the stored procedures used by your BI reports. In the sample database, this user is BIReporting.
2. Assuming that you have deployed all your reports, data sources, and shared datasets, open Report Manager using the URL for your report server. This brings you to the browser screen shown in Figure 12-3.

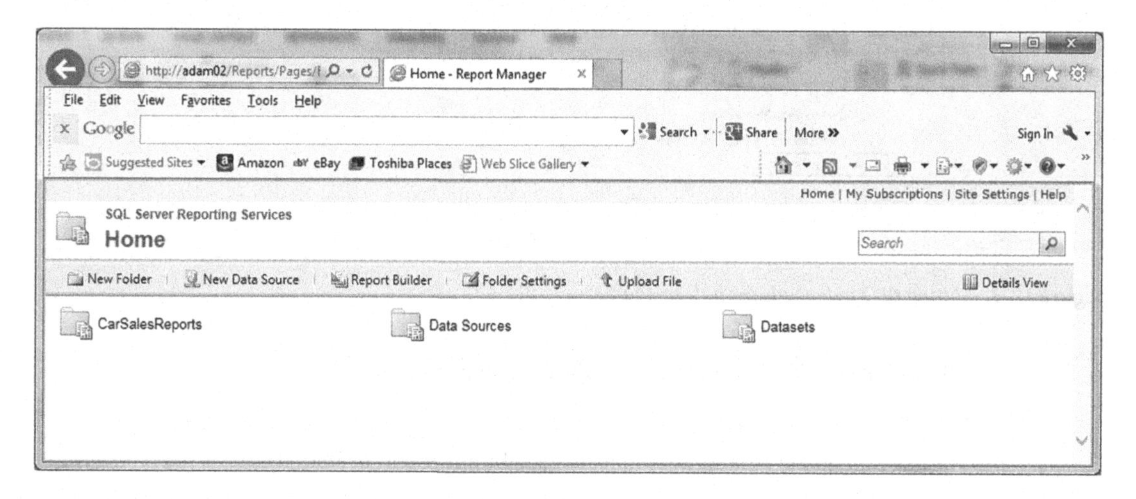

Figure 12-3. The Report Manager

3. Click Data Sources. This will show the available data sources.
4. Click the data source that underlies the shared dataset(s) that you want to cache. The Properties browser page will appear for this data source.
5. Click the radio button next to Credentials stored securely in the report server.
6. Enter the user name (BIReporting for the sample report project) and the password (also BIReporting). The browser page will look like Figure 12-4.

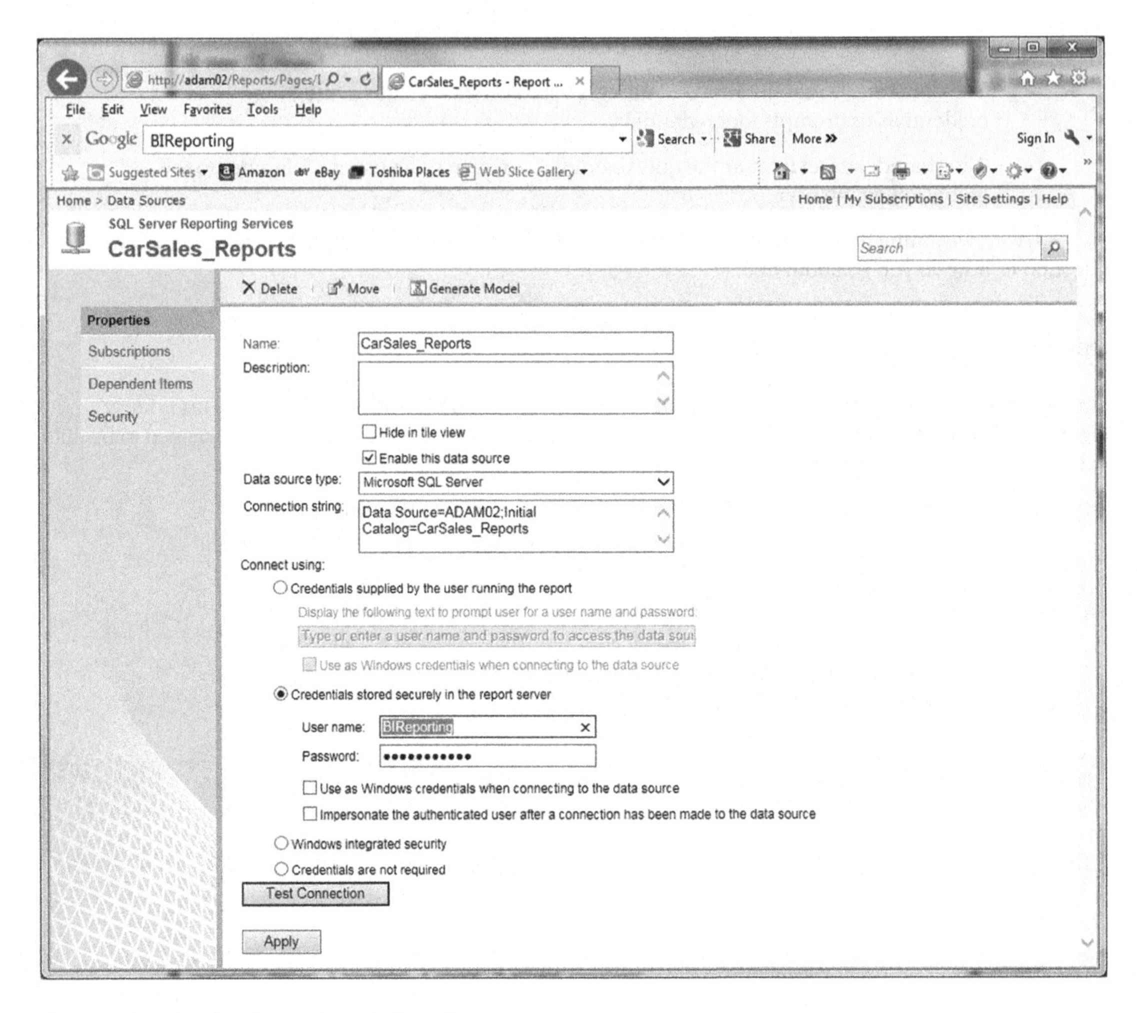

Figure 12-4. *Setting the credentials for a data source*

7. Click Test Connection to make sure that the credentials are valid.
8. Click Apply.

This must be done for all the data sources that are used by any report or shared dataset that you wish to cache.

Enabling Shared Datasets for Caching

Now that you have stored the credentials for the data source, you can enable a shared dataset for caching. Here is how you do this.

1. Open or browse to the Report Manager web page.
2. Click Datasets. Any datasets that you have deployed will be displayed, as shown in Figure 12-5.

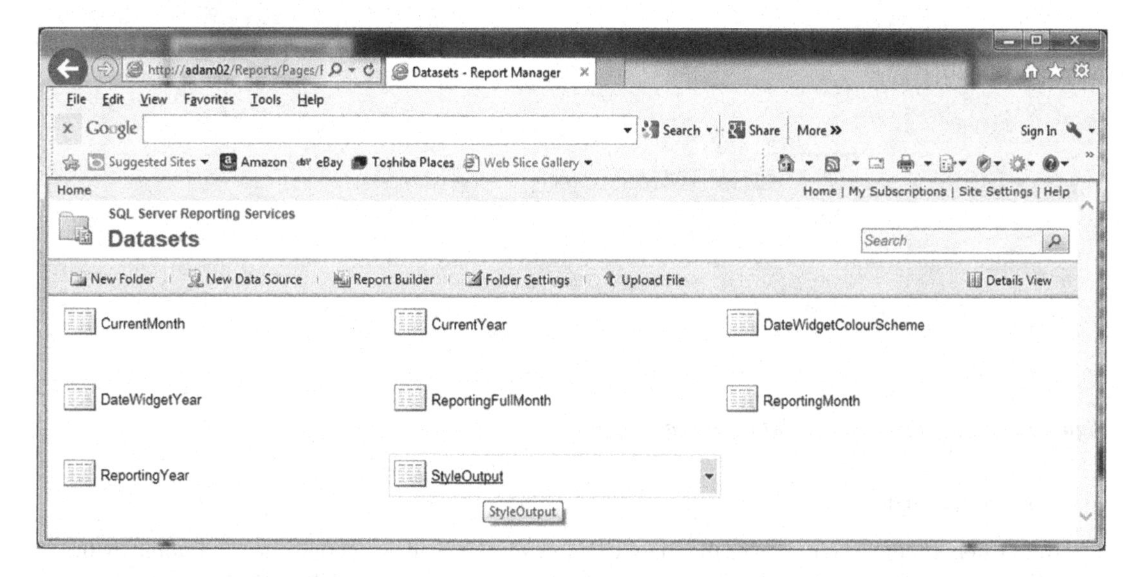

***Figure 12-5.** Available datasets*

3. Click the dataset that you want to modify (StyleOutput, in this example). The Properties page for this dataset will be displayed.
4. Click Caching on the left. The cache details will be displayed.
5. Check Cache shared dataset.
6. Select the button to Expire the cache after this number of minutes.

7. Enter a number of minutes (600, in this example). The browser window will look like Figure 12-6.

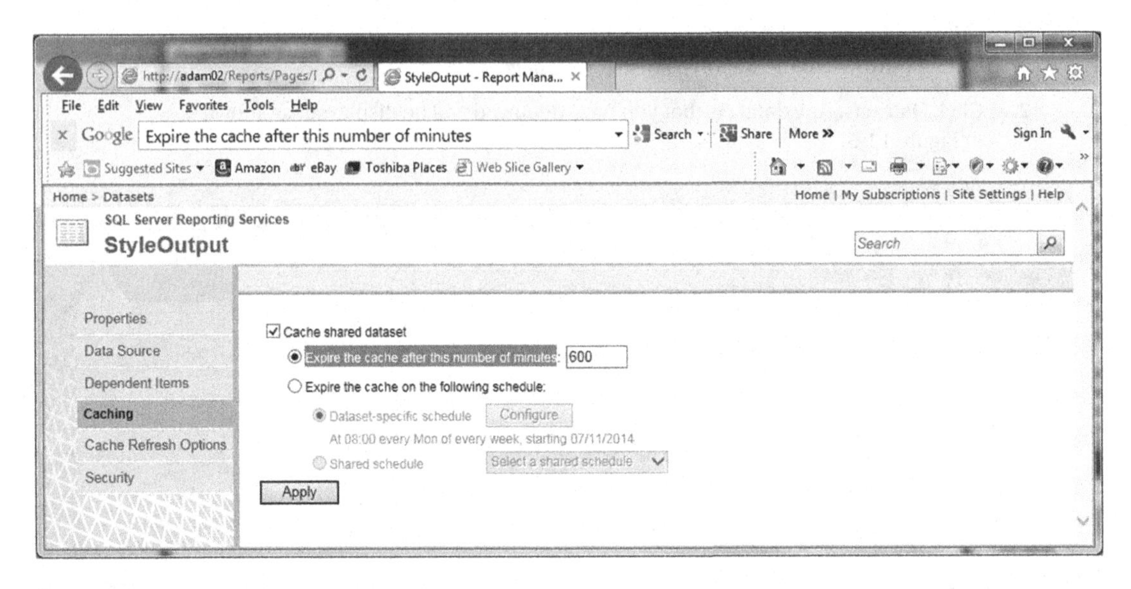

Figure 12-6. *Enabling a shared dataset for caching*

8. Click Apply.

Now, when *any* report is run using the shared dataset, the data will be queried once and stored in the cache for the parameter combination that was used by the query. This data will then be used for any report that uses this dataset. After 10 hours (600 minutes) the cache will be cleared.

■ **Note** If you see the warning icon and the message "Credentials used to run the shared dataset are not stored" you need to store the data source credentials as described previously.

Report Caching

As was the case with datasets, caching can shorten the time needed to deliver a report. This is particularly efficient if the report is large, frequently accessed, or is based on multiple datasets. Indeed, if the report uses several datasets, then report caching is a good way of caching multiple datasets at the same time. The only downside is that the *cached* data for a specific report cannot be read from the cache by other reports, so you need to cache multiple reports independently, even if they use the same datasets. As was the case with shared datasets, you need to store the credentials used by the data source(s) used by any report that will be cached. So here is how to cache a report.

1. Open or browse to the Report Manager web page.
2. Browse through the hierarchy of folders until you find a report that you want to cache. I have chosen SmartPhone_SalesAndTargetWithPreviousMonthAndPreviousPeriod from the sample reports.
3. Click the pop-up arrow that appears when you place the pointer over the report. The menu appears as shown in Figure 12-7.

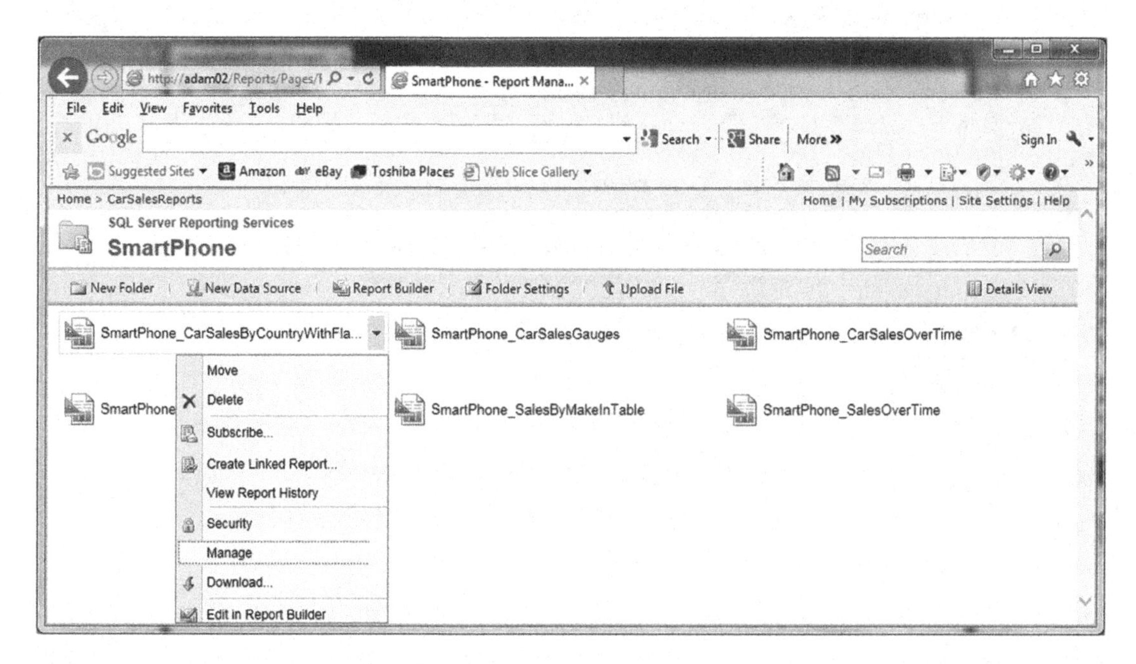

Figure 12-7. The pop-up menu for reports

4. Select Manage. The manage report browser window will appear.
5. Click Processing Options on the left.
6. Select the option Cache a temporary copy of the report. Expire copy of report after a number of minutes.

7. Enter a number of minutes (600, in this case). The browser window will look something like Figure 12-8.

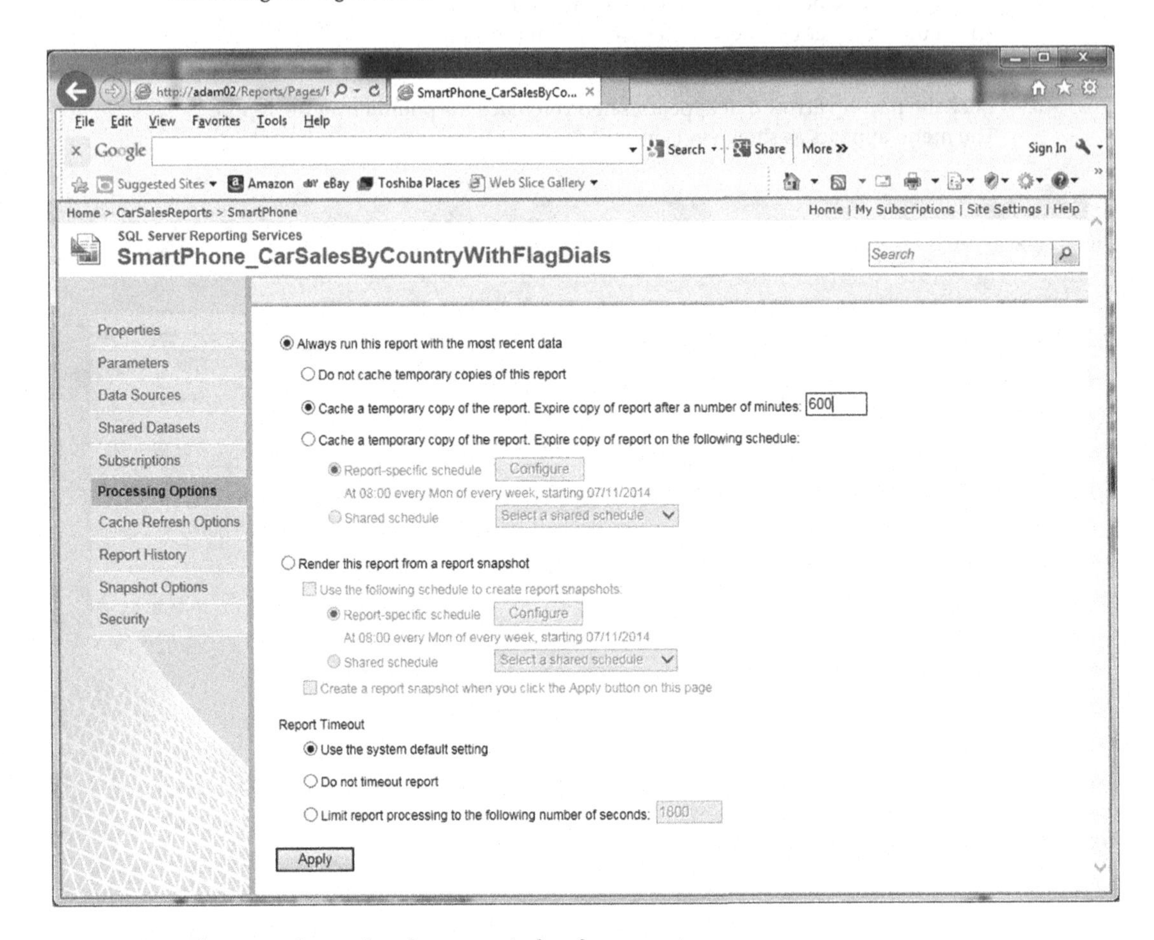

Figure 12-8. *The processing options browser window for a report*

8. Click Apply.

Now, when a report is run, the data for this report will be queried once and stored in the cache for the parameter combination that was used by all the queries in the report. This data will then be used for the next 10 hours (600 minutes) and then the cache will be cleared.

Using Parameters and Filters when Caching Reports and Datasets

The standard thinking when building reports in SSRS is that all filtering and data selection should be carried out at the server. After all, optimized data selection is where relational databases (as well as datawarehousing technologies) excel.

I suspect, however, that the accepted best practice might not always apply when dealing with BI result sets. More often than not, the datasets that you prepare for gauges and charts are narrow and shallow (having only a few columns and a few dozens of rows). This is entirely unlike the potentially wide and very deep record sets that you see in conventional reporting.

To appreciate the difference, let's look at an example. Suppose that you need to return sales and costs per make of vehicle for a chart that will be sent to a tablet. You confidently assume that the user will want to flip between years and months using report parameters.

Now let's consider the dataset. In the sample database, there are only seven makes of car. So each query that is sent back to the database will return at most seven records. This is a trivial amount of data. I realize that we are talking about the sample database that may not correspond to real-world data, but I am trying to illustrate the generic point for the moment. However, these same few records will be queried each time a user wants to see data for a different year or month. This means that the source database will get hit many, many times for a tiny dataset. Even if each query is cached, you will get potentially dozens of queries running one after the other, as each different set of parameters is a fresh database query on a server that could already be overloaded.

Now think what happens if you add the ReportingYear field to your query, and group by year. If there are four years, your dataset will have at the most 28 records. This is nothing for SQL Server or a network to handle. Take the analysis one step further and add the ReportingMonth field to the query and group on this too. You could end up with 336 records. Again, this is a trivial amount of data (especially as the record set is unlikely to be very wide).

In cases like this, you may *not* want to send parameters back to the database for every possible combination of parameters, but prefer to return a dataset that contains all the possible combinations of parameters so that the report can be *filtered* on certain fields instead. I realize that as an SSRS developer you may have gotten used to thinking that SSRS parameters are only applied to SQL queries. This is because in SSRS, parameters are generally passed to the query that returns the data, whereas filters are applied to the data that is returned to the report. However, there is nothing to prevent you from using SSRS parameters to filter *datasets* that have been returned to the report by a stored procedure (or other query). This means that if your query returns only a small dataset, you could be better off using as few parameters as possible in the stored procedure, and setting up SSRS parameters that will be applied as filters to the dataset that is returned by the query. If the resulting dataset is then cached, you will be minimizing the load placed on the database server (as the query is only run once, even if it is more complex than it was before) as well as speeding up the report rendering and consequently improving the user experience. Moreover, you are making your work life easier as you will be caching far fewer datasets than if you were caching a dataset for every possible combination of parameters. Indeed, you may be able to return a single dataset that has no parameters in the underlying query, and consequently only have to cache one dataset instead of dozens.

So, contrary to what you have seen throughout the book, you may find it faster to group by certain fields, rather than use them as parameters passed back to the database and then filter on certain elements in SSRS. As an example of how this can be applied, here is a report that has the same year and month parameters that you have used in nearly all the BI reports in this book, as well as a parameter for the make of car. However, the underlying stored procedure contains no parameters, and all the report parameters merely filter the cached, shared recordset, just as they could filter the recordset for a cached report.

■ **Note** The drawback with this approach is that the dataset must contain all the information that a user could possibly want. So you will need to make sure that every possible combination of selection parameters is contained in the dataset or no data will be displayed. Another risk is that the query underlying the dataset does not map to the queries supplying lists of information to the SSRS parameters. You may, for instance, have a dataset that returns data for 10 years where your selection parameter list contains 15 years. Consequently, you will need to pay close attention to keeping the parameters in synch with the dataset query if you use this approach.

The report (with absolutely no attempt at enhanced presentation, as we are looking at the data here) looks like Figure 12-9.

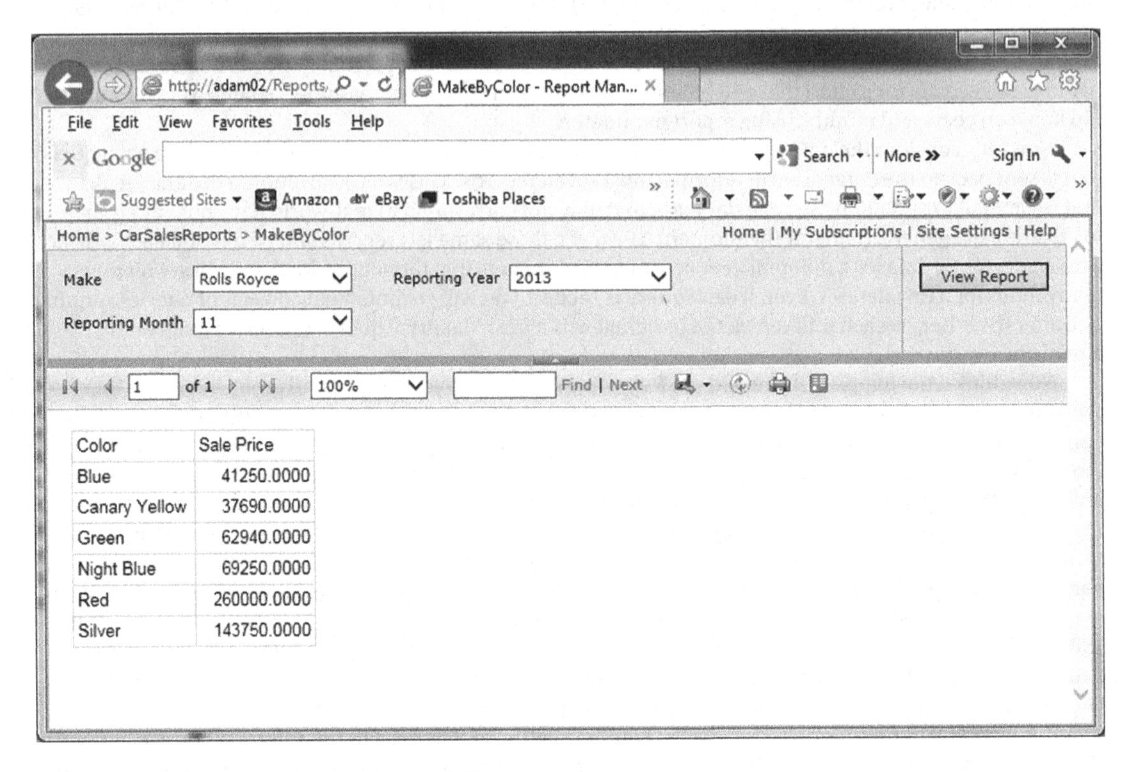

Color	Sale Price
Blue	41250.0000
Canary Yellow	37690.0000
Green	62940.0000
Night Blue	69250.0000
Red	260000.0000
Silver	143750.0000

Figure 12-9. *A report using parameterized filters*

The Source Data

The source data for this report is two stored procedures. The stored procedure Code.pr_MakeSales is a simple query to return the data that will then be filtered. The stored procedure Code.pr_MakeList is used in the report filter to return the list of makes of vehicle.

```
-- Code.pr_MakeSales

SELECT
 Make
,Color
,SUM(SalePrice) AS SalePrice
,ReportingYear
,ReportingMonth
FROM      Reports.CarSalesReporting
GROUP BY  Make, Color, ReportingYear, ReportingMonth

-- Code.pr_MakeList

SELECT DISTINCT Make FROM Reports.CarSalesReporting
```

The stored procedure pr_MakeSales returns just 302 records from the sample data, and this is every combination of make, color, year, and month for all vehicle sales. A quick overview of a few records is given in Figure 12-10.

	Make	Color	SalePrice	ReportingYear	ReportingMonth
1	Aston Martin	Black	95000.00	2012	12
2	Aston Martin	Black	178500.00	2013	3
3	Aston Martin	Black	37000.00	2013	7
4	Aston Martin	Black	195500.00	2014	12
5	Aston Martin	Black	225500.00	2015	3
6	Aston Martin	Black	362500.00	2015	4
7	Aston Martin	Black	79500.00	2015	9
8	Aston Martin	Blue	120000.00	2012	1
9	Aston Martin	Blue	130000.00	2012	8
10	Aston Martin	Blue	69250.00	2013	1
11	Aston Martin	Blue	37000.00	2013	5

Figure 12-10. *Output from the stored procedure Code.pr_MakeSales*

Building the Report

Time, then, to create the report to optimize the use of a cached dataset.

1. Create a new shared dataset named MakeSales. Have it use the data source CarSales_Reports and the stored procedure Code.pr_MakeSales. This dataset will have no filters or parameters.
2. Create a new shared dataset named MakeList. Have it use the data source CarSales_Reports and the stored procedure Code.pr_MakeList. This data set will have no filters or parameters
3. Make a copy of the report __BaseReport.rdl and name it MakeByColor.
4. Add the shared dataset MakeSales and name it MakeSales.
5. Add the shared dataset MakeList and name it MakeList.
6. Add a table to the report and remove the third column. Set it to use the dataset MakeSales.
7. Add the fields Color and SalePrice to the table, in this order.

8. Add a new parameter to the report. Set its properties as follows:

Section	Property	Value
General	Name	Make
	Prompt	Make of Vehicle
	Data type	Text
Available Values	Get values from a query	Selected
	Dataset	MakeList
	Value field	Make
	Label field	Make
Default Values	Specify values	Selected
	Value	Rolls Royce

9. Select the table and display the Properties window.

10. Click the ellipses button for the Filters property. The Tablix Properties dialog will appear with the Filters pane selected.

11. Add three filters by clicking the Add button three times, and set the following filter values:

Expression	Operator	Value
ReportingYear	=	`=Parameters!ReportingYear.Value`
ReportingMonth	=	`=Parameters!ReportingMonth.Value`
Make	=	`=Parameters!Make.Value`

12. The Tablix Properties dialog will look like Figure 12-11. Click OK.

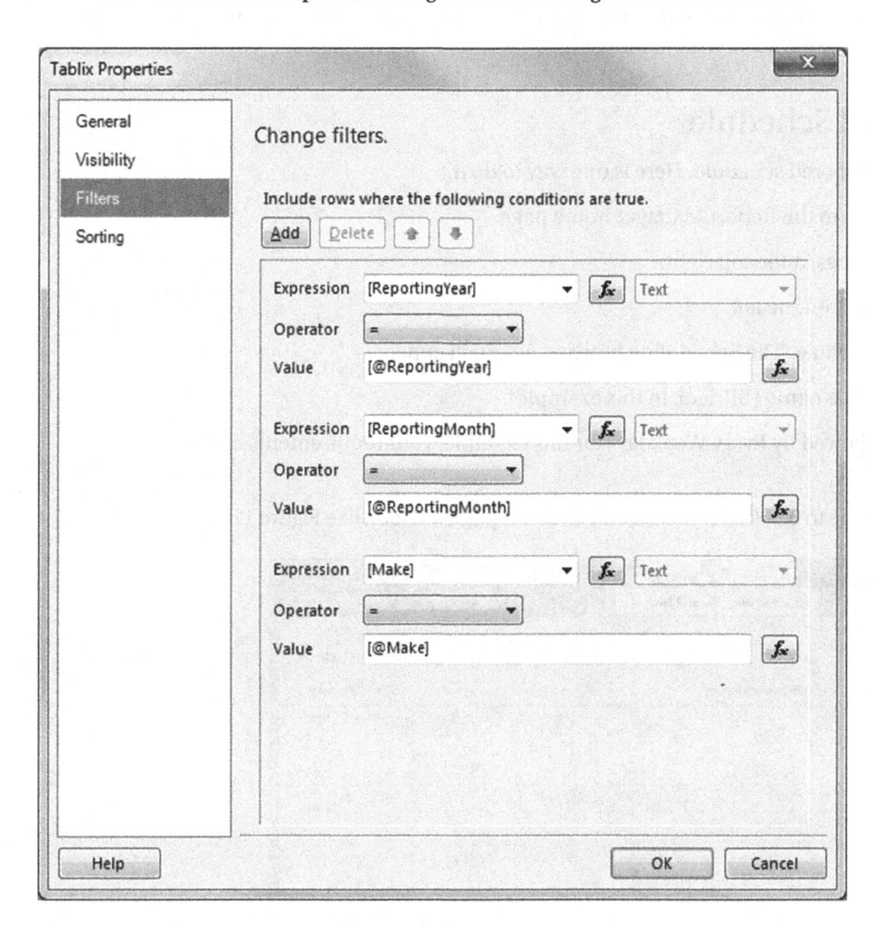

Figure 12-11. *The Tablix Properties dialog when filtering using parameters*

13. Deploy the report and the two shared datasets you created.

14. Set the two shared datasets MakeSales and MakeSales to be cached, as described previously in this chapter.

Now once the dataset has been cached using a cache schedule and you view the report there will be little database activity when you apply a different set of parameters for the year, month, and make. This is because the cached dataset will merely be reused, only with different filters applied.

Cache Schedules

So far, when caching reports and datasets you have manually set the expiration time for the cached object. This is not always the most efficient solution to removing objects from the cache. There are occasions when BI report suites are updated and/or recalculated on a regular basis. So they can be ideally suited to predefined schedules for clearing cached reports and datasets. For instance, if the database containing your reporting data is refreshed overnight (and presuming that users know that the latest data is only available first thing in the morning), you could be best advised to remove all cached objects before running the update process. Any cached reports will be recached with the latest data the first time that they are run.

This is a two-part process. First, you create a shared schedule that will start running at 01:00 every weekday. Second, you apply this to a shared dataset. The only important thing to note is that the SQL Server agent must be running for shared schedules to function.

Creating a Shared Schedule

First, you need to create a shared schedule. Here is one way to do it.

1. Open or return to the Report Manager home page.
2. Click Site Settings at the top right.
3. Click Schedules on the left.
4. Click New Schedule. The Scheduling browser page will appear.
5. Enter a schedule name (BIClear, in this example).
6. Click Day, followed by Every Weekday (for this example; your requirements may be different).
7. Set the start time to 02:00. The Scheduling browser page will look like Figure 12-12.

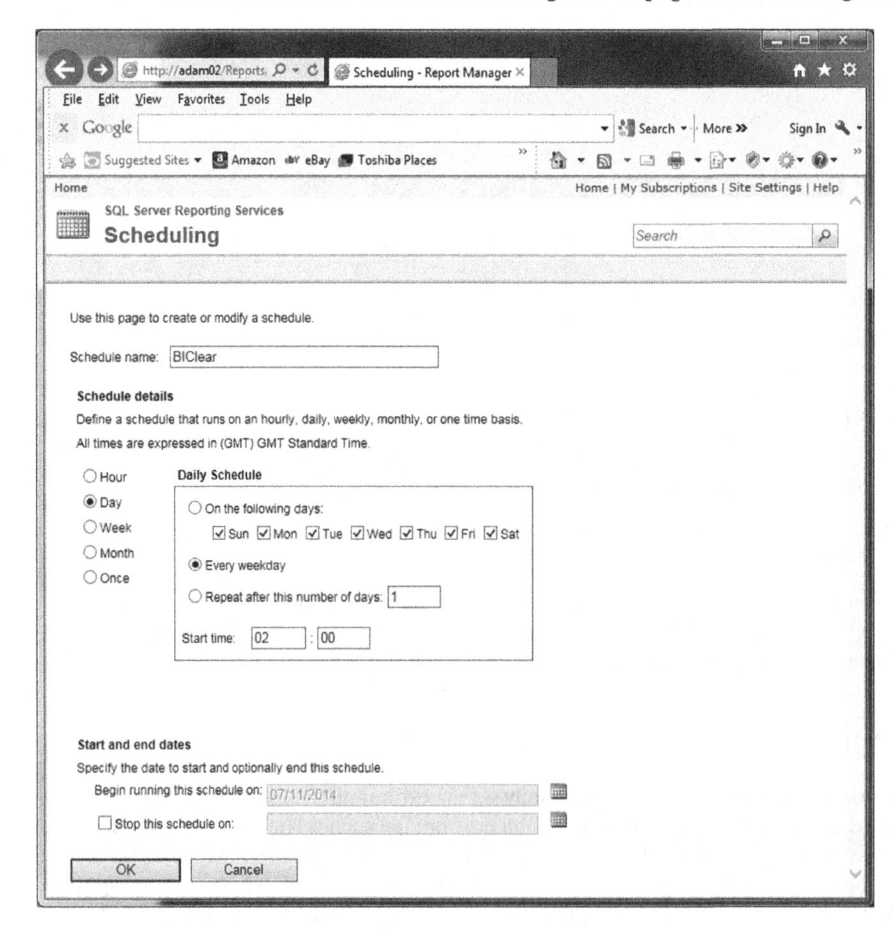

Figure 12-12. *Creating a shared schedule*

8. Click OK. You will return to the Site Settings web page, which will look like Figure 12-13.

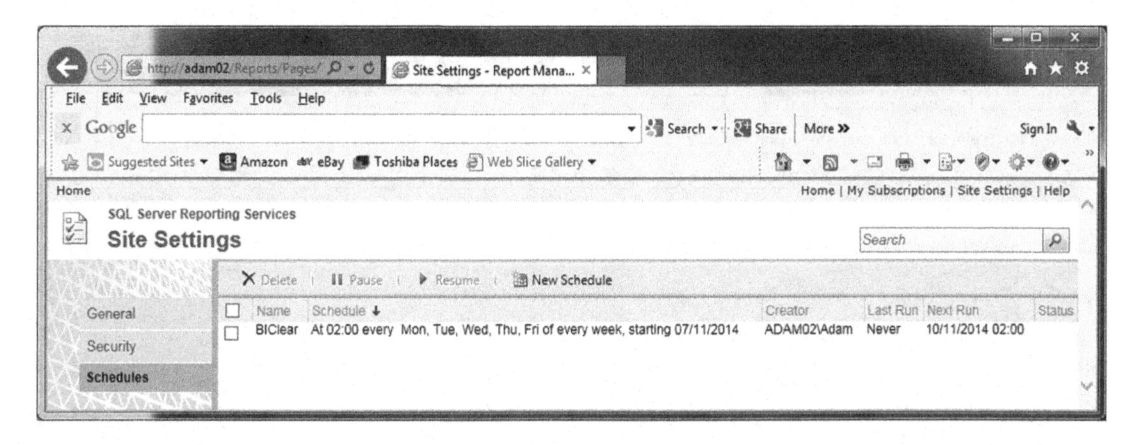

Figure 12-13. Site settings showing a shared schedule

9. Click Home at the top right to return to the Report Manager home page.

Your shared schedule is now active and ready to be applied to any shared reports or datasets. It is worth noting that the site settings page lets you pause, resume, modify, or delete a shared schedule.

You may find yourself expiring the cache and then reloading the report into the cache using a second schedule (shared or specific). In this case, ensure that you leave enough time for the expiry process to run before reloading any reports into the cache.

■ **Note** You can also create and manage shared schedules from SQL Server Management Studio. All you have to do is connect to Reporting Services (and not to a database), expand Shared Schedules, and double-click the schedule you wish to modify.

Applying a Shared Schedule

Applying a shared schedule to a dataset is done like this.

1. Follow steps 1 through 5 from the "Report Caching" section.
2. In the Caching web page, select Expire the cache on the following schedule.
3. Click the Shared schedule radio button.

4. Select the shared schedule (BIClear, in this example) from the list of available schedules. The web page will look like Figure 12-14.

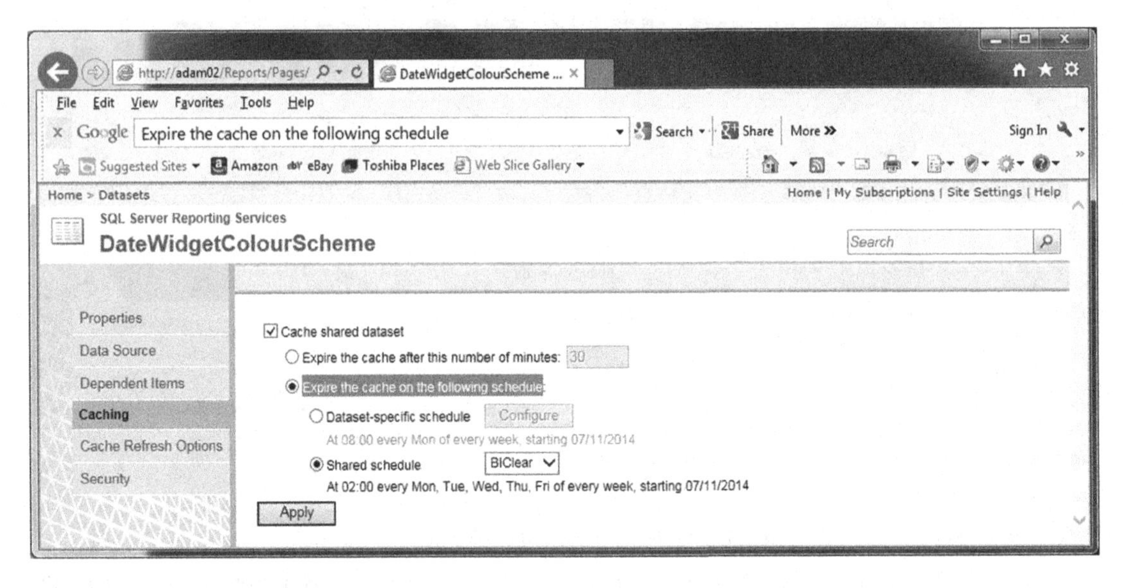

Figure 12-14. *Dataset caching settings used to apply a shared schedule*

5. Click Apply.

It's really that simple. This dataset will now be removed from the cache at 02:00 every weekday.

Pre-Caching Reports and Data Sets

So far, you have seen how to enable reports and datasets for caching, and then how to remove the items from the cache once an initial query (run when a report was loaded) had triggered the first set of data in the cache.

This does not avoid a potentially unfriendly user experience (as the initial query has not yet been run) for the first person who views the report, even if it is cached, or based on cached datasets. Indeed, if the person waiting for the query to run the first time is the CFO, then the unpleasantness could end up being shared with you, the report developer.

So, to avoid sadness and misery all round, you can also refresh the items in a cache (both reports and datasets) on a regular schedule too. This ensures that even the first person who views a report, or that uses a dataset, gets to read the latest version from the cache.

There are two main ways of loading an item into the cache (other than opening the report manually). These are

- Setting a scheduled cache refresh.
- Defining a subscription using the NULL delivery provider.

There is one essential difference between the two. A scheduled cache refresh can only cache a report with one set of parameters, whereas if you define a subscription using the NULL delivery provider, you can preload multiple parameter combinations for the same report. Although you can define several cache refresh settings to allow for multiple report parameter combinations, you may find it best to use cache refresh when you have only a single parameter combination (for example, when setting the current year/month in a dashboard) and to use a subscription and the NULL delivery provider when you have multiple parameters to set when caching a report.

That's the theory. Now it's time to see the two approaches in action and apply them to a couple of the reports you built in earlier chapters.

Cache Refreshing

Here is how you can set cache refresh for a report. I will be using a dashboard for this example, as it is a prime example of a report that only needs one parameter combination: the current year and month. This is because most users only want to see the current data. A dashboard is also a top candidate for caching because it is based on multiple complex datasets that can take a while to run when the report is loaded the first time.

1. Set up a shared schedule (as described previously). Name it Loadcache and set the run time to 06:00 every weekday.
2. Navigate to the Properties page for a shared dataset or report (as you have done previously) and click Cache Refresh Options on the left. If there are any existing cache refresh plans, they will be listed.
3. Click New Cache Refresh Plan. The Cache Refresh Plan web screen will appear.
4. Enter a description for the cache refresh plan. For this example, use "Daily Cache" here.
5. Set any parameters required by the report (or use the report default). As this is a dashboard I will use the default because it will return data for the current month and year.

6. Select Shared Schedule, and choose the Loadcache shared schedule that you created in step 1. The web page will look like Figure 12-15.

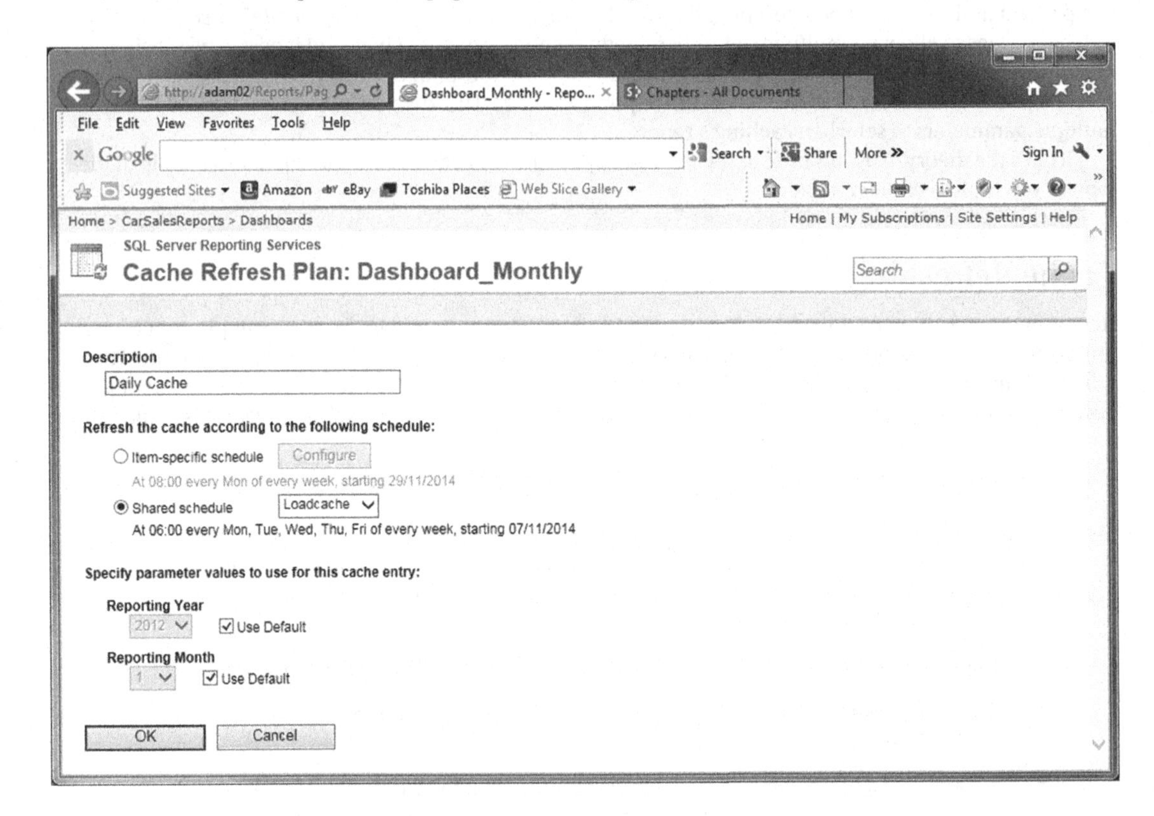

Figure 12-15. *Defining a cache refresh plan*

7. Click OK. The cache refresh options web page will reappear, with the new cache refresh plan listed.

Once you have created a cache refresh plan, you can modify it or even delete it from either the Report Manager or SSMS.

■ **Note** You do not need to remove any existing cached objects from the cache if you are refreshing them.

Subscriptions to Load the Cache

Another option that you could use to load the Reporting Services cache is to set up a subscription. Although this approach is a little more long-winded than using cache refresh, it does have one real advantage: you can define a *set* of parameters for the underlying queries. This means that you can preload *some*, but not necessarily all, of the permutations of parameters that are possible for a report. This allows you to cache frequently-used combinations of parameters, and not overload the SSRS cache with parameter combinations that are more rarely required by users. Also, when compared to preparing datasets that will be filtered in the report, this approach has the advantage of letting you circumvent the "all or nothing" approach of prepared datasets that have to include all the available data that a user might want, as we saw previously.

The set of parameter combinations that you will use will be an SQL snippet. It can be a hard-coded set of values, but it is more likely to be a database query that returns a collection of parameters.

■ **Note** Using the NULL delivery provider is only available if you have a SQL Server Enterprise license.

---

1. In Report Manager, navigate to the folder containing the report that you want to cache. Click the pop-up menu arrow for the report and select Manage. Then click Subscriptions in the left pane, followed by New Subscription at the top of the list of current subscriptions.
2. Enter a description for this data-driven subscription and select Null Delivery Provider from the pop-up list specifying how recipients are notified. The browser window will look like Figure 12-16.

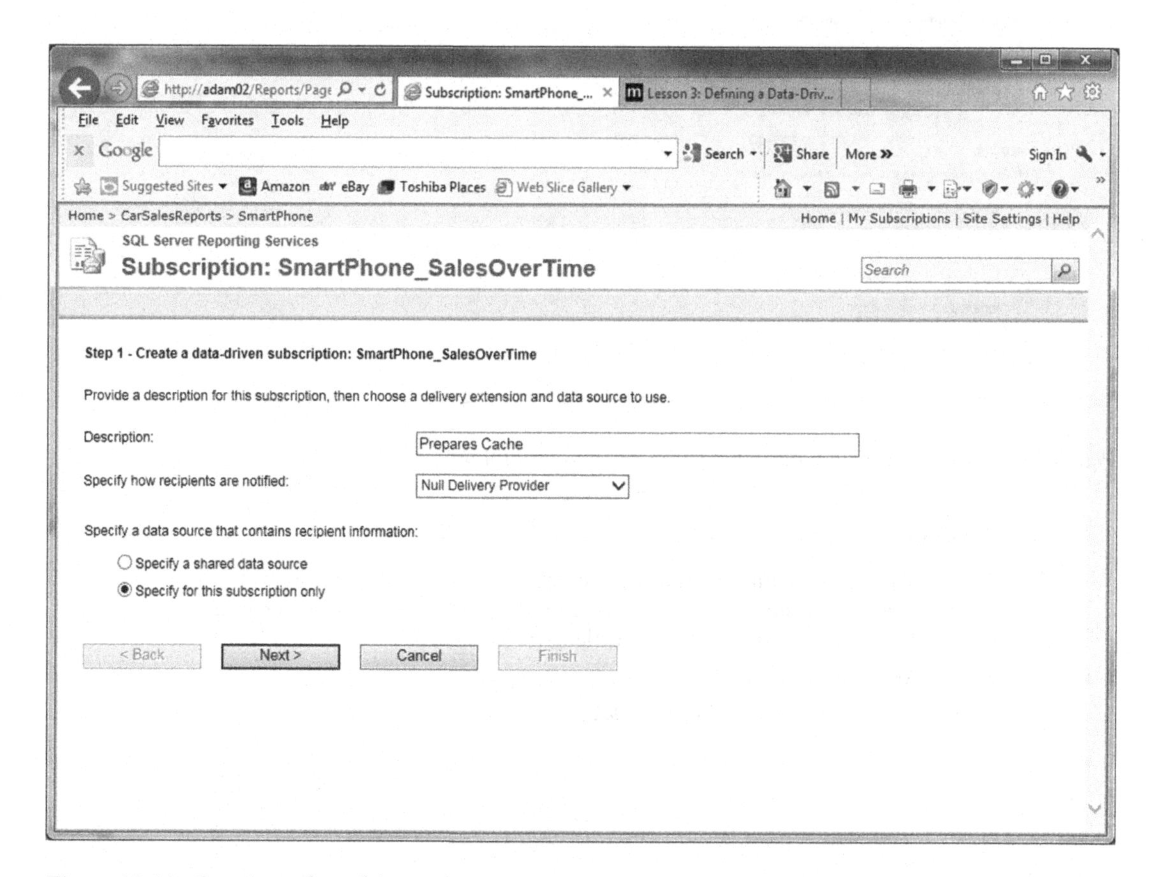

Figure 12-16. *Creating a data-driven subscription*

3. Click Next.
4. Leave the data source type as SQL Server (in this example, anyway) and enter the data source string to connect to the server. You can find this in SSDT if you open a shared data source that you are already using.
5. Enter the credentials for the SQL Server user that you have defined previously. In this example, it is BIReporting as both user name and password. The browser window will look like Figure 12-17.

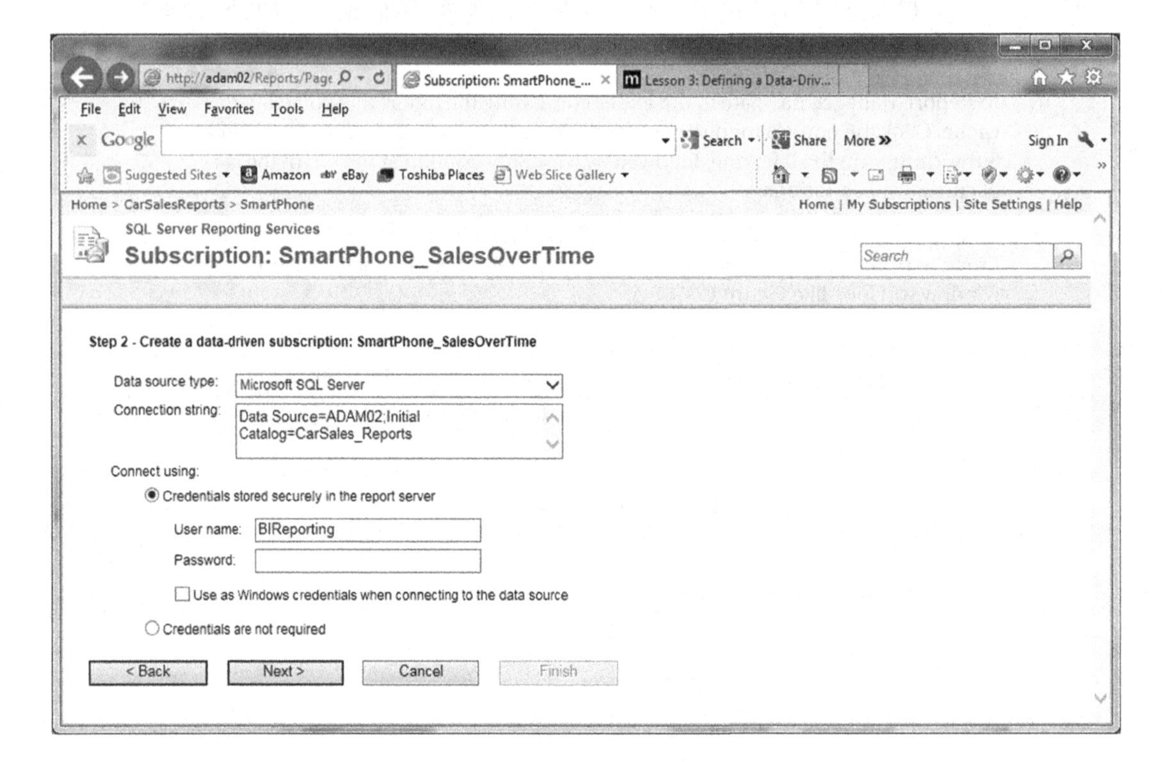

Figure 12-17. *Assigning a data source for a data-driven subscription*

6. Click Next.
7. Enter the SQL snippet that will return the combination of parameters that you want to apply to this subscription. In this example, you are returning all the year and month combinations that are in the dataset. The SQL is

```
SELECT DISTINCT ReportingYear, ReportingMonth
FROM            Reports.CarSalesData
```

8. Click Validate to ensure that the SQL and the security settings are valid. The browser window will look like Figure 12-18.

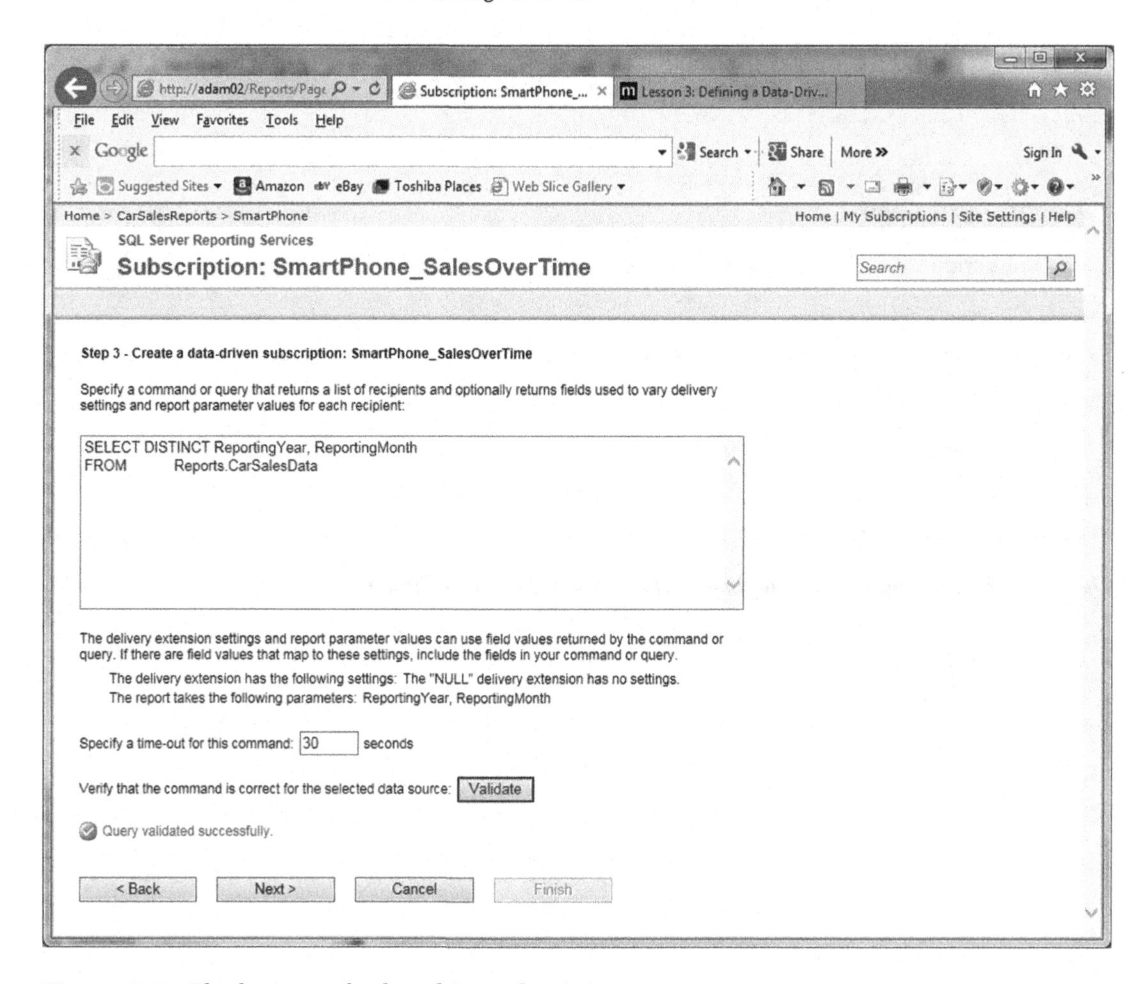

***Figure 12-18.** The data query for data-driven subscription*

9. Click Next.

10. The NULL destination for a data-driven subscription window will be displayed. The browser window will look like Figure 12-19.

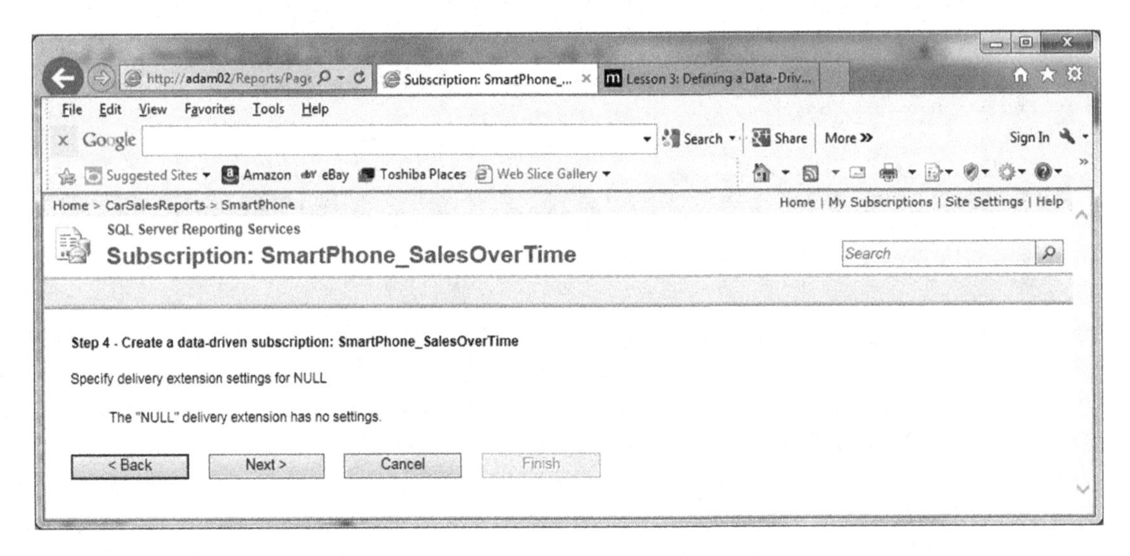

Figure 12-19. *Applying the NULL destination for a data-driven subscription*

11. Click Next.

12. For all the parameters in the report, either select a field from the database (returned by the SQL query that you provided earlier) or enter a fixed value. The browser window will look like Figure 12-20.

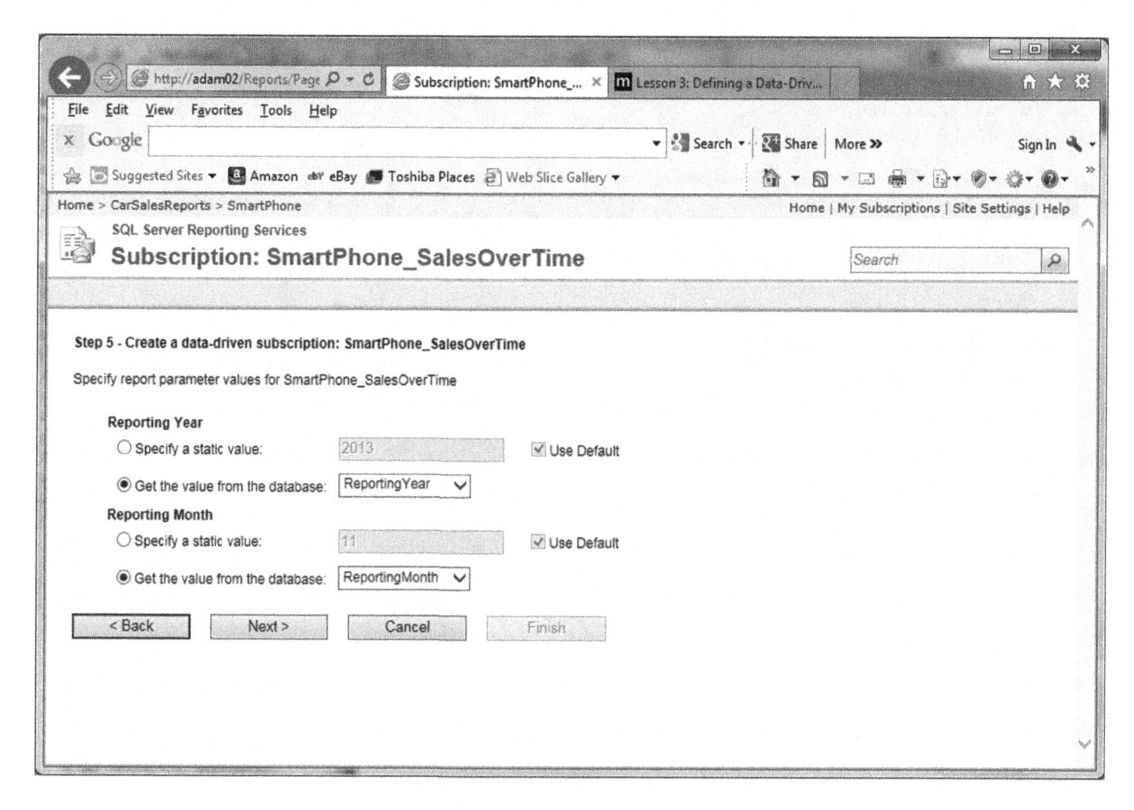

Figure 12-20. *Setting parameters for a data-driven subscription*

13. Click Next.

14. Apply or create a schedule just as you did previously when caching a report. The browser window will look like Figure 12-21. In this case, I am using a shared schedule. You saw how to create these earlier in this chapter.

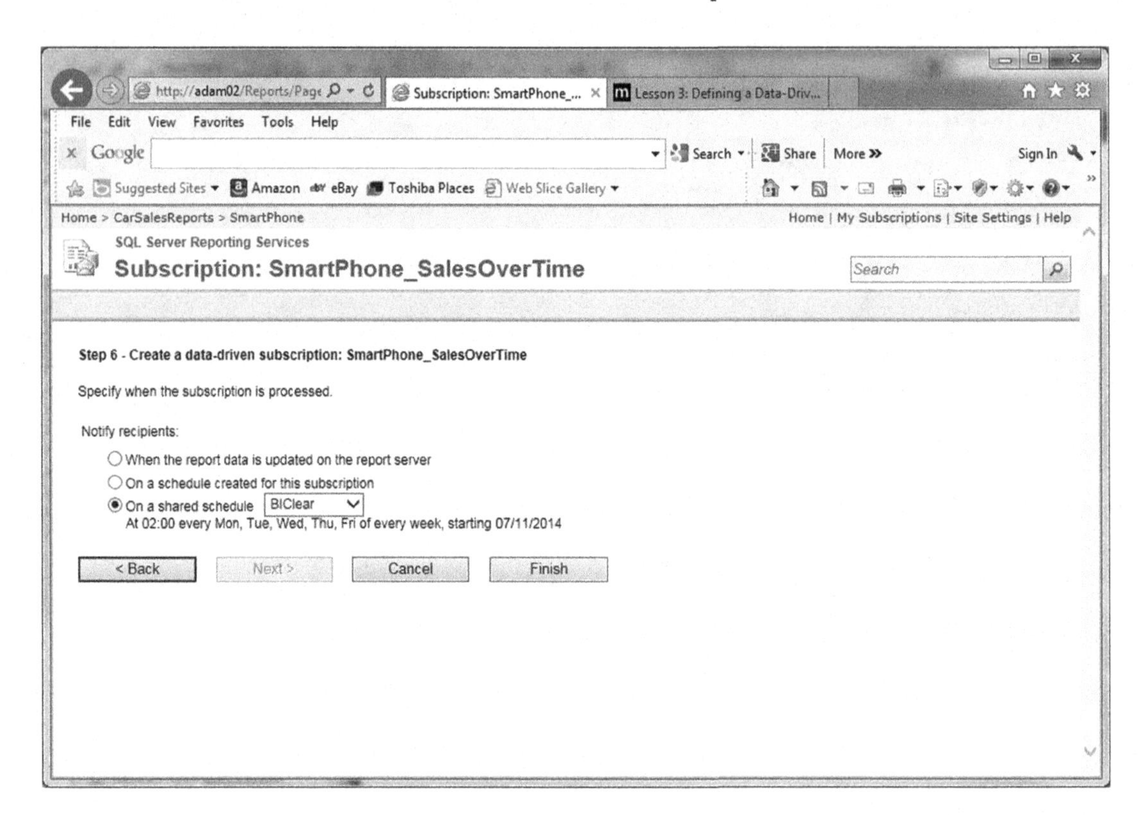

Figure 12-21. *Applying a schedule to a data-driven subscription*

15. Click Finish. Your data-driven subscription has been created and will be added to the list of existing subscriptions, as shown in Figure 12-22.

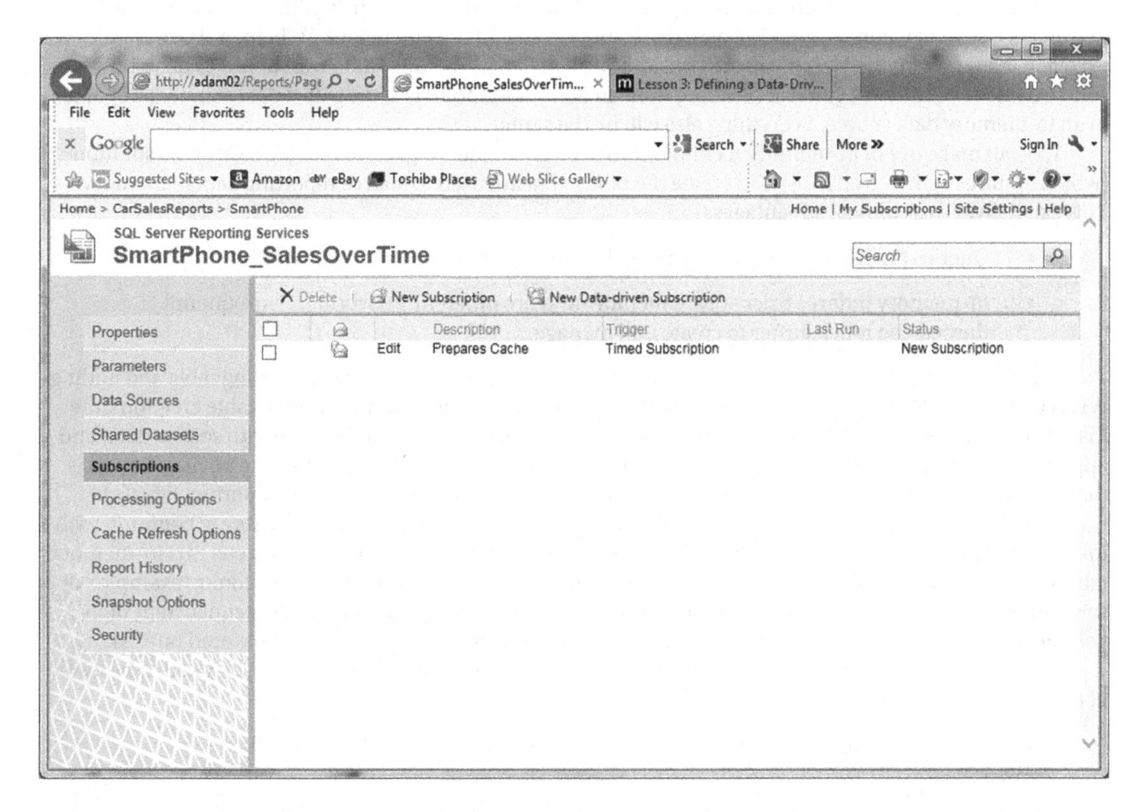

Figure 12-22. *The list of available data-driven subscriptions*

When the schedule runs, the report will be cached using all the parameter combinations that were available in the SQL snippet you applied in step 8.

■ **Note** Rerunning a subscription to apply caching using the NULL delivery provider will *not* update a cached report with latest version of the data when the cache has not expired. Consequently, you must *always* expire cached reports before reloading the cache to ensure that the latest versions are available.

In-Memory Data Warehouses

One technology that can make dataset caching redundant (in some circumstances, at least) is to load the data for your BI output into an SSAS tabular warehouse. Although you can configure a SSAS tabular warehouse to spill to disk, it will normally reside in memory. As you probably know, this is only available with SQL Server 2012 or 2014, and in either case is an Enterprise-version only feature. As the source data for your dashboards and mobile reports will be in memory there will be no disk access for reporting data. The simple consequence is speed. In many cases, you may see a considerable reduction in the time taken to render reports that use SSAS tabular data warehouses as the source for the data. So you may be able to avoid having to cache reports entirely in some circumstances.

Creating tabular data warehouses is a vast subject, and one that needs, if anything, a separate book. Fortunately, several volumes already exist on this subject, so I advise you to look at them or the many excellent resources on the Web if you are considering using an in-memory data warehouse to host the source data for your BI. All you will be doing from an SSRS standpoint is setting up a data source that points to an in-memory data source. Everything else will be the same.

You can make use of in-memory technology to accelerate querying *without* using SSAS. One solution (which requires SQL Server 2014) is to create the tables that are used to source reporting data in memory. This can deliver two colossal advantages:

- Queries on in-memory tables can be blisteringly fast.
- In-memory indexes exist only in memory and are inherently covering. Consequently, they can be much easier to create and manage.

Creating an in-memory table for reporting is as simple as creating any other reporting table and adding WITH (MEMORY_OPTIMIZED=ON) to the DDL. You also must include any indexes at table creation time in the DDL that you write. Then you can carry out a SELECT ... INTO to load the table with source data, and append delta data as and when you need to, just as you would with a persisted table. The obvious caveat is that you need to be sure that your server has sufficient available memory to store the reporting table(s).

The Microsoft documentation claims that 5-to-20 times performance improvements are common with in-memory tables. My tests with reporting tables certainly bears this out. Moreover, you can mix in-memory tables with standard "on-disk" tables. So if you cast your mind back to the vertical partitioning example you saw earlier in this chapter, the core data could be placed in an in-memory table to allow concurrent high-volume reads, while the more rarely-used data in the second vertical partition could be stored on-disk.

Should you think about implementing an in-memory table or tables for reporting, you need to be aware of the following limitations:

- In-memory tables are only available in the Enterprise version of SQL Server.
- You cannot create geospatial or LOB columns.
- In-memory tables cannot be truncated, so you must delete data, which is slower.
- You cannot add clustered indexes or columnstore indexes.

There is no guarantee that an in-memory approach will always be better than a "classic" optimized BI database or datawarehouse because it will depend on the data and environment. So you will have to test and compare. If you are dealing with a tiny SQL server-based relational database as your data source, any increase in perceived report rendering could be minimal. With larger data warehouses you could see a significant difference. However, I can only advise that you experiment, at the very least. You could be in for a pleasant surprise.

Conclusion

This chapter showed some of the ways you can speed up the delivery of your business intelligence to your users, and hopefully taught you some new techniques and solutions too. The ideas that you saw here are not exhaustive; there are many other ways to accelerate the perceived speed with which your charts, gauges, and other visualizations appear in front of your users. The key point is to remember that getting the information to your users fast is often as important as creating the dashboards and tables themselves. So I urge you not to stop once your report is finished, but to go that extra mile to reduce rendering times to the bare minimum.

If there is one thing to take away when you are thinking about speeding up BI delivery, it is that you may well need to apply a range of approaches. So be prepared to look for complementary techniques rather than a single "magic bullet" approach. And above all, test and compare the solutions that you apply. This way you will have the satisfaction of having delivered the output that they requested rapidly, cleanly, and efficiently. This is, after all, what BI is all about.

We have reached the end of this book. I hope that you have enjoyed reading it, and that you will be able to apply the content to your own business intelligence delivery. I also hope that you will have fun producing your BI with SSRS, and that your users will come to appreciate what you have delivered.

Happy development with Reporting Services!

APPENDIX A

Sample Data

If you wish to follow the examples used in this book, and I hope you will, you will need some sample data to work with. All the files referenced in this book are available for download and can easily be installed on your local PC.

This appendix explains where to obtain the sample files, how to install them, and what they are used for.

Creating the Sample Data Directory Structure

First, you should create a directory and four subdirectories on a disk that has at least 100MB of free space to hold the sample data. Once you have created the directory/folder structure, it should like Figure A-1.

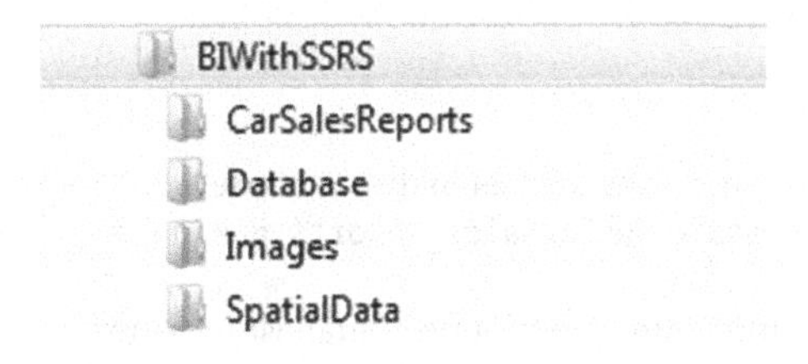

Figure A-1. *The sample data directory structure*

The directories contain the elements outlined in Table A-1.

Table A-1. *The Sample Data Directory Structure*

Folder	Contents
BIWithSSRS	The root directory, which will only contain subdirectories and (temporarily) the compressed source files
CarSalesReports	The directory that contains the SSDT report project and all the report, dataset, and data source files
Database	The database backup file and potentially the database files once the database has been restored
Images	Any images used in reports
SpatialData	Spatial data used for maps

Downloading the Sample Data

The sample files used in this book are currently available on the Apress site. You can access them as follows:

1. In your web browser, navigate to the following URL: www.apress.com/978-1-4842-0533-4.
2. Scroll down the page and click the Source Code/Downloads tab.
3. Click the link Download now, and save the following files into the directory BIWithSSRS:
   - CarSales_Reports.zip
   - CarSalesReports_Application.zip
   - Images.zip
   - SpatialData.zip
4. Uncompress the files to the following directories that you created previously:
   a. CarSales_Reports.zip ⇨ BIWithSSRS\Database
   b. CarSalesReports_Application.zip ⇨ BIWithSSRS\CarSalesReports
   c. Images.zip ⇨ BIWithSSRS\Images
   d. SpatialData.zip ⇨ BIWithSSRS\SpatialData

Sample Database

To follow the examples in in the book, you will need the sample SQL Server dataset in the database CarSales_Reports. This database is available in the sample data as the file CarSales_Reports.Bak in the directory BIWithSSRS\Database.

Before you can load this database you will need access to a functioning SQL Server database instance. If you need to, you can download and install the free SQL Server 2014 Express version, which is currently available at www.microsoft.com/en-in/download/details.aspx?id=42299&WT.mc_id=rss_alldownloads_devresources.

Once installed, you will need to follow these steps to restore the database backup.

1. Open SQL Server Management Studio Express.
2. Open a new query window by clicking New Query in the toolbar.
3. Run the following script:

```
USE [master]
RESTORE DATABASE [CarSales_Reports] FROM  DISK =
N'C:\BIWithSSRS\Database\CarSales_Reports.bak'
WITH  FILE = 1,  NOUNLOAD,  STATS = 5
,
   MOVE 'CarSales_Reports' TO 'C:\BIWithSSRS\Database\CarSales_Reports_Data.mdf',
   MOVE 'CarSales_Reports_Log' TO 'C:\BIWithSSRS\Database\CarSales_Reports_Log.ldf'
GO
```

The database will be restored, and can be used in the examples.

Index

H

I

J

K, L

M

T

U

V

W, X

Y, Z